Evolving

THE THEORY AND PROCESSES OF ORGANIC EVOLUTION

Francisco J. Ayala
University of California, Davis

James W. Valentine
University of California, Santa Barbara

The Benjamin/Cummings Publishing Company
Menlo Park, California

Cover photo: Drosophila engyochracea, a Hawaiian vinegar fly. Photo courtesy H. T. Spieth.

Photos pp. 5, 19, 170, 228, 397, 414, 415, The Bettmann Archive; 68, Owen Franken/Stock, Boston; 172 *middle,* Fred Bodin/Stock, Boston; 169 *top,* © Heather Angel/Biofotos; 169 *bottom,* Animals Animals/© Marty Stouffer Productions; 172 *left,* 172 *right,* 174, 195 *top left,* 195 *bottom left,* Animals Animals/© Leonard Lee Rue III; 195 *top right,* Animals Animals/© M. Krishnam; 195 *bottom right,* Animals Animals/© Denise Hendershot

Sponsoring editor: Jim Behnke
Production editor: Tom Belina
Editor: Tom Belina
Cover designer: Michael Rogondino
Artists: Roger W. Myers, Ginny Mickelson, Linda A. Hawke

Library of Congress Cataloging in Publication Data

Ayala, Francisco Jose, 1934–
    Evolving : the theory and processes of organic evolution.

    Bibliography: p.
    1. Evolution.   I. Valentine, James W., joint author.   II. Title.
QH366.2.A97   575   79-788

ISBN 0-8053-0310-3
ABCDEFGHIJKL-MA-782109

The Benjamin/Cummings Publishing Company, Inc.
2727 Sand Hill Road
Menlo Park, California 94025

To Mitzi and Cathryn

Contents

3   HEREDITARY VARIATION   **60**

4   PROCESSES OF EVOLUTIONARY CHANGE   **110**

Preface

Life is a dynamic system of populations in constant change. The earth's environment oscillates and shifts as a result of natural causes and the activities of man; the earth's organisms adapt to these changes through natural selection. Some evolutionary processes lead to the formation of new species and some species become extinct. The biosphere continues to evolve.

Aims and Focus of Text

Understanding of the evolutionary processes has grown remarkably in recent decades. There have been a number of outstanding accounts of new discoveries, but mostly at an advanced level. The purpose of this text is to provide an up-to-date account of the modern understanding of evolution for readers with only a high-school biology background, or its equivalent, in a format compact enough to serve in a one-semester college course.

One major aim is to cover as wide a spectrum of significant evolutionary concerns as possible, including the most recent findings. Since all fields of biology are affected by evolution, we cannot cover everything. Instead, we concentrate on microevolution, macroevolution, systematics, and evolutionary ecology. Some socioethical issues are also considered. We cover both the processes and the history of evolutionary change, although the emphasis is somewhat greater on the former than on the latter. Thus the coverage of biology and paleontology is broad.

Special Features

We have supplemented the basic account of modern evolutionary science with some special topics in order to reach a wider audience. The special topics, set aside from the running text (in sections labeled "Closer Look"), include examinations of the consequences of new ideas in evolution, descriptions of specialized techniques in evolutionary research that are especially promising, and extended treatments of ancillary topics (such as Darwin's biography) that might lead students to further reading. Basic mathematical expressions are included in the text but are not required for an understanding of the concepts they represent, which are explained in plain language as well. A mathematical introduction to natural selection is included as an appendix which might be used in courses emphasizing basic population genetics.

The topics set aside in the "Closer Look" sections and in the appendix can be omitted without affecting the students' comprehension of the rest of the book. Students with some knowledge of genetics may skip our elementary treatment of the subject, presented in Chapter 2. Depending on the level or purpose of the course, instructors may want to emphasize some chapters and omit, or cover only lightly, other chapters.

A list of questions for discussion and recommendations for additional reading follow each chapter. The questions, while rooted in the textual material, are designed to elicit creative answers; indeed some of them have no well-corroborated answers. They are not examination questions, but ones that lead to reflection and further inquiry. The recommended additional readings include some references that can be read immediately after the chapters in which they are listed, as well as some works (termed moderately advanced or advanced) that would be best read after completion of this book or after the less advanced readings. Technical terms are defined as they first appear in the text, but a glossary is included as additional help with unfamiliar terms. The list of references at the end of the book documents primary sources, particularly of newer ideas, and of illustrations.

Acknowledgments

The order of authorship, much to the disgust of one of us, was determined alphabetically. We are indebted, in a very fundamental way, to the many scientists whose discoveries have provided the knowledge in which this book is rooted. Because of the introductory character of the book, only a few of them are mentioned by name. We thankfully acknowledge Mss. Candy Miller, Lorraine Barr, and Elizabeth Toftner, who congenially toiled through the tedium of manuscript typing, proofreading, and indexing.

Francisco J. Ayala
James W. Valentine

1

The Evolution of Evolutionary Science

The millions of diverse living species we find around us in the modern world are all descended from a common ancestor that lived in the remote past (Fig. 1.1). The processes that have brought this diversity about are collectively called evolution. During the nineteenth century it was common to speak of the "theory" of evolution in the sense that there was some likelihood that it was incorrect and that evolutionary descent had not occurred. Today the proposition that evolution has occurred is no longer in doubt; the evidence in favor of evolution is overwhelming. Actually, evolution is not a simple theory at all, but rather is an entire field of study, a branch of science. Within this branch of science there are numerous theories that have been scientifically corroborated, although much still remains to be learned.

Science and Evolution

In science, as in other domains, the solution to one problem generally reveals many new problems; an increase in knowledge may settle old controversies, but it also raises new questions that we were formerly too ignorant to pose. Our knowledge of evolution is rapidly increasing at present, and this new knowledge brings with it many unanswered questions. It is thus a time of great excitement for evolutionists, for they can see real growth in their ability to explain evolutionary processes, while at the same time they find the new questions extremely challenging. It is our hope to convey a feeling for this challenge and excitement. We will describe our present understanding of evolution, indicate the major scientific challenges that we face, and suggest the directions in which scientific speculations are leading in the search for answers.

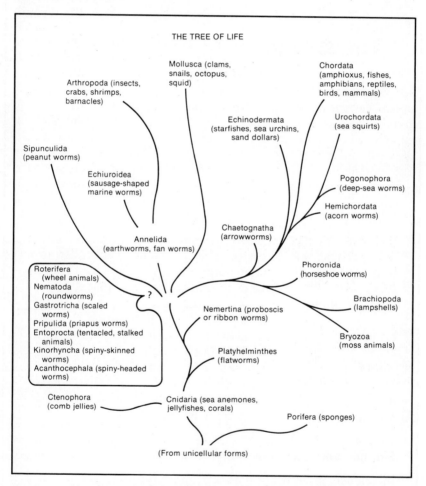

THE TREE OF LIFE

Mollusca (clams, snails, octopus, squid)

Arthropoda (insects, crabs, shrimps, barnacles)

Chordata (amphioxus, fishes, amphibians, reptiles, birds, mammals)

Echinodermata (starfishes, sea urchins, sand dollars)

Urochordata (sea squirts)

Sipunculida (peanut worms)

Echiuroidea (sausage-shaped marine worms)

Pogonophora (deep-sea worms)

Hemichordata (acorn worms)

Chaetognatha (arrowworms)

Annelida (earthworms, fan worms)

Phoronida (horseshoe worms)

Roterifera (wheel animals)
Nematoda (roundworms)
Gastrotricha (scaled worms)
Pripulida (priapus worms)
Entoprocta (tentacled, stalked animals)
Kinorhyncha (spiny-skinned worms)
Acanthocephala (spiny-headed worms)

?

Nemertina (proboscis or ribbon worms)

Brachiopoda (lampshells)

Bryozoa (moss animals)

Platyhelminthes (flatworms)

Ctenophora (comb jellies)

Cnidaria (sea anemones, jellyfishes, corals)

Porifera (sponges)

(From unicellular forms)

Fig. 1.1. Major types (phyla) of animals, showing the inferred patterns of relationship. Some phyla (such as Arthropoda) are represented today by literally millions of species, while others (such as Phoronida) contain several species only.

To call a speculation scientific may seem contradictory, but in fact speculation is an indispensable part of scientific progress. The scientific method is widely misunderstood, and indeed it is not a subject that has been settled to the satisfaction of philosophers and historians, nor of scientists themselves. However, certain attributes of scientific activity can be identified as of obvious importance; these include observation, interpretation, and testing of the interpretation.

Most research intends to answer or illuminate questions based upon previous observations. Offspring resemble their parents: Why? Dinosaurs were the dominant terrestrial vertebrates for millions of years: Why?

The search for answers can be divided into two parts. The first part is the framing of a possible explanation to create a scientific hypothesis that might account for the observations. The second is the performance of a test to see if the hypothesis stands up. Sometimes scientists grope for long periods of time before arriving at a plausible hypothesis. Often it is difficult for them to explain where their ideas have come from. To propose a new hypothesis is a creative act that often requires considerable ingenuity and originality. During the twentieth century Einstein exhibited such creativity to an astonishing degree; Darwin was similarly astonishing during the nineteenth.

Only hypotheses that expand our knowledge beyond the observed facts have any scientific interest. Any such expansive hypothesis explains data beyond the phenomena we have observed— it predicts facts not yet collected. This provides opportunity to test the hypothesis, for if we proceed to collect the appropriate facts and they are not as predicted, then the hypothesis is *falsified.*

Science is easily separated from nonscientific areas of knowledge, such as logic or theology, by this process of testing by attempted falsification. This is the key attribute of the scientific method and probably the attribute that is least understood by nonscientists. As pointed out by philosopher Karl Popper, a scientific hypothesis can never be proved, no matter how much it is tested; at best, it has not been falsified. When predictions of a hypothesis are verified by observation, so that the hypothesis is not falsified, it is still not *proved,* because there may be many other hypotheses compatible with the observations; we can never be sure that we have thought of everything.

When facts agree with a hypothesis, we feel that the hypothesis is supported because it has passed a test and survived a potential falsification. A hypothesis that has passed many attempts to falsify it is said to be *corroborated.* Such a hypothesis is a most robust and useful one and will ordinarily be adopted. Nevertheless it is only the most likely explanation available and does not necessarily represent the whole truth.

Our scientific understanding of nature is founded on bodies of well-corroborated hypotheses that have more explanatory power than other suggested hypotheses. Together, these bodies of hypotheses form a sort of superhypothesis or model of the way nature works. Such a model is called a *paradigm*. Any given scientific discipline, such as the study of evolution, has a paradigm that is generally accepted by most scientists. As Thomas Kuhn has pointed out, this is the paradigm that is presented in college classrooms and in textbooks, the one that most scientists have learned and the one that forms the basis of their thinking and underlies their research. Scientists usually seek to extend the paradigmatic explanations into new areas, to explain new data, and to resolve observations that do not seem, at least at first, to jibe with the accepted paradigm.

Scientific research does not automatically lead to immutable knowledge. Rather it leads to a body of explanations capable of being attacked, but accepted because the explanations have withstood all attempts at falsification. When scientists say a hypothesis is "proved," they mean that it has proved to be robust by withstanding many attempts to falsify it. But we cannot be certain that it will always be so.

Creationism

When evolution was first suggested in order to explain many puzzling facts of the world of life, it was opposed by many people because it contradicted the religious teachings in which they believed. At first the scientific paradigm of evolution was much less extensive than today. Corroboration of many of the assumptions on which it was founded was lacking, so that even some scientists were unsure that the idea would stand up.

With the passage of time, however, the evolutionary paradigm has grown and solidified until at present it appears to be unshakeable. Indeed, many people who are deeply religious find no difficulty in reconciling their beliefs with scientific findings on evolution.

A major perceived difficulty is the doctrine of creation, which holds that the world was created by a deity and did not evolve as a result of the inherent properties of matter. Creationism and allied religious beliefs are impossible to cast into a scientific mold. There can be no scientific examination of creationism because the hypothesis cannot be falsified, even in principle. A deity with the power to create the earth and its life, or the entire universe, is essentially an omnipotent entity beyond the capacity of the scientific method to discover. Any results of divine activity are clearly beyond the scientific method to detect if the divinity wished to conceal them. Lack of evidence to demonstrate divine creation can always be interpreted as owing to the will of the creator. Since the hypothesis of creationism is not falsifiable, it lies entirely beyond the purview of science.

Lamarckism

The earliest ideas of organic evolution are attributable to classical Greek thinkers. Later philosophers and even theologians also mentioned the idea. In the eighteenth and early nineteenth centuries a number of naturalists supported the concept of evolution, frequently in order to explain the close resemblances among many groups of species, such as catlike or deerlike animals. The common resemblances could be interpreted as due to a common ancestry.

The great French naturalist Jean Baptiste Lamarck (1744–1829) was the most famous evolutionist of his day (Fig. 1.2). He argued that the patterns of resemblance of organisms arise

Fig. 1.2. Jean Baptiste Lamarck, 1744–1829. A great naturalist, Lamarck proposed (incorrectly as it turned out) that variations arising from the use or disuse of organs could be inherited, leading to improved adaptation and to evolution over long periods of time.

through evolutionary modifications—animals that resemble each other most closely, such as lions and tigers, have close common ancestors. In common with many other naturalists of his time, Lamarck recognized that animals were adapted by natural characteristics to certain modes of life. However, he believed that over a period of time environmental change could directly call forth adaptive responses in animals, such as changes in their forms, which became inheritable.

Many of the facts and observations that suggested evolution to Lamarck also impressed other eighteenth and nineteenth century naturalists. But the Lamarckian explanation of the evolutionary process was too general to be convincing. It was all too easy to assert that environmental change created evolutionary change, but what exactly was the process, and how could this proposition be tested? It was Charles Darwin who would provide the answers.

Darwinism

Charles Darwin (1809–1882) is justly celebrated as being responsible for our basic understanding of the evolutionary process. His explanation was more subtle than Lamarck's. He observed that while inheritance is obviously conservative, in that members of the same family resemble one another, it is not perfectly so; the offspring of the same parents are not identical. Moreover, some of the variation observed among offspring is itself inheritable. Breeding of dogs, horses, and other domestic animals demonstrated that. Breeders eliminate undesirable traits and enhance desirable qualities in animal populations by selecting for breeding only those individuals with the desired characteristics.

Darwin reasoned that in nature individuals with qualities permitting them to be better adjusted to their environments, to be more fecund, or otherwise to be superior to other individuals of their kind, would tend to leave more offspring. Thus these qualities, which are frequently inheritable, will be proportionately increased in the succeeding generations.

Furthermore, as Thomas Malthus had pointed out, generally more individuals are born than survive to breed. Therefore there is a constant winnowing of populations, and those that survive tend to be those better adapted to environmental conditions. Nature is in fact breeding populations in such a way that better adaptation to the prevailing environment is a common result. This was called *natural selection* by Darwin.

Darwin was not the only person to conceive of this process; A. R. Wallace (1823–1913) developed the idea independently some time after Darwin had thought of it and published it jointly with Darwin in a short paper read (in the absence of both men) to the Linnaean Society of London in 1858. Other workers had also suggested the process of natural selection, or had come very close to doing so, in earlier publications. But Darwin was the only one to develop a comprehensive paradigm of evolution based on natural selection, beginning with his famous book *On the Origin of Species by means of Natural Selection,* which appeared in 1859.

Darwin had a wide knowledge of animal and plant distributions across the world, gained primarily as a naturalist on HMS *Beagle,* which circumnavigated the globe in the years 1831–1836. From this experience, and from a long program of reading and discussions within the small and relatively close-knit scientific community of his time, Darwin was able to weave together an interlocking set of hypotheses explaining resemblances among organisms, their patterns of distribution, and their fossil records—a set of hypotheses making sense and providing coherence to a wide body of observation and experience that had accumulated by the mid-nineteenth century.

Ironically, the weakest part of Darwin's paradigm was the treatment of heredity, although this was a critical difference between Darwin's ideas and Lamarck's. The basic rules of heredity were discovered by Gregor Mendel during Darwin's lifetime. They remained unknown to Darwin, however, and indeed Mendel's work was not known to science generally until the turn of the century. Darwin's theory depended chiefly upon two processes, first hereditary variation, and then preservation and increase of favorable variations. Lamarckism required that organisms respond directly through some unexplained internal process to environmental change. Hereditable variation is an observed fact, while the internal forces implied by Larmarckism are unknown. Lamarckian doctrine is thus more a description of adaptation than an explanation of it. Nevertheless, Darwin himself thought that Lamarckian responses to environmental changes might play a role in evolution.

The rationality, scope, and coherence of Darwin's paradigm, and its great explanatory power, eventually converted scientists to accept the theory of evolution by natural selection. The evolutionary paradigm of Lamarck, never developed to the extent of Darwin's and never widely accepted, nevertheless commanded some serious support well into the twentieth century. With the development of modern genetics and molecular evolution, however, Lamarckism has been definitively falsified, while the notion of natural selection has retained its explanatory ability.

CLOSER LOOK 1.1 Charles Darwin

Charles Robert Darwin was born in Shrewsbury, England on February 12, 1809 (coincidentally on the same day that Abraham Lincoln was born). He was the son of a successful physician and the grandson of another physician and noted natural historian, Erasmus Darwin (1731–1802), who had done some work on the notion of evolution. Charles Darwin grew up in Shrewsbury, where he attended a rather old-fashioned private school, and eventually enrolled as a medical student at the University of Edinburgh. He soon found that medicine was not to his liking, and after many deliberations it was decided that he would become a clergyman. He accordingly obtained a place in Christ's College, Cambridge, and graduated in 1831. From his letters and papers, and from the accounts of his friends, Darwin much enjoyed the life at Cambridge, especially the riding and hunting; he was not an exceptional scholar. Nevertheless, he did have an enthusiasm for natural history that had been noticeable at Edinburgh, and while at

Charles Darwin in 1840. By this time he had worked out the main points of his paradigm of evolution. He is 31.

Cambridge he pursued interests in geology and biology. Indeed the first published words that can be attributed to Darwin are collecting records of beetles from his Cambridge days.

Darwin's love for general natural history led him into friendship with the Professor of Botany at Cambridge, John Stevens Henslow. Shortly after Darwin graduated, but before he had decided on a particular future employment, Henslow recommended him for a post as an unpaid naturalist to accompany HMS *Beagle,* bound for a circumnavigation of the world with particular attention to surveying the coasts of South America. That this post was eventually awarded to Darwin now seems like a long chance. Others were asked before him, but refused; his father strongly opposed the idea, but was eventually persuaded to approve; and when Darwin finally wrote to accept (after refusing initially) he had a competitor. The deciding factor seems to have been that he was a charming and likeable person and impressed the captain of the *Beagle,* Robert FitzRoy. The unpaid naturalist was to be a companion to FitzRoy, otherwise doomed to isolation in lonely splendor from his crew during the years of voyage. To win the approval of the captain was thus an essential test. The experience of the voyage was to bring Darwin's attention to the problem of the origins of the marvelous adaptations that he observed in animals and plants across the world.

HMS *Beagle* left Devonport, England on December 27, 1831 and sailed to the Cape Verde Islands, where Darwin did

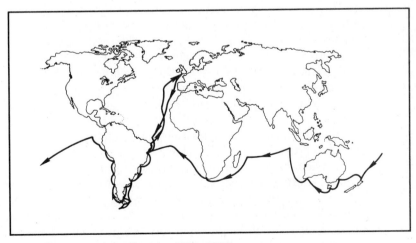

The route of the *Beagle,* 1831–1836.

his first collecting. The next important stop was Brazil, and for the next three-and-a-half years the ship would sail the coasts of South America while FitzRoy accomplished the surveying. During this period, Darwin was frequently able to disembark for extended trips ashore, visiting such diverse environments as the tropical jungles of Brazil, the grasslands of Argentina, the desolation of Tierra del Fuego, and the heights of the Andes Mountains. Finally the ship called at the Galápagos Islands, where Darwin collected the famous birds now known as Darwin's finches, together with many other native organisms. The *Beagle* then completed her voyage, calling at a few other places before sailing into Falmouth, England in October, 1836.

Darwin's experience of a rich sample of the world's organisms and environments stood him in good stead. He saw that species were closely adapted to their environments and also understood the lessons of geology that environments change more or less constantly. It therefore seemed to follow that species must change as well. During the two years or so after the voyage he pieced together a complex of evidence and hypotheses that formed a theory of evolution. It was in September, 1838 that he read Malthus on populations and realized that natural selection, which he already considered to have a role in evolution, must act importantly in populations where more young are born than can survive to breed. Darwin already believed that man was as much a product of evolution as other species. Nevertheless he did not publish any of these views. He married in 1839 and in the 1840s became ill, moving from London to a rural setting in Kent where the Darwins became a large family. Throughout these years he contributed a steady stream of papers and books to science and became increasingly well known and admired, chiefly because of the results of the *Beagle* voyage. He wrote essays on evolution in 1842 and 1844, the latter essentially a book and containing much that appeared later in *The Origin of Species*. Still he did not publish this material, plunging instead into a study of barnacles that lasted eight years.

Publication of his evolutionary views was finally forced on him when another naturalist, Alfred Russell Wallace, sent Darwin a short paper that set forth the same views, including an account of natural selection, developed entirely independently. The two men jointly published a short paper in 1858, and then the following year Darwin published *The Origin of Species,* much reduced from a projected great book on evolution. The book has been continuously in print ever since; the last printing to include changes by Darwin himself was the 1876 issue of the sixth edition. Twelve years after the appearance of *The Origin of Species,* Darwin published *The Descent of Man,* which included a solid mass of additional work on human evolution. Despite continued ill health Darwin published copiously, working until his death on April 19, 1882. In all he published sixteen books, contributed to nine more, and had 116 publications in serials of various kinds, of which about thirty were important contributions. He is buried in Westminster Abbey.

Darwin's scientific methods have often been misrepresented, especially by humanist biographers, but within the last twenty years a much clearer appreciation of his approach has emerged from studies of his voluminous notebooks, letters, and other papers. He created not a simple new hypothesis, but as Howard Gruber has put it, a whole new point of view, what we call here a whole paradigm of evolution. One cannot do this by gathering myriad facts at random, or by simply having a brilliant idea. One must develop a remarkable judgment as to what important facts are and must work laboriously to understand and explain their significance, seeking continuously for new insights, correcting when possible old mistakes, enlarging the web of observation and inference, searching for harmony and mutual support—for generality—among the confusion of data.

This at least is much of the story of Darwin's work; he was convinced early, by observation and logic, of the fact of evolution, and he proceeded to test the predictions and ramifications of this conviction. The ramifications are still being pursued by thousands of scientists, and important advances in understanding evolution still occur regularly. That Darwin could have taken the paradigm as far as he did certainly required all of the courage, patience, intelligence, and great good sense that are his traditional attributes.

Mutationism

The difficulties associated with heredity continued to plague the Darwinian paradigm after the rediscovery of the laws of inheritance in 1900. When it became understood that inheritance was particulate, and that hereditary novelty arose spontaneously through the sudden alterations of hereditary particles (*genes*) known as *mutations,* still another evolutionary paradigm was proposed. This new paradigm held that mutation, the only known source of genetic novelty, was the driving force of evolution. Natural selection was unimportant, perhaps operating to weed out of a population unsuitable mutations. According to this paradigm, evolutionary advance is made when a valuable mutation appears, replacing its less valuable predecessor. This mutation hypothesis of evolution sucessfully competed with Lamarckian and Darwinian ideas in the early decades of this century. The leading mutationists included H. de Vries, W. Bateson, and T. H. Morgan, all outstanding geneticists and scientists.

Arguments that led to the downfall of mutationism were developed in the late 1920s and early 1930s. They were essentially mathematical arguments based on the fact, so obvious today, that evolution within populations of organisms proceeds by the spread of favorable characters, and thus of the favored genes that underlie them. Even if a new mutation appears that is immediately favored, its spread in the population depends upon the size of the population, the number of generations, the rate at which the same mutation reappears, the degree to which the mutation favors its possessors, and other factors. The interplay of such factors determines the rate and direction of change in the frequency of genes. Evolution, it seemed, could be redefined as a change in gene frequency, since such changes must underlie the changes in form and habit that we see in organisms. The genes need not be newly arisen mutations, but so long as their frequencies increase or decrease, often owing to environmental changes, then we can say that evolution is occurring.

Furthermore, it became apparent that since individuals in a sexually reproducing population are not all genetically identical, the total number of genes available for inheritance by the next generation is very much larger than the number of genes carried by any single individual—all the genes carried by *every* individual that can reproduce are available. All the genes in a population at any one time form a reservoir of all the genetic variability of the population—the *gene pool.* It is the shifting frequencies of genes in this pool that constitute evolution at the most basic level.

The theory of how gene frequencies change was worked out principally by the Russian S. S. Chetverikoff, the American Sewall Wright, and the Britons R. A. Fisher and J. B. S. Haldane. Much of their work was completed before the mid-1930s. Yet as their work unfolded, their fellow evolutionists had only a limited perception of the significance of the findings. The weight of authority of the master geneticists of previous decades, and of their disciples, prevented any immediate, radical change in the mutationist paragigm. Textbooks of the mid-1930s continued to explain evolutionary advance as flowing essentially from the chance occurrence of novel mutations. Many authorities thought that Darwinism had been falsified.

The Synthetic Theory

A revolutionary change in attitude towards Darwinism can be traced to the publication in 1937 of *Genetics and the Origin of Species,* by Theodosius Dobzhansky (Fig. 1.3). Dobzhansky incorporated the work of Chetverikoff, Wright, Fisher, and Haldane with a wealth of experimental evidence to show that natural selection is the single most effective agent of evolutionary change, just as Darwin had proposed. In effect, Dobzhansky showed that Darwinism and the findings of genetics were compatible. New species usually arise through the accumulation of different genes within

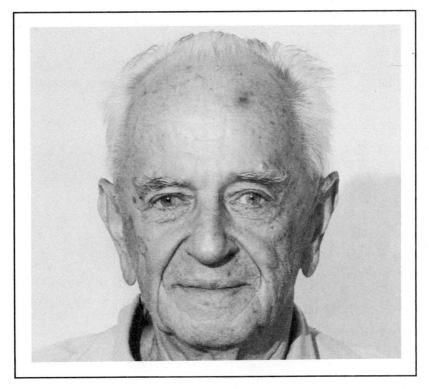

Fig. 1.3. Theodosius Dobzhansky, 1900–1975. A prolific experimentalist who made many contributions to microevolution, he displayed genius as a synthesizer, combining experimental and theoretical results to create the foundations of the synthetic paradigm of evolution.

reproductively isolated populations of some parent species. These populations become so different that they cannot breed back to the parental population and thus can be recognized as distinct species.

Following Dobzhansky's book a number of major contributors, among them George Simpson, Ernst Mayr, and Ledyard Stebbins (Fig. 1.4), established that this revitalized Darwinian model explains a wide range of morphological, paleontological, biogeographic, and ecological facts about plants and animals alike. The Darwinian paradigm became infused with the discoveries of genetics and of other branches of biology, emerging with vastly expanded information content and explanatory powers. This new paradigm is often called the *synthetic theory* of evolution because it resulted from a synthesis of the findings of several scientific disciplines. It is this paradigm that has formed the basis of the textbook accounts of evolution during the last three decades.

Present Trends in Evolutionary Research

Despite the success of the synthetic paradigm, many problems and issues in evolution remained unsolved. As work continued on these matters, the science of genetics underwent a revolution of its own, beginning with the discovery in 1953 by James Watson and Francis

Fig. 1.4. Three of the principal architects of the synthetic paradigm of evolution. *Left,* Ernst Mayr (1904–   ), a zoologist who approached the paradigm from the standpoint of taxonomy and speciation. *Center,* George Gaylord Simpson (1902–   ), a paleontologist who studied macroevolutionary patterns and evolutionary rates. *Right,* G. Ledyard Stebbins (1906–   ), a botanist who investigated evolutionary processes in plants, especially their genetic and reproductive patterns.

Crick of the structure of DNA (deoxyribonucleic acid), the genetic material. The genes themselves—segments of DNA molecules—are codes that spell out the composition of certain organic compounds (polypeptides) that form enzymes and other substances determining the development of an organism from a fertilized egg. Since then, understanding of the nature and control of gene activity has burgeoned, to the point that we now have a new paradigm of genetics, a molecular genetic paradigm. This new knowledge has not immediately changed the synthetic paradigm of evolution in any fundamental way, though it has certainly enriched it with new understanding. What it has done is to provide evolutionists with many new ways of studying evolution, on the level of the molecules that participate directly in the evolutionary process—the genes and their primary products. Using new molecular techniques of study, investigators are now attacking major unresolved problems and issues in the synthetic paradigm. For example, one important question concerns the amount and kind of genetic differences that accompany the evolution of new species from old. Another question concerns the significance of the regulation of gene activity, through special regulatory genes, in the evolution of new types of organisms.

A second scientific revolution has recently occurred in the earth sciences, one that is as fundamental in its way as is the molecular revolution in the biological sciences. This revolution is based on the demonstration that the major structural features on the face of the earth, such as the continents and ocean basins, are dynamic features. Oceans grow and shrink, while continents break up into fragments or coalesce into large land masses and move across the earth in ever-changing geographic patterns. Indeed such motions are occurring today (Fig. 1.5). They are slow by some standards, amounting to several centimeters per year, but over millions of years of geological history they have profoundly altered the configuration of the land and the seas, of mountains and deserts, causing major changes in the earth's environments.

Since natural selection adapts organisms to changed environmental conditions, the history of these environmental changes should be reflected in the evolutionary history of life. Thus, this revolution in the earth sciences has opened the way for explanations of many features of evolutionary history that have puzzled students of the fossil record since the beginnings of evolutionary science.

A number of important issues in environmental science relating to the stability and endurance of natural communities are of much interest owing to the pervasive environmental effects of human activity. Important evidence relating to these questions is

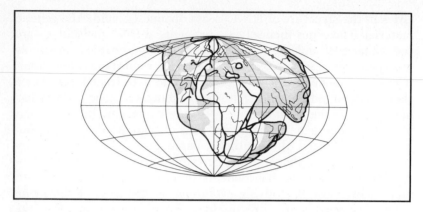

Fig. 1.5. Map of the earth as it was thought to have appeared about 300 million years ago according to Alfred Wegener (1880–1930). Wegener was a pioneer of the notion of continental drift, first publishing his ideas in 1915. His interpretation of continental positions and of the timing of drift were wrong in detail, but his general concept, that continents have undergone large-scale horizontal movements over the earth's crust, has proven to be correct. The supercontinent depicted here was actually formed (in slightly different configuration) about 225 million years ago.

now forthcoming from the study of evolutionary ecology and of evolutionary patterns and trends in the fossil record.

New discoveries and insights continually arise in the fields that bear upon evolutionary studies, and they continually enrich the evolutionary paradigm. From all these sources it is now possible to present a general account of the history of life, from its origin to the emergence of mankind and of human cultural patterns. Furthermore we are able to begin to address social problems to which evolutionary studies are relevant. We are developing a better understanding of why certain human attributes—intelligence, human sexuality, racial diversity—have evolved, and we are opening studies on new areas of human concern, such as the extent to which our activities are genetically conditioned, and our present and potential ability to correct genetic defects in children.

QUESTIONS FOR DISCUSSION

1. In what ways are scientific hypotheses research tools that lead to new understanding of the properties of matter and of the flux of history?

2. Why do creation hypotheses not form research tools that lead to new understanding?

3. What are the chief distinctions between the evolutionary ideas of Lamarck and Darwin?

4. Define natural selection.

5. Why are there so many kinds of animals? Why aren't there fewer—or many more?

Recommended for Additional Reading

Darwin, C. R. 1839. *Journal of researches into the geology and natural history of the various countries visited during the voyage of H.M.S. "Beagle," etc.* London: Henry Colburn (second issue).

The voyage of the *Beagle,* a still readable account of the scientifically most famous circumnavigation of the world, available in cheap editions.

Darwin, C. R. 1859. *The origin of species by means of natural selection, or the preservation of favoured races in the struggle for life.* London: John Murray.

Perhaps the single greatest scientific work ever written, this account of evolutionary processes and their results is still widely available in cheap editions, and still makes interesting reading.

Dobzhansky, T. 1937. *Genetics and the origin of species.* New York: Columbia University Press.

The classic that synthesized the findings of genetics with natural selection; there are later editions. A more up-to-date treatment of evolutionary genetics by the same author is cited in a later chapter. Advanced.

Mayr, E. 1942. *Systematics and the origin of species.* New York: Columbia University Press.

Simpson, G. G. 1944. *Tempo and mode in evolution.* New York: Columbia University Press.

Stebbins, G. L. 1950. *Variation and evolution in plants.* New York: Columbia University Press.

These three works established the power of the synthetic paradigm of evolution. They employ a neo-Darwinian approach creatively to form explanations for the facts of animal systematics, paleontology, and botany, respectively. All these authors have now written more up-to-date treatments, which are cited in later chapters.

2

Basic Genetics

When people speak of biological evolution, they have in mind the transformation of some kinds of organisms into other kinds by gradual change through long periods of time, such as the evolution from fish to mammal or from ape to human. Underpinning the evolutionary process are hereditary changes. Evolution occurs because the genetic constitution of organisms changes from one generation to another. Nonhereditary changes, such as losing a leg or developing strong muscles, have no evolutionary significance because they disappear with the individual. Evolution is a cumulative process that is made up of hereditary changes passed from generation to generation. In order to understand evolution, we must therefore understand biological heredity—the process by which the characteristics of individuals are passed on to their progenies. The study of biological heredity is *genetics;* in this chapter we introduce some fundamental concepts of genetics.

Mendel and the Birth of Genetics

The process of biological heredity was not well understood by Darwin or his contemporaries. Several theories had been advanced by various people, one theory by Darwin himself; but these theories lacked confirmation and later were proved to be wrong. Darwin knew that the weakest point in his paradigm of the evolutionary process was lack of a proper understanding of biological heredity. Yet, a few years after the publication of *The Origin of Species* in 1859, but unknown to Darwin, Gregor Mendel (1822–1884) had discovered the fundamental principles of heredity (Fig. 2.1).

Mendel was a monk in an Augustinian monastery in Brünn, Austria (now Brno, Czechoslovakia), of which he would later become abbot. For many years, he used garden peas (*Pisum sativum*) to investigate how individual traits are inherited. His experiments are masterful examples of scientific research, even by today's standards. He published his results in 1866, but they appeared in the *Proceedings* of the Natural History Society of Brünn, an obscure publication that received little attention from scientists. Mendel's discoveries lay generally unknown until 1900 when three scientists obtained results similar to Mendel's. The three investigators had been working independently, Hugo de Vries in Holland, Carl Correns in Germany, and Erich von Tschermak in Austria. Genetics has developed continuously, and at times explosively, since the rediscovery of Mendel's work in 1900.

Fig. 2.1. Gregor Mendel (1822–1884), the founder of genetics.

There is perhaps no better introduction to the fundamentals of genetics than a description of Mendel's experiments and conclusions. Many scientists before Mendel had attempted to elucidate how biological characteristics are inherited. Mendel succeeded where they had failed owing to his brilliant methodology. Previous investigators had crossed plants or animals and had looked at the overall similarities between the offspring and their parents. But the results were confusing: the offspring resembled one parent in some traits, the other parent in other traits, and neither one in still other characteristics; no precise regularities could be discovered. Mendel saw the need to focus on a single trait at a time, the shape of the seeds for example, rather than on the whole plant. Before starting crosses between plants, he also made sure that they were true breeding rather than hybrids; he obtained many pea varieties from seedsmen and bred them for two years in order to select for his experiments only those whose progenies all resembled the parents in a given trait. Mendel studied the seven characters shown in Figure 2.2. For each character he had plants that differed in clear-cut ways.

Dominance and Segregation

In one experiment Mendel crossed plants producing round seeds with plants producing wrinkled seeds. The results were clear-cut: all the hybrid plants of the first generation (F_1) produced round seeds, independently of whether the round-seed plant had been the female parent or the male parent. The wrinkling seemed to be suppressed under the dominance of the roundness. Additional experiments showed that all seven characters behaved in this way; in each case only one of the two contrasting traits appeared in the F_1 hybrids. Mendel called such traits (round seeds, yellow peas, axial flowers, and so on) *dominant,* and their alternatives (wrinkled seeds, green peas, terminal flowers) he called *recessive.*

Scientists later found that dominance of one trait over another is a common phenomenon but not a universal one. In some cases the F_1 hybrids are intermediate between the two parents. In other cases the traits of both parents are expressed. In the snapdragon, for example, a plant with crimson flowers crossed to a plant with white flowers produces F_1 hybrids all with pink flowers. In humans, normal pigmentation of the skin is dominant over albinism; but the characteristics of blood group A and blood group B are both equally expressed in individuals who have inherited the two, one from each parent.

Mendel then planted seeds from the hybrids and allowed the plants to fertilize themselves. Round and wrinkled seeds appeared

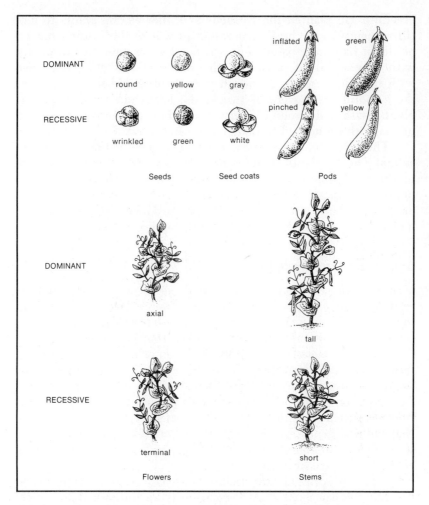

Fig. 2.2. The seven traits studied by Mendel in the garden pea, *Pisum sativum*. He had plants that differed in clear-cut ways with respect to a given trait.

side by side in the same pod in the second generation (F_2) of the cross between round-seed and wrinkled-seed plants. He counted the seeds; 5,474 were round and 1,850 were wrinkled. The ratio was very close to 3:1 (in fact, 2.96:1), and a similar ratio occurred in all the other crosses. For example, there were 6,022 yellow peas and 2,001 green peas (a ratio of 3.01:1) in the F_2 progeny of the cross yellow x green, although the F_1 peas were all yellow. In every case, the dominant trait was about three times as common as the recessive trait.

Mendel was now ready to investigate whether the round-seed plants and the wrinkled-seed plants of the F_2 generation were true breeding. He planted the F_2 seeds and allowed the plants to fertilize themselves. The F_3 generation produced the results that he no doubt had anticipated. The wrinkled seeds gave rise to plants with only wrinkled peas. But the situation was quite different with the round seeds. Although indistinguishable in appearance, these round seeds were of two kinds: about one-third produced plants with only round seeds; the other two-thirds produced plants with round and wrinkled seeds in the ratio 3:1.

These results were repeated for every one of the seven characters. The F_2 plants showing the recessive trait bred true; they produced F_3 plants identical to their parents. However, the F_2 plants showing the dominant trait were of two kinds, one-third bred true, while the other two-thirds produced F_3 progenies in which the dominant and recessive traits appeared in the ratio 3:1.

Genes, the Carriers of Heredity

Mendel advanced the following hypothesis in order to explain the results of his experiments with peas. Contrasting traits, such as the roundness or wrinkling of peas, are determined by "factors" (now called *genes*) that are transmitted from parents to offspring through the *gametes,* or sex cells; each factor exists in alternative forms (now called *alleles*) responsible for one or the other form that the character may take. For each character every organism has two genes, one inherited from the male parent, the other inherited from the female parent. Thus, pea plants have two genes for seed shape; these genes may exist in the form that determines round seeds (allele for roundness), or in the form that determines wrinkled seeds (allele for wrinkling).

It is appropriate at this point to introduce two other terms of the genetics vocabulary. A *homozygote* is an individual in which the two genes for a given character are identical, i.e., an individual with two identical alleles. A *heterozygote* is an individual in which the two genes for a given character are different, i.e., an individual with two different alleles. Thus, the true breeding round-seed plants are homozygous for roundness, the true breeding wrinkled-seed plants are homozygous for wrinkledness, and the F_1 hybrids from the cross round x wrinkled are heterozygous for roundness and wrinkling.

From the phenomenon of dominance, Mendel inferred that in heterozygous individuals, one gene (allele) is dominant, the other recessive. From the segregation in the progeny of hybrids (heterozygotes), Mendel concluded that the two genes for each trait do not fuse or blend in any way, but rather remain distinct throughout the life of the individual and segregate in the formation of gametes so that half the gametes carry one gene and the other half carry the other gene. For example, the F_1 hybrid plants from the round x wrinkled cross produce gametes of two kinds in equal amounts. One kind carries the allele for roundness, the other kind carries the allele for wrinkling. This hypothesis accounts for the results observed in the F_2 and following generations, as shown in Figure 2.3.

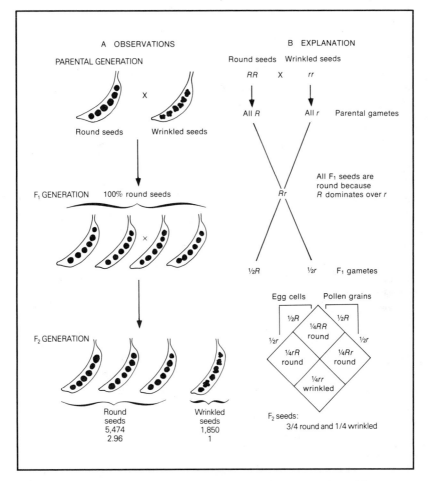

Fig. 2.3.   *A.* Mendel crossed pea plants having round seeds with plants having wrinkled seeds. In the first generation (F_1) all seeds are round. When F_1 plants are self-crossed or crossed to each other, the second generation (F_2) seeds are, approximately, three-quarters round (5,474 seeds) and one-quarter wrinkled (1,850 seeds). *B.* Explanation proposed by Mendel. *R* and *r* represent the factors *(genes)* determining roundness and wrinkledness, respectively.

It is customary to represent pairs of genes with letters, dominant alleles with capital letters, recessives with lowercase letters. For example, the allele for roundness may be represented as R and the allele for wrinkling as r. The genetic makeup of Mendel's plants is, then, as follows. The true breeding round plants are RR, the true breeding wrinkled plants are rr. The F_1 hybrids are Rr, having received the R allele from one parent and the r allele from the other parent. These plants produce only round seeds because the R allele is dominant and the r allele recessive. The Rr plants produce two kinds of gametes, R and r, in equal amounts. When Rr plants fertilize themselves (or are crossed with other Rr plants), they produce three kinds of progeny:

(1) One-quarter are RR, namely the true breeding round-seed plants. These plants result from the union of an R male gamete (the frequency of R among male gametes is ½) and an R female gamete (the frequency of R among female gametes is also ½, and the frequency of RR plants is, therefore, $½ \times ½ = ¼$).

(2) One-half are Rr, namely the round-seed plants that in the F_3 generation produce both round and wrinkled seeds. The Rr plants result either from the union of an R male gamete and an r female gamete (each one of these gametes occurs with a frequency of ½ and thus the frequency of these plants is $½ \times ½ = ¼$), or from the union of an r male gamete and an R female gamete (again each of these gametes occurs with a frequency of ½ and thus the frequency of these plants is $½ \times ½ = ¼$). In the F_3 generation, the Rr plants produce round and wrinkled peas in the ratio 3:1 because they are in fact identical to the F_1 hybrids (Rr) and thus will produce seeds of the same kinds and in the same proportions as the F_1 hybrids.

(3) One-quarter are rr, namely the wrinkled plants, which are true breeding because they are homozygous and thus produce only one kind of gamete. These rr plants result from the union of an r male gamete and an r female gamete; since the frequency of r is ½ among male gametes and also ½ among female gametes, the frequency of rr plants is $½ \times ½ = ¼$.

In summary, the main elements of Mendel's hypothesis, and those that have become the foundation of genetics, are: (a) each character is determined by two genes, one inherited from each parent; and (b) the two genes do not fuse or blend but segregate in the production of gametes so that half the gametes receive one gene and the other half receive the other gene. Something else besides this fundamental discovery made possible the development of genetics, namely Mendel's method of *genetic analysis*. Mendel made crosses between organisms differing in a given character; from the ratios observed in the progenies of crosses between the alternative forms of the trait, he inferred the existence of hereditary factors. This method of genetic analysis would thereafter be the way in which geneticists would ascertain the presence of genes since these could not be observed directly.

Mendel tested his hypothesis in various ways. One kind of test, much employed by geneticists, is called *backcross* or *testcross*. It consists of crossing F_1 hybrid individuals to the recessive parent (Fig. 2.4). If Mendel's hypothesis was correct, such a cross would produce offspring with half showing the recessive trait and half showing the dominant trait. Consider the F_1 hybrid from the cross round x wrinkled. The hybrid peas, although all round in appearance, give rise to plants (*Rr*) producing two kinds of gametes, *R* and *r*, in equal proportions. When these gametes join with *r* gametes coming from the wrinkled (*rr*) parent, there should be two kinds of peas, half round (*Rr*) and half wrinkled (*rr*). This is precisely what Mendel obtained. Studies with many different organisms have confirmed the validity of Mendel's hypothesis.

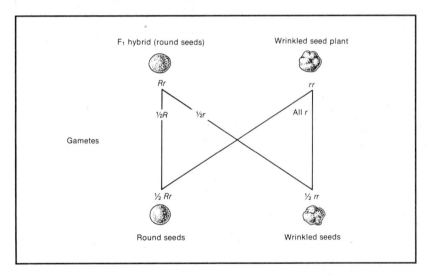

Fig. 2.4. Testcross or backcross. A hybrid F_1 plant is crossed with a wrinkled-seed plant. If Mendel's hypothesis is correct, the F_1 hybrid plant will be *Rr* and produce gametes half of which carry the *R* factor while the other half carry the *r* factor. When these gametes associate at random with the *r* gametes produced by the wrinkled-seed parent, half the progeny should be *Rr* (round seeds) and the other half should be *rr* (wrinkled seeds). In this as well as in other testcrosses, Mendel in fact observed that as predicted half the progeny exhibited the dominant trait and the other half exhibited the recessive trait.

Independent Assortment

Mendel's experiments described so far concern the inheritance of alternative expressions of a single character. To describe what happens when two characters are simultaneously considered, Mendel formulated the Principle of Independent Assortment, which says that genes for different characters are inherited independently of one another. The Principle of Independent Assortment is far from universally valid, but studying the exceptions led to major advances in genetics in the twentieth century.

Mendel derived the Principle of Independent Assortment from the results of crosses between plants having different traits with respect to two or three separate characters. One experiment involved crossing plants having round and yellow seeds with plants having wrinkled and green seeds. The F_1 generation offered no surprises. Mendel already knew that round dominates over wrinkled and that yellow dominates over green. As expected, all peas in the F_1 generation were round and yellow.

The interesting results came in the F_2 generation. Mendel considered two alternative possibilities: (1) that traits derived from one parent are transmitted together, and (2) that they are transmitted independently of each other. With characteristic lucidity, Mendel formulated the expectations from these alternative possibilities. If (1) is true, there would be only two kinds of seeds in the F_2 generation, round-yellow and wrinkled-green, and they should appear in the ratio 3:1 according to the Law of Segregation. However, if (2) is correct, there should be four kinds of seeds: round-yellow (two dominant traits), round-green (dominant-recessive), wrinkled-yellow (recessive-dominant), and wrinkled-green (two recessive traits), which should appear in the proportions 9:3:3:1 (Fig. 2.5).

Mendel grew plants from the F_1 round-yellow hybrid seeds and allowed them to self-fertilize. The peas in the F_2 generation were of four kinds: 315 round-yellow, 108 round-green, 101 wrinkled-yellow, and 32 wrinkled-green. This was close enough to the proportions 9:3:3:1 expected from the second alternative, and Mendel concluded that the factors determining different characters are inherited independently. (Note that the results of this experiment also confirm the Law of Segregation, since the proportions 9:3:3:1 are calculated on the assumption that the Law of Segregation is true. Moreover, it is easy to see that the expected 3:1 ratio was obtained for each character separately. In the F_2 generation there are 423 round versus 133 wrinkled seeds, and 416 yellow versus 140 green seeds.)

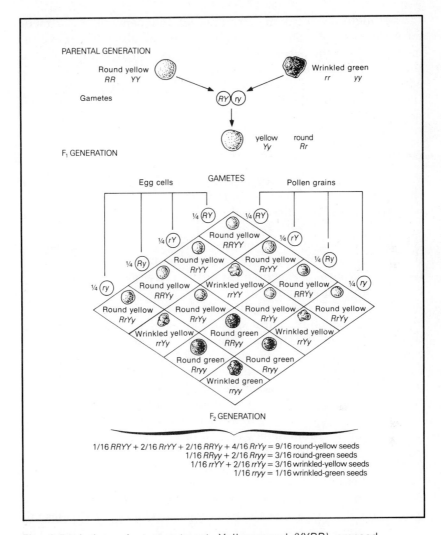

PARENTAL GENERATION

Round yellow
RR   *YY*

Wrinkled green
rr   *yy*

Gametes   RY   ry

yellow   round
Yy   *Rr*

F₁ GENERATION

GAMETES

Egg cells   Pollen grains

¼ RY   ¼ RY

¼ rY   Round yellow *RRYY*   ¼ rY

¼ Ry   Round yellow *RrYY*   Round yellow *RrYY*   ¼ Ry

¼ ry   Round yellow *RRYy*   Wrinkled yellow *rrYY*   Round yellow *RRYy*   ¼ ry

Round yellow *RrYy*   Round yellow *RrYy*   Round yellow *RrYy*   Round yellow *RrYy*

Wrinkled yellow *rrYy*   Round green *RRyy*   Wrinkled yellow *rrYy*

Round green *Rryy*   Round green *Rryy*

Wrinkled green *rryy*

F₂ GENERATION

1/16 *RRYY* + 2/16 *RrYY* + 2/16 *RRYy* + 4/16 *RrYy* = 9/16 round-yellow seeds
1/16 *RRyy* + 2/16 *Rryy* = 3/16 round-green seeds
1/16 *rrYY* + 2/16 *rrYy* = 3/16 wrinkled-yellow seeds
1/16 *rryy* = 1/16 wrinkled-green seeds

Fig. 2.5. Independent assortment. Yellow-round *(YYRR)* crossed with green-wrinkled *(yyrr)* produces F₁ hybrids which are all yellow-round *(YyRr)*. The F₁ hybrid plants produce four kinds of gametes, each with a frequency of one-quarter. Random association between female and male gametes of the four kinds produces F₂ seeds in sixteen possible combinations. (The frequency of the seeds in each square is one-sixteenth.) When the F₂ seeds are grouped according to their phenotypes, there are nine squares with round-yellow seeds, three squares with round-green seeds, three squares with wrinkled-yellow seeds, and one square with wrinkled-green seeds; thus, the four kinds of seeds are expected in the ratios 9:3:3:1. Mendel obtained 315:108:101:32 seeds of the respective kinds in good agreement with the expectations.

Mendel confirmed the Principle of Independent Assortment for various combinations of two characters and also with an experiment in which the parents differed simultaneously with respect to three characters—all three characters were inherited independently of one another. However, it would be seen early in the twentieth century that the Principle of Independent Assortment is not universally valid.

Genes and Chromosomes

While Mendel was crossing peas and Darwin was formulating the theory of evolution, other scientists were looking at cells. In the nucleus of cells they discovered certain bodies, called *chromosomes* ("colored bodies"), with remarkable properties. As the nineteenth century was coming to a close, scientists had discovered that all the cells of an organism have the same number of chromosomes, with one important exception—the sex cells (gametes) have only half as many chromosomes as the body cells.

It was discovered that body cells divide by a process called *mitosis,* during which the chromosomes in a cell are first doubled and then divided into the two daughter cells. This process explains the constant number of chromosomes in all body cells of an organism. However, scientists found that gametes are formed by a special process of cell division called *meiosis.* In meiosis each cell divides twice but the chromosomes are doubled only once; thus the resulting gametes only have half as many chromosomes as body cells. The number of chromosomes typical of an organism is restored at fertilization when two gametes (a male sex cell and a female sex cell) are joined to form the fertilized egg (*zygote*) (Fig. 2.6).

To illustrate this, we can represent the number of chromosomes in a gamete as N. A zygote will have $2N$ chromosomes, half obtained from each of the two gametes. The zygote divides mitotically, producing two cells, each with $2N$ chromosomes. The cells divide mitotically again and again during development, so that the organism consists of many cells each with $2N$ chromosomes. However, the gametes are produced by meiosis and thus carry only N chromosomes, half the number of chromosomes in the body cells. When two gametes unite at fertilization, the $2N$ number of chromosomes is restored again. Thus, the $2N$ number of chromosomes is maintained generation after generation. The number of chromosomes characteristic of an organism varies considerably among species; some have as few as two, others have several hundred chromosomes.

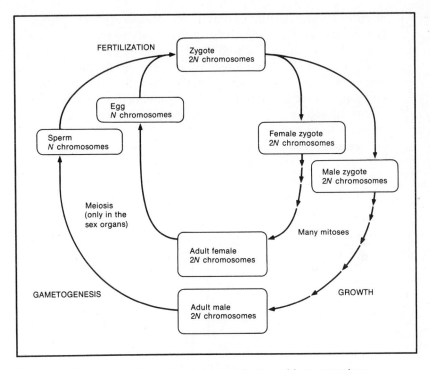

Fig. 2.6. The cycle of growth and reproduction. Most organisms develop from zygotes having two sets of chromosomes (2N), one inherited from each parent. The zygote divides into two cells, and each one of these into two others, and so on, by means of a process known as mitosis, which ensures that each cell in the body has the same two sets of chromosomes as the zygote. The sex cells are an exception; by means of a process, known as meiosis, two cell divisions take place with only one duplication of the chromosomes, thus sex cells have only one set of chromosomes (N). When fertilization takes place two sex cells, each with one set of chromosomes, unite and thus produce a zygote with 2N chromosomes.

CLOSER LOOK 2.1   Mitosis and Meiosis

The genetic information is contained in the chromosomes, which are threadlike bodies in the nucleus of each cell. Each chromosome consists of a centromere with two attached arms. All body cells have the same number of chromosomes; sex cells have half as many chromosomes as body cells. Mitosis (plural, mitoses) is the process of cell division by which one cell results in two daughter cells, each with the same number of chromosomes as the parental cell. Meiosis (plural, meioses) is a special kind of cell division that leads to the formation of sex cells and during which the number of chromosomes per cell is halved. The characteristic number of chromosomes of a body cell is restored when a male sex cell and a female sex cell join at fertilization. In the diagrams below only two pairs of chromosomes $(2N = 4)$ are represented, although most organisms have more.

In most cells, chromosomes exist in pairs. The two members of a pair are called homologous; chromosomes that are not members of the same pair are called nonhomologous.

Mitosis	Meiosis

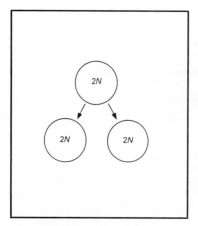

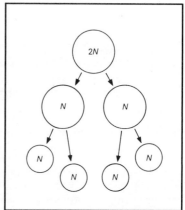

Two daughter cells with the same number of chromosomes (2N) as the parental cell are formed. During mitosis each cell divides once, but the chromosomes are previously doubled, resulting in no change in the number of chromosomes. Four stages are recognized during mitosis: prophase, metaphase, anaphase, and telophase.

Sex cells are formed, each with half as many chromosomes (N) as the other body cells. During meiosis each cell divides twice, but the chromosomes are doubled only once, which accounts for the reduction in chromosome number. The stages in the two meiotic divisions are designated with the same names as in mitosis: prophase, metaphase, anaphase, and telophase.

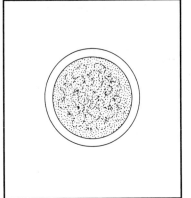

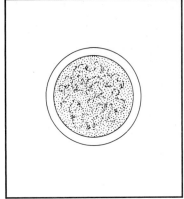

Interphase. This is the regular nondividing condition of a cell. There are $2N$ chromosomes in each cell, which are not easily distinguishable because they exist as very thin long threads intermingled with each other.

Interphase. The regular stage preceding meiosis is similar to the interphase of mitosis.

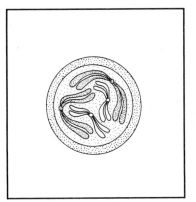

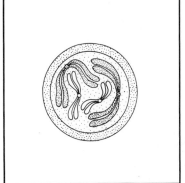

Previous to the start of the mitotic stages, the chromosome materials are doubled. The two halves of each double chromosome remain united at the centromere.

Previous to the start of the meiotic divisions, the chromosome materials are doubled. The two halves of each double chromosome remain united at the centromere.

Mitosis	Meiosis

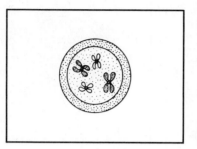

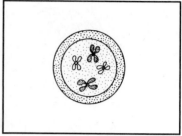

Prophase. The chromosomes shorten.

Prophase. The chromosomes shorten.

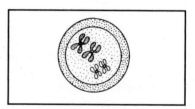

The two homologous chromosomes of each pair align.

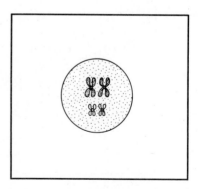

Metaphase of first meiotic division. The chromosomes (homologous chromosomes still aligned) move to the center. The nuclear membrane has disappeared.

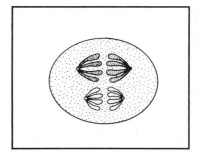

Anaphase of first meiotic division. The centromeres, followed by the chromosome arms, separate toward opposite poles.

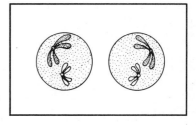

Telophase of first meiotic division. Two cells are formed, each with half the number of chromosomes (*N*), but each chromosome is doubled.

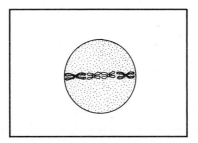

Metaphase. Unpaired chromosomes align in the center. The nuclear membrane has disappeared.

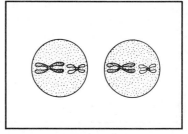

Metaphase of second meiotic division. Unpaired chromosomes align in the center.

Mitosis	Meiosis
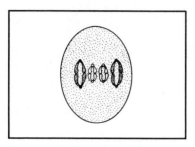	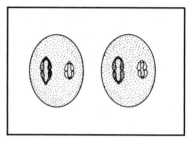

Anaphase. The centromeres divide. The centromeres, followed by the chromosome arms, migrate towards opposite poles.

Anaphase of second meiotic division. The centromeres divide. The centromeres, followed by the chromosome arms, migrate toward opposite poles.

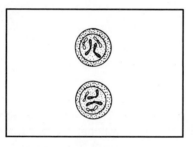	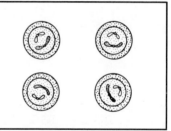

Telophase. The chromosomes have fully separated. Nuclear membranes are formed and the cell divides. Two new cells, each with the same number of chromosomes (2N) as the original one, have formed.

Telophase of second meiotic division. The chromosomes have fully separated. Nuclear membranes are formed and each cell divides. From the two divisions of meiosis a total of four cells result, each with half the number of chromosomes (N) as the original cell before meiosis started.

In 1902, two investigators—W. S. Sutton in the United States and Theodor Boveri in Germany—independently suggested that genes are contained in chromosomes. Their argument was based on the parallel behavior between chromosomes at meiosis and fertilization on the one hand, and genes on the other hand. If genes are contained in chromosomes, Mendel's results can be explained. The existence of two genes for a given character, one inherited from each parent, parallels the transmission of chromosomes that are received one of each kind from each parent. The two genes for a given trait segregate in the formation of the gametes because the two chromosomes of each pair go to different gametes during meiosis. Independent assortment will occur when genes for different traits are in different pairs of chromosomes, because nonhomologous chromosomes assort themselves in the gametes independently of the parent from which they come. (The two chromosomes of a pair are called *homologous;* chromosomes that are not members of the same pair are called *nonhomologous.*)

Figure 2.7 shows the chromosomal basis of the Law of Segregation and of the 3:1 ratios observed by Mendel. The cross between round-seed plants and wrinkled-seed plants produces F_1 hybrids carrying different alleles in two homologous chromosomes. These hybrids produce two kinds of gametes in equal proportions; half the gametes carry the chromosome with the roundness (*R*) allele, the other half carry the chromosome with the wrinkling (*r*) allele. At fertilization, these gametes combine at random producing zygotes in the proportions found by Mendel.

Figure 2.8 shows how two pairs of genes will assort independently if they are contained in nonhomologous chromosomes. The cross round-yellow x wrinkled-green produces hybrids (*RrYy*) with different alleles in the two homologous chromosomes of each pair. There are two possible orientations of the chromosomes at metaphase of the first meiotic division. One orientation leads to gametes carrying the genes in the parental combinations (*RY* and *ry*); the other orientation results in gametes with genes combined differently than in the parents (*Ry* and *rY*). Because the two orientations are equally probable, the joint result is four kinds of gametes with equal probabilities. At fertilization, when the four kinds of gametes (*RY, ry, Ry,* and *rY*) combine at random, two at a time, the 16 combinations shown in Figure 2.5 are possible. These 16 combinations reduce to four classes of peas in the expected proportions 9:3:3:1.

Linkage and Crossing Over

Mendel observed independent assortment for the seven characters he studied in peas because the seven genes that he studied happened to be each in a different nonhomologous chromosome. The sweet pea, *Pisum sativum,* has seven pairs of chromosomes and

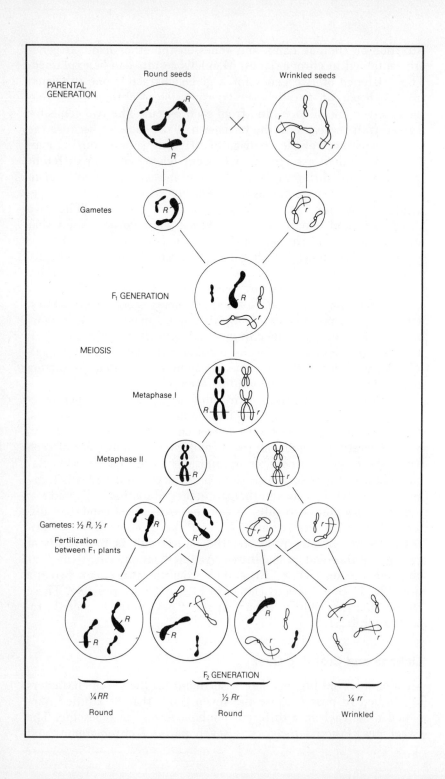

PARENTAL GENERATION

Round seeds

Wrinkled seeds

Gametes

F_1 GENERATION

MEIOSIS

Metaphase I

Metaphase II

Gametes: ½ R, ½ r

Fertilization between F_1 plants

F_2 GENERATION

¼ RR

Round

½ Rr

Round

¼ rr

Wrinkled

Fig. 2.7. Chromosomal basis of Mendel's Law of Segregation. A cross between a round-seed plant and a wrinkled-seed plant: only two pairs of chromosomes are shown, one of them carrying the gene for either roundness or wrinkling. Meiosis results in gametes all carrying the gene (allele) for roundness, *R*, in the case of round-seed plants, but for wrinkling, *r,* in the case of wrinkled-seed plants. The F₁ plants have one chromosome carrying *R* and the other carrying *r;* meiosis in F₁ plants produces gametes with *R* and gametes with *r* in equal proportions. Random union of these gametes at fertilization produces the observed 3:1 ratio of round versus wrinkled seeds.

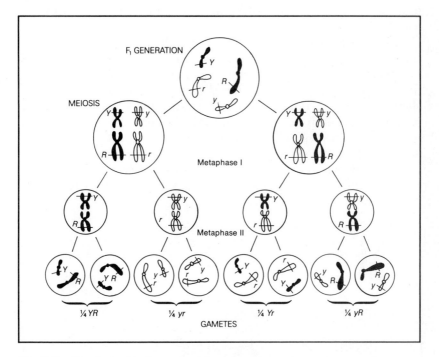

Fig. 2.8. The Law of Independent Assortment as a consequence of independent association of nonhomologous chromosomes at meiosis. The cross round-yellow **x** wrinkled-green produces hybrids (*RrYy*) carrying either *R* or *r* in two homologous chromosomes and *Y* or *y* in the other two homologous chromosomes. In metaphase I of meiosis, the chromosomes coming from the same parent may align on the same side (as on the left of the diagram) or on different sides (as on the right). In the first case, the resulting gametes have the same gene combinations (*YR* and *yr*) as were present in the parental generation; in the second case, the alternative combinations appear (*Yr* and *yR*). The final results are four kinds of gametes each with a frequency of one-quarter; random combination among these four kinds of gametes produces the 9:3:3:1 ratios observed by Mendel (as shown in Fig. 2.5).

thus it is a remarkable coincidence that no two characters were determined by genes in the same chromosome.

Genes that are in the same chromosome are said to be *linked*. However, it is not the case that linked genes are always kept together in the formation of the sex cells. This is because homologous chromosomes exchange parts during meiosis, specifically in late prophase when the two homologous chromosomes of each pair first align with one another (Fig. 2.9). The exchange of genes between homologous chromosomes is called *crossing over*. As a consequence of crossing over, linked genes may be transmitted to the progeny in combinations different from those in which they are present in the parents.

Drosophila melanogaster, the vinegar fly (often inappropriately called fruit fly), is one of the favorite organisms used in genetic experiments. *Drosophila* flies are easily and inexpensively reared in the laboratory; a new generation is produced in a short time, just about two weeks; and as many as several hundred progeny can be obtained from each mated pair of flies. We shall illustrate the phenomenon of crossing over with an experiment involving two characters, body color and wing length, in *Drosophila melanogaster.*

Normal body color (yellowish gray) is determined by a gene that is dominant over the gene for black body; normal wings are

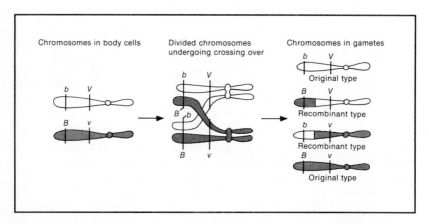

Fig. 2.9. Crossing over. On top, two homologous chromosomes are represented, one carrying the alleles *b* and *V*, the other carrying the alleles *B* and *v*. During late prophase the duplicated homologous chromosomes are aligned with each other and may exchange parts. Four kinds of chromosomes, and therefore four kinds of gametes, result whenever crossing over takes place; two of them have the original combinations (*bV* and *Bv*) while the other two have complementary combinations (*bv* and *BV*).

determined by a gene that is dominant over the gene producing very short ("vestigial") wings. The genes for body color and for wing length are linked; they are both in one of the four pairs of chromosomes that exist in these flies, the one known as chromosome II. We can represent the gene for normal body color as *B*, which is dominant over the gene for black body color, *b;* and the gene for normal long wings as *V*, which is dominant over the gene for vestigial wings, *v*.

Thomas H. Morgan, who received the Nobel Prize for his genetic research, crossed females having black body color and long wings (*bbVV*) with males having normal body color and vestigial wings (*BBbb*). The F_1 flies were heterozygous for both genes (*BbVv*) but had normal body color and wings, because gray body is dominant over black and long wing is dominant over vestigial (Fig. 2.10). Heterozygous females (*BbVv*) of the F_1 generation were, then, crossed to males homozygous for the two recessive genes (*bbvv*). In this cross, only the males contibuted gametes carrying recessive genes; and thus all genes, whether recessive or dominant, contributed by the heterozygous females showed in the following generation. Four types of progeny were obtained in the following proportions:

1. Gray vestigial   41.5 percent
2. Black long   41.5 percent
3. Gray long   8.5 percent
4. Black vestigial   8.5 percent

If the genes for body color and wing length were completely linked, only two kinds of flies would have appeared: gray-vestigial and black-long. If the two genes assorted independently, four kinds of flies would be expected, but in equal proportions (Fig. 2.11). Four kinds of flies are indeed observed, but those representing the parental gene combinations are considerably more common than the flies representing alternative combinations. The gene for body color and the gene for wing length are linked, but not completely linked. The nonparental classes of gametes, called *recombinant* gametes, are due to chromosomal exchanges during meiosis (Fig. 2.9).

The phenomenon of crossing over has made possible an important achievement of genetics, namely the construction of genetic maps (Figs. 2.12 and 2.13). The principle used in genetic mapping is simple. If genes are linearly ordered along chromosomes, the farther apart they are in the chromosomes the more likely it is that crossing over will occur between them (Fig. 2.14). In crosses such

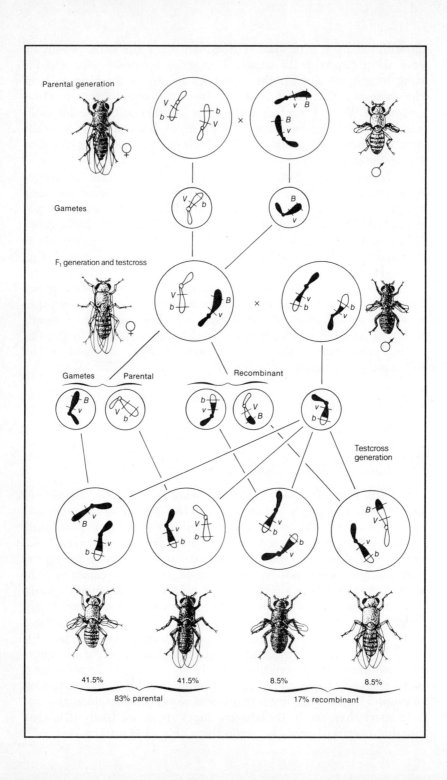

Parental generation

Gametes

F₁ generation and testcross

Gametes   Parental   Recombinant

Testcross generation

41.5%   41.5%   8.5%   8.5%

83% parental   17% recombinant

Fig. 2.10. Recombination in the vinegar fly, *Drosophila mel-anogaster*. The parental flies carry different alleles for two traits, body color and wing size; the genes controlling these traits are on the same chromosome. The F₁ females (*BbVv*) produce four kinds of gametes, two with the parental combinations (*Bv* and *bV*) and two with recombined genes (*bv* and *BV*). This is made manifest by a backcross between F₁ females and males homozygous for the two recessive alleles (*bbvv*). The frequency of the flies in the backcross generation of flies having a trait from one parent and a trait from the other parent reflects the frequency of recombination between the gene for body color and the gene for wing size.

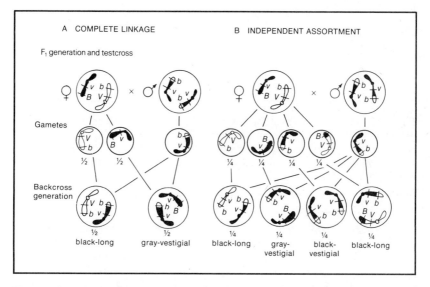

Fig. 2.11. Results from the backcross shown in Figure 2.10 in two hypothetical cases. *A.* If the genes for body color and wing length are completely linked, only two kinds of flies should appear in the backcross generation. *B.* If the genes for body color and wing length assort independently, four kinds of flies should appear in the backcross generation, but all four kinds in equal frequencies. The results shown in Figure 2.10 indicate that the genes for body color and wing length are not completely linked, because four kinds of flies were observed in the backcross generation; those results show moreover that the genes are linked (i.e., do not assort independently), because in the backcross generation the recombinant flies are less common than the parental flies.

as the one described above and shown in Figure 2.10, the larger the number of the recombinant flies the farther apart are the genes involved. The position that a gene has in a chromosome is known as its *locus* (Latin for "place").

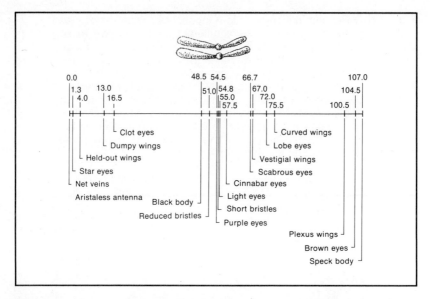

Fig. 2.12. Genetic map of the second chromosome of *Drosophila melanogaster*. The positions of several genes along the chromosome are determined by the frequency of crossing over. Closely located genes recombine with each other less frequently than far apart genes.

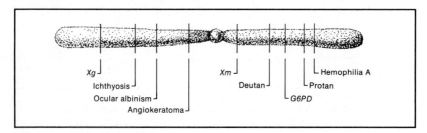

Fig. 2.13. Genetic map of the human X chromosome (the chromosome involved in the determination of sex). *Xg, Xm,* and *G6PD* are genes determining the synthesis of three different proteins. Deutan and protan are two forms of color blindness. About 100 more genes are known to belong to the X chromosome, but the location of most of them is not known.

CLOSER LOOK 2.2 Genetic Mapping

In the cross described in the text between *Drosophila melanogaster* females heterozygous for black body and vestigial wings (*BbVv*) and males homozygous for the two recessive genes (*bbvv*), the proportion of recombinant progeny (gray-long and black-vestigial flies) is 17 percent. We may say that the gene locus for black body color and the gene locus for vestigial wings are 17 "crossing-over units" apart. (*Gene locus,* plural *loci,* refers to the position of a gene in the chromosome, or simply to the gene itself without specifying which allelic form it has. Thus the gene locus for black body may have either the black body allele or the normal gray body allele).

Consider now the following cross. Female flies with normal body color (*BB*) but cinnabar eye color (*cc*) are crossed with males having black body (*bb*) but the normal red eye color (*CC*). The F_1 progeny will be heterozygous for both genes (*BbCc*) but normal in body and eye color because of the dominance relationships. When F_1 females are crossed with males homozygous for both the black body and the cinnabar eye genes (*bbcc*), four types of progeny are obtained in the following proportions:

1. Gray cinnabar   45.8 percent
2. Black red   45.8 percent
3. Gray red   4.2 percent
4. Black cinnabar   4.2 percent

Classes 3 and 4 represent crossing-over classes: the gene loci for body color and for eye color appear in combinations different from the parental ones. The total proportion of crossing-over flies is 8.4 percent, from which we conclude that the gene locus for black body and the gene locus for cinnabar eye are 8.4 crossing-over units apart.

We now have two pieces of information: (1) the genes for black body and for vestigial wings are 17 units apart; and (2) the genes for black body and for cinnabar eye are 8.4 units apart. There are two possibilities concerning the position of the three genes:

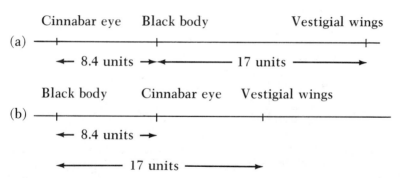

The two alternative possibilities lead to rather different predictions concerning the expected proportion of crossing over progeny in a cross involving the genes for cinnabar eye and vestigial wings. If (*a*) represents the correct position of the three genes, the proportion of crossovers will be greater between cinnabar and vestigial than between black and vestigial. If (*b*) is correct, the opposite will be the case. The second alternative has been demonstrated; the proportion of crossovers obtained experimentally between cinnabar and vestigial is 8.6 percent. We conclude that the gene locus for cinnabar eye color is placed approximately half way between the gene loci for black body and for vestigial wings.

The procedures outlined here have made possible the construction of genetic maps for various organisms, such as those shown in Figures 2.12 and 2.13.

Multiple Alleles

The examples given so far have concerned only two alleles at each gene locus. However, it is often the case that more than two alleles exist at a given gene locus. In 1900 Karl Landsteiner discovered the first human blood groups, known as the ABO blood groups. Blood groups must be taken into consideration when matching donors with recepients in blood transfusion.

There are four common blood groups: O, A, B, and AB. These are determined by three alleles, I^A, I^B, and i. Alleles I^A and I^B are dominant over allele i, but are codominant with each other—an individual heterozygous $I^A I^B$ has blood group *AB* because it has both *A* and *B* antigens in the blood. With three alleles, six two-gene combinations are possible, which result in only four blood groups because of the recessivity of the i allele:

Genotype	Blood group (antigens in blood)
ii	O
$I^A I^A$, $I^A i$	A
$I^B I^B$, $I^B i$	B
$I^A I^B$	AB

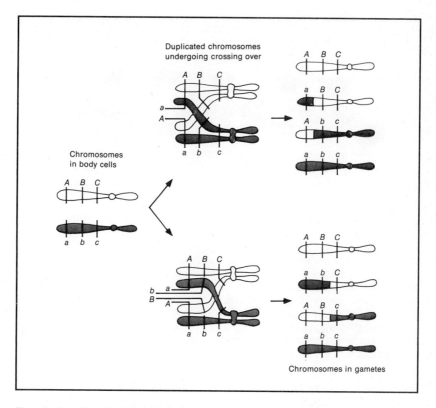

Fig. 2.14. Crossing over between three genes in the same chromosome. The *B* gene is located between *A* and *C* and thus closer to either one than these are to each other. Two cases of crossing over are represented, one, *top,* between *A* and *B,* and the second, *bottom,* between *B* and *C;* crossing over between *A* and *C* has occurred in both cases. In general, the farther apart two genes are the greater the probability of crossing over between them.

Biochemical techniques have shown that multiple alleles exist at many gene loci. One technique used in evolutionary studies is *gel electrophoresis,* which is described in Chapter 3. The genetic constitution of an individual is identified by looking at colored spots that appear in a gel; spots in different positions are due to different alleles. Figure 2.15 shows 12 individuals with different two-gene combinations (genotypes) at a gene locus known as *Pgm* (because it codes for an enzyme called phosphoglucomutase). The 12 individuals manifest three alleles, represented as *S* (for "slow"), *M* (for "medium"), and *F* (for "fast"). These alleles are codominant, and thus heterozygous individuals show two spots. Given three alleles, six different genotypes are possible, each with a different banding pattern because no one allele is dominant:

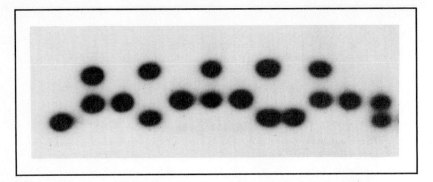

Fig. 2.15. Gel electrophoresis of the enzyme phospho-glucomutase in 12 *Drosophila pseudoobscura* flies. The individual on the extreme left is homozygous for the *S* ("slow") allele and thus shows only one spot on the gel. The next individual is heterozygous for the *M* ("medium") and the *F* ("fast") alleles and thus shows two spots. The genotypes of the 12 individuals, going from left to right are *SS, MF, MM, SF, MM, MF, MM, SF, SS, MF, MM*, and *SM*. With three alleles, six genotypes may exist; all three possible hetero-zygous genotypes are represented in the gel but one of the three possible homozygous genotypes (*FF*) is missing. (Numbers rather than letters are sometimes used to represent different alleles as in Figures 3.8, 3.9, and 3.10.)

Individual	Genotype	Spot pattern
1	SS	One spot, at slow position
2	MM	One spot, at medium position
3	FF	One spot, at fast position
4	SM	Two spots, at slow and medium positions
5	SF	Two spots, at slow and fast positions
6	MF	Two spots, at medium and fast positions

Continuous Variation

The traits studied by Mendel fell into discrete alternative classes: either green or yellow peas, either round or wrinkled seeds, either axial or terminal flowers, and so on. The other examples given so far in this chapter are also discrete alternative traits: either gray or black body, either vestigial or long wings, either O, A, B, or AB blood group. Many other traits in all sorts of organisms fall into discrete, easily separable classes.

There are, however, other traits that do not fall into discrete classes, but rather exhibit *continuous variation*. The height of hu-mans is one such trait: most normal people range from somewhat more than five feet to more than seven feet through all possible in-termediate steps. We could classify people into tall, medium, and short, but that would be rather arbitrary. Other examples of continuously varying characters are the weight of people, the amount of milk produced by cows, and the protein content of corn plants.

Are continuously varying characters hereditary? Do the laws of heredity so far discussed apply to them? The answer to these questions is yes, but some qualifications are needed. Traits may vary more or less continuously for one or both of two reasons: (1) because of environmental effects, and (2) because several gene loci affect the same trait. The possible effects of the environment are obvious in many cases: a well-fed chicken is likely to be bigger than a starved chicken, an athlete who "pumps iron" every day is likely to have larger muscles than a book worm who does little physical exercise.

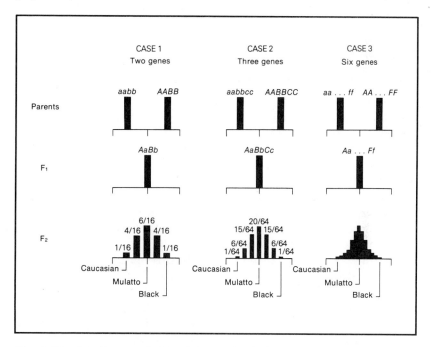

Fig. 2.16.   Inheritance of skin pigmentation in humans. Lowercase letters represent genes determining light skin; capital letters represent genes determining dark skin. It is assumed that alleles at different loci contribute equally to skin pigmentation and that there is no dominance. On the left, two gene loci are assumed to determine the difference in skin pigmentation between Caucasians and Blacks. The F_1 progeny consists of intermediate-colored mulattos. Matings between mulattos in the F_2 generation result in five classes of individuals, most having intermediate-colored skin. As the number of genes involved increases, *middle* and *right*, the number of classes in the F_2 also increases; and more and more individuals have intermediate skin color.

Many traits are continuously varying because they are influenced by many genes, each with a small effect. The simple rule is that the greater the number of gene loci with cumulative effects in a single trait, the more nearly continuous the variation among individuals will be. This is shown in Figure 2.16 using as an example the difference in the intensity of skin pigmentation between Blacks and Caucasians. As the number of gene loci affecting skin pigmentation becomes larger, the number of possible classes intermediate between the two extremes increases. Moreover, a greater and greater proportion of the individuals fall in the intermediate categories. In general, most cases of continuous variation are due to the joint effects of environmental variation and multiple genes.

CLOSER LOOK 2.3   Multiple Genes and Continuous Variation

Many traits vary continuously because they are influenced by many genes, each with a small effect. We shall see how the cumulative effects of several gene loci result in more or less continuous variation using as an example the difference between Blacks and Caucasians in the intensity of skin pigmentation. Differences in skin pigmentation may be objectively measured by the skin reflectance at a certain wavelength. At a wavelength of 685 mμ (millimicrons, a millimicron is one billionth of a meter) the average skin reflectance of Caucasians is about 0.225, and that of Blacks is about 0.425. This difference is largely due to the effects of several gene loci, probably about five.

The larger the number of gene loci cumulatively affecting a trait, the more nearly continuous the variation among individuals (Fig. 2.16). Assume first that the difference in skin pigmentation between Caucasians and Blacks is due to one gene locus with two alleles, b and B, so that an individual bb has a reflectance of 0.225, and each B allele adds a pigmentation of 0.100 over that. The F_1 progeny of a marriage between a Black (BB) and a Caucasian (bb) will be Bb, and have a reflectance of $0.225 + 0.100 = 0.325$, exactly intermediate between the two parents. Matings between such intermediate (mulatto) individuals will produce three kinds of children; BB ¼, Bb ½, and bb ¼. The three types fall into three discrete classes.

Assume now that the difference in skin pigmentation between Blacks and Caucasians is due to two gene loci, each with two alleles, so that a Black has the genotype *AABB* and a Caucasian has the genotype *aabb*. Each capital letter allele adds 0.050 reflectance units over the basic reflectance of 0.225. A marriage between a Black and a Caucasian produces mulatto children (*AaBb*), with a reflectance of 0.225 + 0.050 + 0.050 = 0.325, the same as in the case when only one gene locus is involved. However, when mulattos marry with each other, we have a more complex situation than before. At each locus three genotypes are possible: *AA*, *Aa*, and *aa*, or *BB*, *Bb*, and *bb*, with frequencies ¼, ½, and ¼ in both cases. Assuming independent assortment between the two gene loci, the nine possible combinations and their expected frequencies are:

		Genotypes at the *A* locus		
		AA ¼	Aa ½	aa ¼
Genotypes at	*BB* ¼	$AABB\ ^1/_{16}$	$AaBB\ ^1/_8$	$aaBB\ ^1/_{16}$
the *B* locus	*Bb* ½	$AABb\ ^1/_8$	$AaBb\ ^1/_4$	$aaBb\ ^1/_8$
	bb ¼	$AAbb\ ^1/_{16}$	$Aabb\ ^1/_8$	$aabb\ ^1/_{16}$

The skin reflectance can be calculated using the rules given above. For example, individuals with the genotype *AaBB* will have a reflectance of 0.225 + 0.050 + 0.050 + 0.050 = 0.375 (there are three capital letter alleles in that genotype, and thus we add 0.050 three times). The skin reflectance of the nine genotypes are:

		Genotypes at the *A* locus		
		AA	Aa	aa
Genotypes at	*BB*	0.425	0.375	0.325
the *B* locus	*Bb*	0.375	0.325	0.275
	bb	0.325	0.275	0.225

We see that with respect to skin reflectance, the genotypes fall into five categories with the following frequencies:

Skin Reflectance	Genotypes	Frequencies
0.425	*AABB*	$^1/_{16}$
0.375	*AABb* and *AaBB*	$^1/_8 + ^1/_8 = ^1/_4$
0.325	*AAbb*, *AaBb*, and *aaBB*	$^1/_{16} + ^1/_4 + ^1/_{16} = ^3/_8$
0.275	*Aabb* and *aaBb*	$^1/_8 + ^1/_8 = ^1/_4$
0.225	*aabb*	$^1/_{16}$

With one gene locus, there was only one intermediate category between "pure" Blacks and "pure" Caucasians. With two gene loci there are three intermediate categories;

moreover, these intermediate categories account for seven-eighths of the individuals rather than only one-half, as was the case with one gene locus.

Assume now that five gene loci account for the difference in skin pigmentation between Blacks (*AABBCCDDEE*) and Caucasians (*aabbccddee*), so that each capital letter allele increases skin reflectance by 0.20 units. In the F_2 generation, there are $3^5 = 243$ possible different genotypes (three genotypes at each of five loci, and therefore $3 \times 3 \times 3 \times 3 \times 3 = 3^5$), ranging in frequency from $(\frac{1}{4})^5 = 1/1024$ (for "pure" Black and "pure" Caucasian) to $(\frac{1}{2})^5 = 1/32$ (for the heterozygote at all five loci, *AaBbCcDdEe*). The 243 genotypes associate into eleven classes, having 0, 1, 2, . . . or 10 capital-letter alleles. Most individuals fall into the categories with intermediate skin pigmentation.

The main conclusion concerning continuous variation is that as the number of variable gene loci involved increases, the number of intermediate categories (and the total number of *individuals* in the intermediate categories) increases. Moreover, the differences between neighboring categories become blurred because of environmental effects modifying the expression of a trait even in individuals with the same genotype.

Genotype and Phenotype

The *phenotype* of an organism is its appearance: its morphology, physiology, and ways of life. The *genotype* is the genetic information it has inherited. The distinction between phenotype and genotype must be kept in mind because the relation between the two is not without ambiguity. This is because the phenotype results from complex networks of interactions between genes, and between genes and the environment. The phenotype changes continuously throughout the life of an organism, from the moment of fertilization to its death; however, the genotype remains constant except for the relatively rare occurrence of genetic mutations (to be discussed in Chapter 3).

Individuals having the same phenotype with respect to a given trait do not necessarily have identical genotypes. For example, yellow peas can be either homozygous for the yellow allele or heterozygous for the yellow and green alleles; in the previous discussion on continuous variation, we saw that several genotypes may fall in the same phenotypic class. (Fig. 2.16 and Closer Look 2.3)

Moreover, individuals with identical genotypes may have different phenotypes owing to interactions with other genes and, most importantly, owing to interactions with the environment. Even identical twins may differ in height, weight, and length of life because of different life experiences. Figure 2.17 is a good illustration of environmental effects; cuttings of the same *Potentilla*

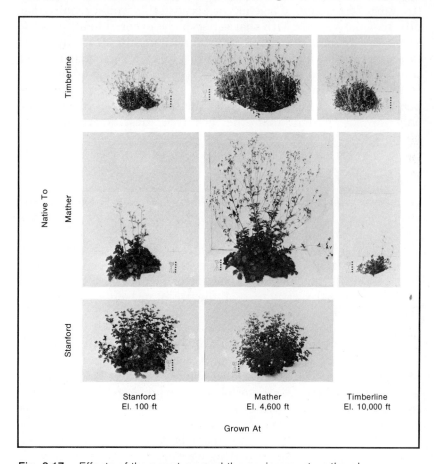

Fig. 2.17. Effects of the genotype and the environment on the phenotype. Cuttings from *Potentilla glandulosa* plants collected at different altitudes were planted together in three different experimental gardens. Plants in the same row are genetically identical to each other, since they have been grown from cuttings of a single plant; plants in the same column are genetically different but have been grown in the same environment. Plants genetically identical (for example, those in the bottom row) may prosper or fail to survive depending on the environmental conditions. Genetically different plants may have quite different phenotypes even when grown in the same environment (for example, the three plants in the first column).

glandulosa plant, although genetically identical to each other, differ in appearance, fertility, length of development, and so on, when grown at different altitudes.

The different phenotypes that a given genotype may have in different environments are expressions of the *range of reaction* (also called "norm of reaction") of the genotype. We never know the entire range of reaction of a genotype because this would require that different individuals with that genotype be exposed to all possible kinds of environments, which are virtually infinite. It is important to recognize that an organism does not inherit a particular set of traits but rather a genotype with a very broad range of reaction. How successfully an individual will survive and reproduce depends directly on its phenotype, while it depends on its genotype only to the extent that the genotype has determined the phenotype.

The Hereditary Materials

Genes are the carriers of heredity. Their existence, location in the chromosomes, and other properties are determined by studying segregation in the progenies of crosses between individuals showing alternative expressions of a trait. In the early 1950s it became established that genes are molecules of deoxyribonucleic acid (DNA). One of the most important scientific discoveries of all time is the double-helix structure of the DNA molecule proposed by James Watson and Francis Crick in 1953, and thoroughly confirmed thereafter (Fig. 2.18).

The DNA molecule consists of two complementary chains made up of long sequences of units called *nucleotides*. There are four kinds of nucleotides, each carrying one of four kinds of nitrogen bases: adenine (A), cytosine (C), guanine (G), and thymine (T). The two chains in the double helix are complementary because there are only two possible kinds of associations between nitrogen bases in different chains, namely A with T and C with G. Using these simple rules of pairing we can determine precisely the sequence of bases (and therefore of nucleotides) in one chain whenever we know the sequence in the complementary chain. For example, if the sequence in one chain is ACCTAGAT, the complementary chain will have the sequence TGGATCTA.

The strict complementarity between the nitrogen bases in different chains accounts for the precise replication of genes. During replication, the two chains of the DNA double helix separate by an unwinding process. Each chain serves as a template for the synthesis of a complementary chain. Two double chains result that are identical to each other and to the parental double chain. Assume, for example, that one double helix starts with the sequence

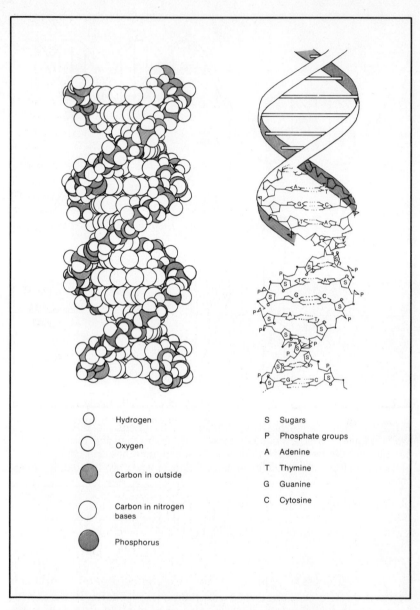

Fig. 2.18. Two representations of the double-helix structure of DNA. *Left,* a "space-filling" diagram. *Right,* additional details. The molecule consists of two complementary chains. The outward backbone of the DNA molecule is made of alternating deoxyribose sugar (S) and phosphate (P), chemically bound to one another. Nitrogen bases connected to the sugars project towards the center of the molecule. There are four kinds of bases: adenine (A), cytosine (C), guanine (G), and thymine (T). The two chains are held together by hydrogen bonds between complementary bases, but A can only pair with T, and C only with G.

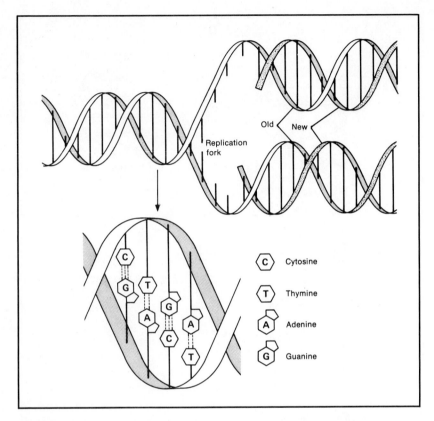

Fig. 2.19. Replication of DNA. The replication process results in two "daughter" double helices, identical to each other and to the parental double helix. Each daughter DNA molecule has one complete chain from the parental molecule and a newly synthesized chain.

ACCTAGAT
TGGATCTA

In the process of replication the two chains separate, and each determines the sequence in a complementary chain. If we represent the nitrogen bases in the newly synthesized chains with bold-face letters, the two resulting double helices will be

ACCTAGAT   **ACCTAGAT**
TGGATCTA and   TGGATCTA

The genetic information is encoded in the sequence of the nitrogen bases in the DNA. The nitrogen bases may be considered as the letters of a genetic alphabet. Sequences of bases make up "words"; genes may be thought of as genetic "sentences." The genetic endowment of an individual may then be thought of as a "book" made up of genetic sentences. Contrary to the strict determination of the nitrogen bases between the two complementary chains, there are no restrictions as to what base may follow any other base along one chain. This makes it possible to have a virtually unlimited number of different DNA molecules. Since there are four different kinds of bases, there are $4 \times 4 = 16$ different combinations of two bases, $4 \times 4 \times 4 = 64$ different chains with a length of three bases, and, in general, 4^n different possible sequences each with a length of n nucleotides. When the length of the chain is in the hundreds or in the thousands of nucleotides, as it is in the case of genes, the number 4^n is staggeringly large.

However, the basic units of information are not the individual bases, but discrete groups of three consecutive bases, called *triplets* or *codons* (because they code for amino acids, as we shall see below). Although there are 64 possible combinations of three bases, there are only 21 different units of information: 61 triplets code for 20 different amino acids, the other three triplets are termination signals. A DNA chain with a length of 900 nucleotides has 300 triplets; the number of potentially different "messages" encoded by chains of that length is $21^{300} = 10^{397}$, a number much greater than the number of atoms in the universe. We can see that there is practically no limit to the number of different genetic messages that can be encoded in long DNA chains. (It should be noted, however, that the triplets of the gene coding for amino acids are not all consecutive. Groups of codons are separated by inserted nucleotides that do not code for any amino acids.)

Transcription and Translation

Genes control the development and metabolism of an organism by determining the synthesis of other molecules, particularly enzymes and other proteins. Proteins consist of long sequences of 20 different kinds of amino acids; one common class of proteins are the enzymes, which control the chemical activities taking place in cells. Two kinds of genes can be identified: *structural genes* are those that determine the sequence of amino acids in a protein; *regulatory genes* are those that control the activity of other genes, for example by turning on and off structural genes.

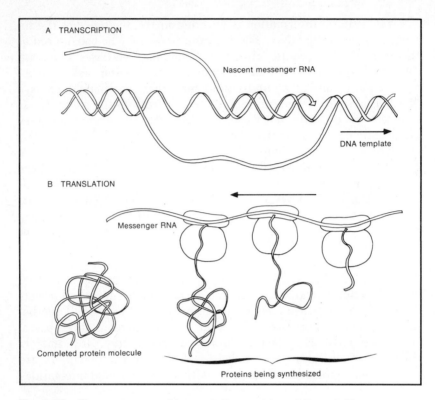

A TRANSCRIPTION

Nascent messenger RNA

DNA template

B TRANSLATION

Messenger RNA

Completed protein molecule

Proteins being synthesized

Fig. 2.20. The processes of transcription and translation. *A.* Transcription. One strand of the DNA helix serves as a template for the synthesis of a complementary molecule of messenger RNA (RNA stands for "ribonucleic acid," a molecule quite similar to a single chain of the DNA double helix). *B.* Translation. The messenger RNA becomes attached to certain bodies, known as ribosomes, that mediate the synthesis of proteins. The sequence of codons in the messenger RNA determines the sequence of amino acids in the protein.

The information contained in the nucleotide sequence of a structural gene determines the synthesis of a particular protein through the processes of *transcription* and *translation* (Fig. 2.20). Transcription is a process by which the information contained in the base sequence of a DNA molecule is converted into a complementary RNA molecule (called *messenger RNA*). Translation is the process by which the information contained in a messenger RNA molecule is converted into the particular amino acid sequence of a protein. The correspondence between nucleotide triplets in messenger RNA and amino acids in proteins is given by the *genetic code* (Fig. 2.21).

The discoveries just summarized make it clear why the inheritance of "acquired characteristics" is not possible. Lamarck

SECOND LETTER

	U	C	A	G	
U	UUU } Phe UUC } UUA } Leu UUG }	UCU } UCC } Ser UCA } UCG }	UAU } Tyr UAC } UAA } Stop UAG }	UGU } Cys UGC } UGA Stop UGG Trp	U C A G
C	CUU } CUC } Leu CUA } CUG }	CCU } CCC } Pro CCA } CCG }	CAU } His CAC } CAA } Gln CAG }	CGU } CGC } Arg CGA } CGG }	U C A G
A	AUU } AUC } Ile AUA } AUG Met	ACU } ACC } Thr ACA } ACG }	AAU } Asn AAC } AAA } Lys AAG }	AGU } Ser AGC } AGA } Arg AGG }	U C A G
G	GUU } GUC } Val GUA } GUG }	GCU } GCC } Ala GCA } GCG }	GAU } Asp GAC } GAA } Glu GAG }	GGU } GGC } Gly GGA } GGG }	U C A G

FIRST LETTER (left) — THIRD LETTER (right)

Fig. 2.21. The genetic code gives the correspondence between the 64 possible codons in messenger RNA and the amino acids (or termination signals). The nitrogen base thymine does not exist in RNA, where uracil (U) takes its place; the other three nitrogen bases in messenger RNA are the same as in DNA: adenine (A), cytosine (C), and guanine (G). The 20 amino acids making up proteins are as follows: alanine (Ala), arginine (Arg), asparagine (Asn), aspartic acid (Asp), cysteine (Cys), glycine (Gly), glutamic acid (Glu), glutamine (Gln), histidine (His), isoleucine (Ile), leucine (Leu), lysine (Lys), methionine (Met), phenylalanine (Phe), proline (Pro), serine (Ser), threonine (Thr), tyrosine (Tyr), tryptophane (Trp), and valine (Val).

believed that evolution—i.e., the gradual adaptation through the generations of organisms to their environments—occurred because characteristics acquired by an individual are hereditary. Through consistent use or disuse, organs may become strengthened or otherwise modified; Lamarck, and even Darwin, believed that modifications caused in an individual by use and disuse could be inherited. In the late nineteenth century and the first half of the twentieth century, Weismann and others demonstrated that acquired characteristics are not transmitted to the progeny. We now know that the relationship between genotype and phenotype is unidirectional. The sequence of nucleotides in DNA is transcribed into RNA, which becomes translated into proteins. The phenotype becomes determined by the interactions between gene products and the environment. But there is no mechanism by which this process can be reversed. Phenotypic modifications resulting from use and disuse, or from other interactions between an organism and the environment, do not change the hereditary information contained in the DNA. The evolution of organisms depends on changes in the DNA, which occur through the processes of gene mutation and chromosomal change, discussed in the next chapter.

QUESTIONS FOR DISCUSSION

1. Many scientists before Mendel attempted to solve the riddle of heredity with little success. What characteristics of Mendel's method made possible his discovery of the basic laws of heredity?

2. What are the fundamental differences between the processes of mitosis and meiosis?

3. With two alleles, say A and a, at one locus, three different genotypes are possible, AA, Aa, and aa, which reduce to only two phenotypes if one allele is dominant over the other. With two alleles at each of two loci, nine genotypes are possible, which reduce to four phenotypes if there is dominance (see Figure 2.5). How many genotypes and how many phenotypes will there be in the case of three loci with two alleles each? Can you work out a general formula in the case of n loci with two alleles each?

4. In view of the distinction between genotype and phenotype, would you say that a person with a high IQ score has superior IQ genes than a person with a lower IQ score? Do you think that environmental influences are likely to play a greater role in human beings than in animals such as a vinegar fly or a mouse? Why? In human beings, do you think that environmental influences have greater effect with respect to behavioral traits, such as IQ, than with respect to physical traits such as eye color? Why?

5. Do you think that a structural gene, i.e., a gene coding for a protein, could also be a regulatory gene?

Recommended for Additional Reading

Bodmer, W. F. and Cavalli-Sforza, L. L. 1976. *Genetics, evolution, and man.* San Francisco: W. H. Freeman.

This is an introductory text; Part I presents in a simple and easily understandable fashion the basic concepts of genetics.

Gardner, E. J. 1975. *Principles of genetics.* 5th ed. New York: John Wiley.

Suzuki, D. T. and Griffiths, A. J. F. 1976. *An introduction to genetic analysis.* San Francisco: W. H. Freeman.

Two standard textbooks of genetics. The book by Suzuki and Griffiths is written in a lively style and emphasizes the methodology used in genetic studies.

Goodenough, U. 1978. *Genetics.* 2nd ed. New York: Holt, Rinehart and Winston.

A textbook with a clear, although advanced, presentation of the recent advances in molecular and microbial genetics.

Strickberger, M. 1976. *Genetics.* 2nd ed. New York: Macmillan.

The most complete genetics text in a single volume.

3

Hereditary Variation

The most obvious unit of living matter is the individual. In unicellular organisms each cell is an individual; multicellular individuals consist of many interdependent cells, many of which may die and be replaced by other cells throughout the life of the individual. The bacterium *Escherichia coli,* the blue-green alga *Euglena pisciformis,* and the ciliate protozoan *Paramecium bursaria* are examples of unicellular organisms. An oat plant, a pine tree, a starfish, and a human being are examples of multicellular organisms.

Populations and Gene Pools

In evolution, the significant unit is not the individual, but the population. A *population* is a community of individuals linked together by bonds of mating and parenthood. In other words, a population is a community of individuals of the same species. The bonds of parenthood that link members of the same population are always present, although mating may be absent in the case of organisms reproducing asexually.

The individual is not the significant unit in evolution because the genotype of an individual remains unchanged throughout its life. Moreover, the individual is ephemeral (although some organisms, such as certain conifers, may live up to several thousand years). On the other hand, a population has continuity from generation to generation. Moreover, the genetic constitution of a population may change—evolve—over the generations. The continuity of a population through time is provided by the mechanism of biological heredity.

A *local population* is a group of individuals of the same species living together in the same territory. The concept of local population is itself easily understandable, but difficulties may arise in applying the concept because the boundaries between populations are often fuzzy (Fig. 3.1). In most species individuals do not exist in homogeneous clusters sharply separated from each other. Moreover, the organisms are not homogeneously distributed within a cluster, even when the clusters are quite discrete, as in the case of organisms living in lakes or on islands: the lakes or the islands may be sharply distinct, but individuals are not evenly distributed within a lake or an island. Animals often migrate from one to another local population, and the pollen or seeds of plants may also move from population to population, which makes local populations of the same species far from independent of each other.

The most inclusive community of individuals linked by bonds of mating and parenthood is the *species* (discussed in Chapter 6). As a rule, the genetic discontinuities between species are absolute; sexually reproducing organisms of different species are kept from interbreeding by isolating mechanisms. Species are independent evolutionary units: genetic changes taking place in a local population can be extended to all members of the species, but they cannot be transmitted to members of a different species.

The concept of the *gene pool* is a useful notion in the study of evolution. The gene pool is the sum total of the genotypes of all individuals in a population. For diploid organisms, the gene pool of a population with n individuals consists of $2n$ complete genomes. One *genome* consists of all the genetic information (genes and chromosomes) received from each parent. Thus, in the gene pool of a population of n individuals there are $2n$ genes for each gene locus and $2n$ chromosomes of each kind. The exceptions are the X chromosomes, and the genes in them (called sex-linked genes), which exist in a single dose in the heterogametic sex. Sex is often determined by the presence of either two or only one chromosome of a certain kind, commonly called the X chromosome. The sex carrying only one sex chromosome is called the *heterogametic sex,* which is the male in most animals, but the female in birds, but-

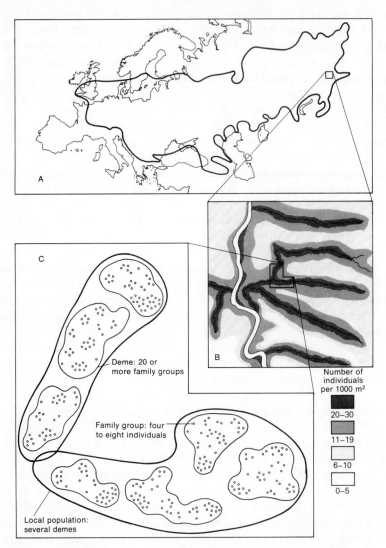

Fig. 3.1. Geographic distribution of *Lacerta agilis. A.* This lizard exists over a broad area encompassing large parts of Europe and western Asia, but its distribution is far from homogeneous. *B. Lacerta agilis* has greater density along streams and rivers than in the intermediate areas. *C.* The lizards occur in small family groups consisting of a few individuals each. Demes consist of about 20 to 40 family groups. Local populations are made up of several demes each. Within a deme matings occur rather freely between members of different family groups, but less than 4 percent of all matings take place between individuals from different demes within the same local population. Fewer than 0.01 percent of all matings involve individuals from different local populations. (Data courtesy of A. B. Yablokov.)

terflies, and others. In humans as well as in many other organisms, individuals of the heterogametic sex (males) have a sex chromosome, called Y, besides the one X chromosome. Y chromosomes carry few genes, although some genes affecting the fertility of the heterogametic sex may be located there.

The genetic constitution of a population changes as evolution occurs. Evolutionary change may therefore be described at the genetic level as change in the gene pool of populations. The number of gene loci and their position in the chromosomes do not necessarily remain constant in evolution. Therefore, evolution does not consist only of changes in the frequencies of alleles and combinations thereof, but also in changes in the amount and in the organization of the genetic material.

Genetic Variation and Rate of Evolution

The existence of genetic variation is a necessary condition for evolution. If, for example, for a certain gene locus all individuals of a population are genetically identical, i.e., homozygous for exactly the same allele, evolution cannot take place at that locus since the allelic frequencies cannot change from generation to generation. On the other hand, if in a different population there are two alleles at the same locus as before, evolutionary change can take place at that locus: one allele may increase in frequency at the expense of the other allele.

Darwin was quite aware that a necessary condition for evolution is the occurrence of hereditary variation. For Darwin this was an incontrovertible fact, although he did not know the processes by which hereditary variation arises. Darwin argued that some hereditary natural variations may be more advantageous than others for the survival and reproduction of their carriers. Organisms having advantageous variants are more likely to survive and reproduce than organisms lacking them. As a consequence, useful variations will spread through the generations while harmful or less useful ones will be eliminated. This is the process of natural selection, which plays the leading role in evolution.

It may be shown mathematically that the more genetic variation there is in a population, the greater the opportunity for evolution. This seems intuitively obvious; the greater the number of variable gene loci, and the more alleles there are at each variable locus, the greater the possibilities for change in the frequency of some alleles at the expense of others. Moreover, it has been shown experimentally that there is a positive correlation between the amount of genetic variation in a population and the population's rate of evolution.

One experiment was performed with *Drosophila serrata,* a species living in eastern Australia, New Guinea, and New Britain. Laboratory populations were set in cages, with food provided in constant amounts, where the flies lived and reproduced in an isolated microcosm. Evolution can be studied in such "population cages" by finding out how a population changes over many generations. Two types of experimental populations were established in the experiment. *Single-strain* populations had founders descended from flies collected in a single locality (either Popondetta, New Guinea, or near Sydney, Australia); a *mixed* population was established by crossing flies collected in two localities (flies from Sydney crossed with flies from Popondetta). The mixed population had greater initial genetic variation than the single-strain populations, since it was started by mixing two of the latter.

The experimental populations were set up in such a way that there was intense competition among the flies for food and space. Such competition stimulates rapid evolutionary change. The adaptation of a population to the experimental environment was measured by the number of individuals in the population.

The results are shown in Figure 3.2. The experimental populations were kept for 500 days, which amounts to about 25 gener-

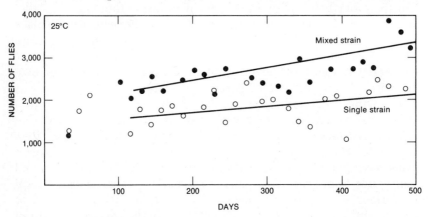

Fig. 3.2.   Correlation between amount of genetic variation and rate of evolution in laboratory populations of *Drosophila serrata* flies. The graph shows the change in number of flies during approximately 25 generations. Both laboratory populations increased in numbers throughout the experimental period, but the average rate of increase was substantially greater in the mixed-strain population (which is genetically more variable) than in the single-strain population. Increases in the number of flies over the generations reflect evolution promoting the adaptation of the populations to the experimental environment.

ations. Two results deserve notice. First, on the average the mixed population (Sydney x Popondetta) had more flies than the single-strain populations from either Sydney or Popondetta (the Sydney population is not shown in Figure 3.2). Second, and even more relevant, the number of flies increased at a faster rate in the mixed population than in the single-strain populations. As illustrated in Figure 3.2 the single-strain Popondetta population and the mixed population both became better and better adapted to the experimental environment as the generations proceeded. However, the rate of evolution was greater in the mixed population than it was in the Popondetta population.

Table 3.1
GENETIC VARIATION AND RATE OF EVOLUTION IN POPULATIONS OF *DROSOPHILA SERRATA*

The rate of evolution is measured by the rate at which the number of flies increases over the generations, as a consequence of the gradual adaptation of the populations to the experimental environment. (Data from Ayala, 1965.)

	Mean number of flies in population	Mean number of flies increased per generation
Single-strain (Popondetta)	1862 ± 79	31.5 ± 13.8
Mixed (Popondetta x Sydney)	2750 ± 112	58.5 ± 17.4

Table 3.1 gives the main results of the experiment shown in Figure 3.2. The average number of flies is 1,862 in the Popondetta population. The average number of flies is 2,750 in the mixed population. The evolution of the experimental populations is shown by the gradual increase in the number of flies in each population. The average number of flies in the Popondetta population increased at a rate of 31.5 flies per generation (each generation lasts about three weeks); the Sydney population increased at about the same rate. The Sydney x Popondetta population, however, increased in size at a rate of 58.5 flies per week, nearly double the rate of evolution of the single-strain populations. The mixed population was started by mixing two single-strain populations, and thus from the start it had much more genetic variation than either one of them. The larger initial amount of genetic variation made possible a faster rate of evolution in the mixed population.

Classical and Balance Models of Population Structure

A population's potentiality for evolving is determined by the genetic variation present in the population. How much genetic variation is there in natural populations? This question was much debated by geneticists for many years. Two hypotheses of the genetic structure of populations were advanced, called the *classical* and the *balance* model, one arguing that there is very little, and the other that there is a lot of, genetic variation (Fig. 3.3).

The classical model was championed by geneticist and Nobel laureate H. J. Muller (1890–1967) and his followers. It proposes that the gene pool of a population consists, at virtually all loci, of a wild-type allele with a frequency approaching one; the rest of the gene pool is made up of deleterious alleles arisen by mutation and kept at very low frequencies by natural selection. A typical indi-

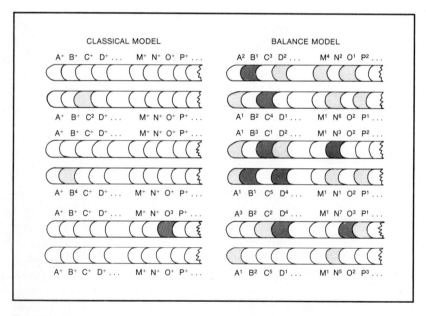

Fig. 3.3. Two models of the genetic structure of populations. The hypothetical genotypes of three typical individuals are shown according to each model. Capital letters symbolize gene loci, and each number represents a different allele; the wild-type allele postulated by the classical model is represented by a plus sign. According to the classical model, individuals are homozygous for the wild-type allele at nearly every locus, although they may be heterozygous for the wild allele and a mutant at an occasional locus (C in the first individual, O in the third). According to the balance model, individuals are heterozygous at many gene loci.

vidual would be homozygous for the wild-type allele at nearly every locus, but at a few loci it would be heterozygous for a mutant and the wild allele. The "normal," ideal genotype would be an individual homozygous for the wild-type allele at every gene locus.

According to the classical model, evolution occurs because rarely, but occasionally, a beneficial mutant arises. Since such a beneficial allele is better for the organisms than the preexisting wild allele, the beneficial mutant will gradually increase in frequency by natural selection and become the new wild-type allele, while the former wild-type allele will be eliminated or reduced to very low frequency.

The proponents of the balance model of population structure include the great evolutionist Theodosius Dobzhansky (1900–1975) and many others. According to the balance model, there is generally no single wild-type or "normal" allele. Rather, the gene pool of a population consists at many loci of an array of alleles in various frequencies. Consequently, individuals are heterozygous at a large proportion of gene loci. There is no "normal" or ideal genotype. Rather, populations consist of an array of genotypes that yield a satisfactory fitness in most environments encountered by the population.

According to the balance model, evolution occurs by gradual change in the frequencies and kinds of alleles at many loci. The pervasive genetic *polymorphisms* (variations in form) found in natural populations are maintained by natural selection (to be discussed in Chapters 4 and 5); alleles do not act in isolation. Rather, the fitness conferred by an allele depends on what other alleles exist in the genotype, and of course on the environment. Gene pools are seen as coadapted systems: the set of alleles favored at one locus depends on the sets of alleles existing at other loci. The balance model accepts, however, that some mutants are unconditionally harmful to their carriers. These deleterious alleles are kept at low frequencies by natural selection and play only a secondary, negative role in evolution.

Some Evidence of Abundant Genetic Variation

It is now known that most natural populations possess a great deal of genetic variation as proposed by the balance model of population structure. Evidence has accumulated over the years showing that genetic polymorphisms are widespread.

Individual variation is obvious whenever organisms of the same species are carefully examined. Human populations, for example, exhibit variation in facial features, skin pigmentation, hair

Fig. 3.4. Variation in facial features, height, and other traits is apparent in human populations.

color and shape, body configuration, height and weight, blood groups, and so on (Fig. 3.4). We notice human differences more readily than variation in other organisms, but morphological variation has been recorded in many species, such as color and pattern in snails, butterflies, grasshoppers, ladybird beetles, mice, and birds; plants often differ in flower and seed color and pattern, as well as in growth habit. It is not immediately clear, however, how much of this morphological variation is due to genetic variation rather than to environmental effects, although gradually it has become established that genetic differences underlie morphological variation in many instances.

Geneticists have moreover discovered that there is much more genetic variation in organisms living in nature than is apparent through observation alone. This genetic variation has been revealed by inbreeding, i.e., by matings between close relatives, which increases the probability of homozygosis (being a homozygote). In this way recessive genes become expressed. Inbreeding has shown, for example, that virtually every *Drosophila* fly has allelic variants that in homozygous condition result in abnormal phenotypes, and that plants carry many gene variants that when homozygous result in abnormal or no chlorophyll. Inbreeding has also shown that organisms carry alleles that in homozygous condition affect their fitness, i.e., modify their fertility and survival probability.

A most convincing source of evidence indicating that genetic variation is pervasive comes from artificial selection experiments.

In artificial selection the individuals chosen to breed the next generation are those that exhibit the greatest expression of the desired characteristic. For example, if we want to increase the yield of wheat, every generation we choose the wheat plants with the greatest yield and use their seed to produce the following generation. If the selected population changes over the generations in the direction of the selection, it is clear that the original organisms had genetic variation with respect to the selected trait.

The changes obtained by artificial selection are often staggering. For example, the egg production in a flock of White Leghorn chickens increased from 125.6 eggs per hen per year in 1933 to 249.6 eggs per hen per year in 1965 (Fig. 3.5). Selection is often successfully practiced in opposite directions. Selection for high

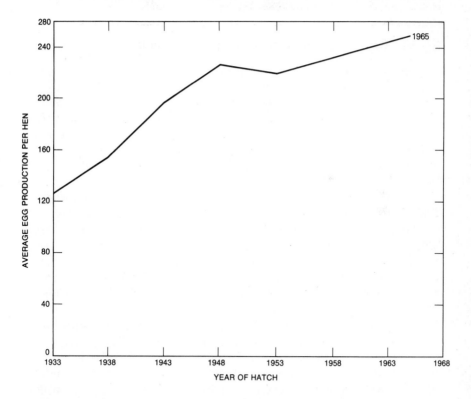

Fig. 3.5. An example of artificial selection: egg production per hen per year in a flock of White Leghorn chickens. In the formation stock the average production was 125.6 eggs. Thirty-two years later, selection had increased productivity to 249.6 eggs, double the initial number. The success of selection indicates that the flock had considerable genetic variability with respect to egg production. The economic significance of doubling the number of eggs laid in a year is obvious.

protein content in a variety of corn increased the protein content from 10.9 to 19.4 percent, while selection for low protein content reduced it from 10.9 to 4.9 percent. Figure 3.6 shows six parallel experiments for the body weight of mice at six weeks of age: selection for higher as well as for lower body weight succeeded in every case, while the unselected control lines showed no significant changes. Artificial selection has been successful for innumerable commercially desirable traits in many domesticated species, including cattle, swine, sheep, poultry, corn, rice, and wheat, as well as in many experimental organisms such as *Drosophila*, where artificial selection has succeeded for more than 50 different traits. The fact that artificial selection succeeds virtually every time it is tried indicates that genetic variation exists in populations for nearly every characteristic of the organisms.

The Problem of How to Measure Genetic Variation

The evidence mentioned in the previous section indicates that genetic variation is pervasive in natural populations, and hence that there is ample opportunity for evolutionary change. But we would like to go one step further and find out precisely how much variation there is. For example, what proportion of all gene loci are polymorphic (variable) in a given population, or what proportion of all gene loci are heterozygous in a typical individual of the population?

Until recently evolutionists were unable to answer such questions because the traditional methods of genetic analysis involve a methodological handicap when trying to solve this problem. Consider the process of finding out what proportion of the genes are polymorphic in a population. We cannot study every gene locus because we do not even know how many gene loci there are, not to mention the enormity of the task. The solution, then, is to look at only a sample of gene loci. Studying samples is what pollsters of public opinion do. If the sample is random, i.e. not biased and thus representative of the population, the values observed in the sample can be extrapolated to the whole population. Pollsters do quite well this way. For example, based on a sample of about 2,000 individuals they are able to predict with fair accuracy in what way many millions of Americans will vote in a presidential election.

In order to ascertain how many gene loci are polymorphic in a population we need to study a few genes that are an unbiased sample of all the gene loci. With the traditional methods of genetics this is impossible, because the existence of a gene is ascertained by examining the progenies of crosses between individuals showing

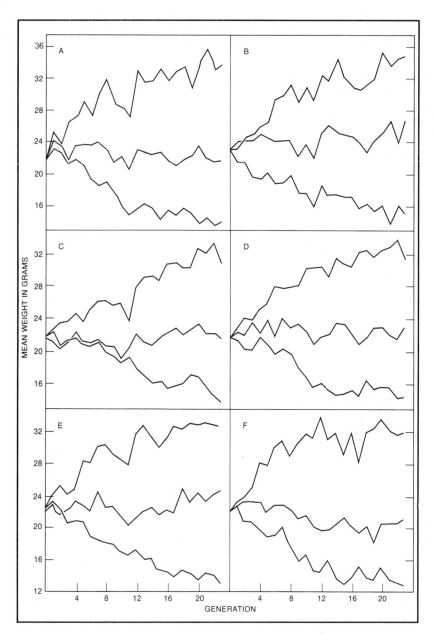

Fig. 3.6. Selection for body weight in mice when six weeks old. Six separate experiments were carried out. In each experiment one line was selected for greater weight, a second for lesser weight, and a third was left unselected. Selection was successful in every case, indicating the presence of genetic variation affecting body weight.

different forms of a given character; from the proportions of individuals in the various classes we infer whether one or more genes are involved. Therefore, the only genes we know to exist are those that are variable. Thus, there is no way of obtaining an unbiased sample of the genome because invariant genes cannot be included in the sample.

The way out of this dilemma became possible with the discoveries of molecular genetics. It is now known that the genetic information encoded in the nucleotide sequence of the DNA of structural genes is translated into a sequence of amino acids making up proteins. We can select for study a series of proteins without previously knowing whether or not they are variable in a population—a series of proteins that represent an unbiased sample of all the structural genes in the organisms. If a protein is found to be invariant among individuals, we can infer that the gene coding for the protein is also invariant. If the protein is variable, we know that the gene is variable, and we can measure how variable it is, i.e., how many variant forms of the protein exist, and in what frequencies.

Quantifying Genetic Variation

Since the early 1950s biochemists have known how to obtain the amino acid sequence of proteins. Therefore, one conceivable way to measure genetic variation in a natural population would be to pick up a fair number of proteins, say thirty, chosen without knowing whether or not they are variable in the population, so that they would represent an unbiased sample. Then, each of the thirty proteins could be sequenced in a number of individuals, say one hundred, to find out how much variation, if any, exists for each one of the proteins. The average amount of variation per protein found in the one hundred individuals for the thirty proteins would be an estimate of the amount of variation in the gene pool of the population.

Unhappily, obtaining the amino acid sequence of a protein is a very demanding task so that several months, or even years, are usually required to sequence each protein. Hence it is not feasible to sequence 3,000 proteins (thirty in each of one hundred individuals) for estimating genetic variation in each population we want to study. Fortunately, there is a technique, *gel electrophoresis* (see Closer Look 3.1), that makes possible the study of protein variation with only a moderate investment of time and money. Since the late 1960s, estimates of genetic variation have been obtained for natural populations of many organisms using gel electrophoresis (Figs. 3.7, 3.8, 3.9, and 3.10).

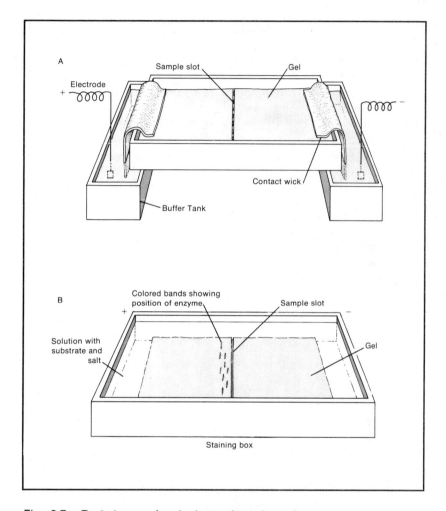

Fig. 3.7. Techniques of gel electrophoresis and enzyme assay used to measure genetic variation in natural populations. *A*. A tissue sample from each organism is homogenized to release the proteins in the tissue. The homogenate supernatants (liquid fractions) are placed in a gel made of starch, agar, polyacrylamide, or some other jellylike substance. The gel with the tissue samples is then subjected, usually for a few hours, to an electric current. Each protein in the samples thus will migrate in a direction at a rate that depends on the protein's net electrical charge and molecular size. *B*. After removing the gel from the electric field, it is treated with chemical solution containing a substrate specific for the enzyme to be assayed, and a salt. The enzyme catalyzes the reaction from the substrate to its product and this product then couples with the salt giving colored bands at the positions where the enzymes had migrated. The genotype at the gene locus coding for the enzyme can be determined for each individual from the number and position of the bands in the gels.

CLOSER LOOK 3.1   Gel Electrophoresis

The apparatus and procedures employed in gel electrophoresis for studying genetic variation in natural populations are shown in Figure 3.7. Tissue samples from organisms are individually "homogenized" (i.e., ground) in order to release the enzymes and other proteins. The liquid part (supernatant) of each homogenate is placed in a gel made of starch, polyacrylamide, or some other jellylike substance. The gel with the tissue samples is then subjected to an electric current for a given length of time. Each protein in the gel migrates in a direction and at a rate that depends on the protein's net electric charge and molecular size. After the gel is removed from the electric field, it is treated with a solution that contains a specific substrate for the enzyme to be assayed, and a coloring salt that reacts with the product of the reaction catalyzed by the enzyme. At the place in the gel to which the specific enzyme had migrated, a reaction will take place which can be symbolized as follows:

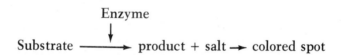

The method is most useful because the genotypes of the individuals in a sample can be simply inferred from the patterns observed in the gels. Figure 3.8 shows a gel that has been treated to reveal the position of the enzyme phosphoglucomutase; the gel contains the homogenates of 12 *Drosophila* flies. The gene locus coding for this enzyme may be represented as *Pgm*. The first and third individuals in the gel, starting from the left, have enzymes with different electrophoretic mobility, and thus with different amino acid sequences, which in turn implies that they are coded by different alleles. We can represent the alleles coding for the enzymes in the first and third individuals as Pgm^{100} and Pgm^{108}, respectively. (The superscripts indicate that the enzyme coded by allele Pgm^{108} migrates 8 mm farther in the gel than the enzyme coded by Pgm^{100}; this way of representing alleles is a common practice in electrophoretic studies, although letters such as S, M, and F are sometimes used to represent different alleles, as was done in Figure 2.15).

Because the first and third individuals each exhibit only one band we infer that they are homozygotes, with genotypes $Pgm^{100/100}$ and $Pgm^{108/108}$, respectively. The second individual in Figure 3.10 exhibits two colored spots; one of the spots has the same migration as that of the first individual and thus is coded by allele Pgm^{100}, while the other spot has the same migration as that of the third individual and thus is coded by allele Pgm^{108}. That is, the second individual is heterozygous with genotype $Pgm^{100/108}$.

Sometimes a protein, such as the enzyme malate dehydrogenase, shown in Figure 3.9, consists of two units (called polypeptides); heterozygotes will then exhibit three colored spots. We can represent the locus coding for malate dehydrogenase as Mdh. The first individual in Figure 3.9 shows only one band and thus is inferred to be homozygous, with genotype $Mdh^{94/94}$. The second individual is homozygous $Mdh^{104/104}$. A heterozygous individual has two kinds of polypeptides, which we can represent as a and b, coded respectively by alleles Mdh^{94} and Mdh^{104}. Three associations of two different units are possible, namely aa, ab, and bb. These correspond to the three colored spots which we see in the third individual of Figure 3.9.

There are proteins that consist of four and even more units; the patterns of heterozygous individuals will then show five or more colored bands, but the principles used to infer the genotypes from the electrophoretic patterns are similar to those just presented. The patterns shown in Figures 3.8 and 3.9 manifest the existence of two alleles at each locus. An invariant locus will be manifested by a colored spot that is the same in all individuals. On the other hand, more than two alleles are often found, as in Figure 3.10, which shows patterns of the enzyme acid phosphatase in *Drosophila*.

Electrophoretic techniques show what the genotypes of the individuals in a sample are: how many are homozygous, how many are heterozygous, and for what alleles. In order to obtain an estimate of how much variation exists in a population, about 20 or more gene loci are usually studied. The information obtained for all the loci needs to be summarized in a simple way that expresses how variable a population is and permits comparing one population with another. This can be accomplished in a variety of ways, but two measures of genetic variation are commonly used. One is

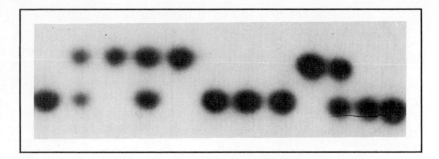

Fig. 3.8. An electrophoretic gel stained for the enzyme phosphoglucomutase. The gel contains tissue samples from each of 12 flies of *Drosophila pseudoobscura*. Flies with only one colored band are inferred to be homozygotes; flies with two bands are inferred to be heterozygotes. Enzymes with different migration are different in amino acid sequence and thus are coded by different alleles. The first individual on the left exhibits only one enzyme band and thus is inferred to be homozygous for the allele Pgm^{100}. The third individual also exhibits only one band, which migrates further than the band of the first individual; the third individual is then homozygous for allele Pgm^{108}. The second individual exhibits two bands; it is heterozygous for alleles Pgm^{100} and Pgm^{108}. Therefore, the genotypes of all 12 individuals are, from left to right: $Pgm^{100/100}$, $Pgm^{100/108}$, $Pgm^{108/108}$, $Pgm^{100/108}$, $Pgm^{108/108}$, $Pgm^{100/100}$, $Pgm^{100/100}$, $Pgm^{100/100}$, $Pgm^{108/108}$, $Pgm^{100/108}$, $Pgm^{100/100}$, and $Pgm^{100/100}$.

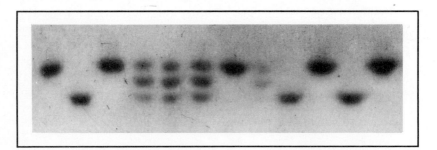

Fig. 3.9. An electrophoretic gel stained for the enzyme malate dehydrogenase. The gel contains tissue samples from each of 12 flies of *Drosophila equinoxialis*. Flies with only one colored band are, as in Figure 3.8, inferred to be homozygotes; but the heterozygotes exhibit three bands because malate dehydrogenase is a dimer (consisting of two units) enzyme. If the gene locus coding for this enzyme is represented as *Mdh*, the genotype of the second and ninth flies is inferred to be $Mdh^{94/94}$; the genotype of the first fly is inferred to be $Mdh^{104/104}$; flies fourth, fifth and sixth all have the heterozygous genotype $Mdh^{94/104}$, and so on. As in Figure 3.8, the numbers representing alleles refer to different amounts of migration of the enzymes coding for them.

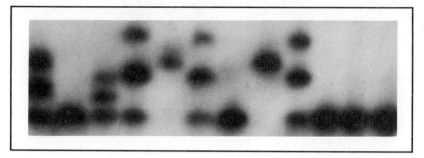

Fig. 3.10. An electrophoretic gel stained for the enzyme acid phosphatase. The gel contains tissue samples from each of 12 *Drosophila equinoxialis* flies. Acid phosphatase is a dimer enzyme and hence heterozygotes show three bands. Four different alleles (88, 96, 100, and 106) are manifested in the gel. The first fly on the left has the genotype $Acp^{88/100}$, the second $Acp^{88/88}$, the third $Acp^{88/96}$, the fourth $Acp^{88/106}$, the fifth $Acp^{100/100}$, and so on.

the amount of *polymorphism* in a population, which is simply the proportion of all gene loci studied that are polymorphic (i.e., in which more than one allele has been observed). The other is the *heterozygosity* of the population, which is the average frequency of heterozygous individuals per locus; first, one finds out how many individuals are heterozygous at each locus studied, and then these values are averaged over all the loci studied. Heterozygosity is a more precise measure of genetic variation than polymorphism and it is generally preferred by evolutionists, although it is often useful to have both measures.

CLOSER LOOK 3.2   Measures of Genetic Variation

There are many ways of expressing numerically how much variation exists in a natural population, but some measures are more precise than others. One measure is the *proportion of polymorphic loci,* or simply the *polymorphism,* in a population. Assume that using electrophoretic techniques we look at 30 gene loci in a certain species of shrimp and find no variation whatsoever at 18 loci, but some variation at 12 gene loci. We can say that $^{12}/_{30} = 0.40$, or 40 percent, of the loci are polymorphic in that population, or that the degree of polymorphism in that population is 0.40. Assume that we examine three other populations of the same shrimp species and that the numbers of polymorphic loci found in each of the populations, out of 30 loci studied in each, are 16, 14, and

14. We then can calculate the *average* degree of polymorphism over the four populations as follows:

	Polymorphic loci	Total number of loci	Polymorphism
Population 1	12	30	12/30 = 0.40
Population 2	16	30	16/30 = 0.53
Population 3	14	30	14/30 = 0.47
Population 4	14	30	14/30 = 0.47
Average			56/120 = 0.47

We conclude that, on the average, 47 percent of the loci exhibit genetic variation in populations of this shrimp species.

The amount of polymorphism is a useful measure of genetic variation in certain cases, but there are two reasons why it is not a very good measure. First, the value obtained depends on how many individuals are examined. For example, assume that in the first shrimp population mentioned above we examined 100 individuals. If we had examined more individuals we might have found variation at some of the 18 gene loci that appeared invariant; alternatively, if we had examined fewer individuals, some of the 12 variable loci might have appeared invariant. Scientists deal with this problem by establishing a *criterion of polymorphism;* for example, one criterion often used is that a gene locus is considered polymorphic only when the most common allele has a frequency no greater than 0.95. Then, as more individuals are examined, additional variants may be found, but on the average the proportion of polymorphic loci will not change. Of course, the degree of polymorphism will vary when we use different criteria. For example, if the criterion used is that the frequency of the most common allele be no greater than 0.98, it is possible that some loci will be considered polymorphic that are not with the 0.95 criterion (e.g., a locus at which there are two alleles with frequencies of 0.97 and 0.03).

There is a second reason why the degree of polymorphism is not a very good measure of genetic variation, namely that a gene locus slightly polymorphic counts the same as a very polymorphic one. Assume that at a certain locus there are only two alleles with frequencies of 0.95 and 0.05, while at another locus there are 20 alleles, each with a frequency of 0.05. It is obvious that more genetic variation exists at the second locus than at the first; yet both will count equally under the 0.95 criterion of polymorphism.

A better measure of genetic variation is the *average frequency of heterozygous individuals,* or simply the *heterozygosity* of the population. That is calculated by obtaining first the frequency of heterozygous individuals at each locus and then averaging these frequencies over all loci. For example, using the *Drosophila* samples shown in Figures 3.8, 3.9, and 3.10, we proceed as follows.

	Number of heterozygotes	Frequency of heterozygotes
Fig. 3.8	3	3/12 = 0.25
Fig. 3.9	5	5/12 = 0.42
Fig. 3.10	5	5/12 = 0.42
Heterozygosity (average of the above)		1.09/3 = 0.36

We conclude that the heterozygosity is 36 percent. Of course, in order for an estimate of heterozygosity to be valid it is necessary to examine more than three loci, but the procedure is as indicated. If several populations of the same species are examined, one may calculate first the heterozygosity in each population and then obtain the average over the various populations.

The heterozygosity of a population is the measure of genetic variation preferred by evolutionists. However, the observed heterozygosity does not reflect well the amount of genetic variation in populations of organisms that reproduce by self-fertilization, as some plants do, because most individuals will be homozygous even though different individuals will have different alleles if there is variation in the population. This difficulty can be overpassed by calculating the *expected* heterozygosity, calculated from the allelic frequencies *as if* the individuals in the population were mating with each other at random (see Chapter 4).

Genetic Variation in Natural Populations

Electrophoretic techniques were first applied to estimate genetic variation in natural populations in 1966, when three studies were published, one dealing with humans, the other two with *Drosophila* flies. Numerous populations of many organisms have been surveyed since that time and many more are studied every year. As an example, we shall examine the results of a study of *Euphausia superba,* a shrimplike crustacean that lives in the ocean around Antarctica and is a main food source of whales and other marine animals.

A total of 36 gene loci coding for enzymes were studied in 126 individuals. No variation at all was detected at 15 loci, while at each of the other 21 loci two or more alleles were found. The allelic frequencies at the 21 variable loci are given in Table 3.2.

Table 3.2
GENIC VARIATION IN *EUPHAUSIA SUPERBA*
Allelic frequencies at 21 variable gene loci. A locus is considered polymorphic when the frequency of the most common allele is no greater than 0.990. (Data from Ayala, Valentine, and Zumwalt, 1975.)

Gene locus	96	98	100	102	106	110	Frequency of hetero-zygotes	Is locus poly-morphic?
Acp-1			0.996			0.004	0.008	no
Ao-1		0.012	0.960	0.028			0.081	yes
Ald-1		0.012	0.988				0.024	yes
Ald-2		0.169	0.831				0.274	yes
Aph		0.004	0.996				0.008	no
Est-1	0.138		0.850	0.012			0.291	yes
Est-4		0.012	0.988				0.024	yes
Est-5		0.028	0.972				0.065	yes
G6pd		0.008	0.992				0.016	no
Got		0.402	0.594	0.004			0.449	yes
Hk-1	0.028		0.969	0.004			0.063	yes
Hk-2		0.004	0.996				0.008	no
Idh			0.996	0.004			0.009	no
Lap		0.004	0.996				0.008	no
Mdh-2	0.020		0.980				0.039	yes
Mdh-3	0.004	0.133	0.864				0.236	yes
Me-2		0.007	0.993				0.014	no
Odh		0.039	0.957	0.004			0.087	yes
Pgi	0.020		0.787	0.178	0.016		0.323	yes
To-2			0.988	0.012			0.024	yes
Xdh	0.004		0.996				0.008	no

No genetic variation was detected at 15 other gene loci not shown in the table. The average frequency of heterozygotes for the 36 loci, which include the invariant ones, is 0.057. There are 13 polymorphic loci using the criterion indicated at top of the table; therefore the polymorphism in this population is $13/36 = 0.361$.

The criterion of polymorphism used in Table 3.2 is that the most common allele has a frequency no greater than 0.990. With this criterion, 14 of the gene loci in the table (but of course none of the 15 invariant loci not included in the table) are polymorphic; therefore the proportion of polymorphic loci is $13/36 = 0.361$, or 36.1 percent. The average heterozygosity for all 36 gene loci is 0.057, or 5.7 percent.

A great amount of genetic variation exists in most natural populations studied. Table 3.3 summarizes the results for 125 animal species and eight plant species in which a fairly large number of gene loci have been surveyed. Among animals, invertebrates have more genetic variation than vertebrates, although there are exceptions. For the species in Table 3.3, the average heterozygosity is 13.4 percent for invertebrates and 6.0 percent for vertebrates. The average heterozygosity in man is 6.7 percent, very similar to the vertebrate average. Plants have great amounts of genetic variation; the average for eight species is 17 percent.

Table 3.3
GENIC VARIATION IN NATURAL POPULATIONS OF SOME MAJOR GROUPS OF ANIMALS AND PLANTS (After Selander, 1976, and other sources.)

Organisms	Number of species studied	Average number of loci studied per species	Proportion of polymorphic loci per population*	Proportion of heterozygous loci per individual
Invertebrates				
Drosophila	28	24	0.529	0.150
Wasps	6	15	0.243	0.062
Other insects	4	18	0.531	0.151
Marine	14	23	0.439	0.124
Land snails	5	18	0.437	0.150
Vertebrates				
Fish	14	21	0.306	0.078
Amphibians	11	22	0.336	0.082
Reptiles	9	21	0.231	0.047
Birds	4	19	0.145	0.042
Mammals	30	28	0.206	0.051
Average values				
Invertebrates	57	21 .8	0.469	0.134
Vertebrates	68	24 .1	0.247	0.060
Plants	8	8	0.464	0.170

*The criterion of polymorphism is not the same for all species.

The amount of genetic variation in most organisms is staggering. Consider man, with an average heterozygosity of 6.7 percent. If we assume that there are 100,000 structural gene loci in man, which is probably approximately correct a human individual would be heterozygous at 6,700 structural genes. Such an individual can potentially produce $2^{6700} = 10^{2017}$ different kinds of gametes. (An individual heterozygous at one locus can produce two different kinds of gametes, one with each allele; an individual heterozygous at n gene loci has the potentiality to produce 2^n different gametes. The number 10^{2017} is unity followed by 2,017 zeros.) Even if we assume that the number of structural gene loci in man is only 30,000, a person would be heterozygous at $30,000 \times 0.067 = 2,010$ gene loci, and could potentially produce $2^{2010} = 10^{605}$ different kinds of gametes. Such a number of gametes, however, will never be produced by any individual, nor by the whole of mankind, since that number is much greater than the number of atoms in the known universe, which is estimated as 10^{70}, a very small number by comparison. It follows that no two human gametes are ever identical, and that no two human individuals (except those derived from the same zygote such as identical twins) that exist now, have existed in the past, or will exist in the future are genetically identical. Such is the genetic basis of human individuality. And the same can be said of other organisms that reproduce sexually: no two individuals developed from separate zygotes are ever likely to be genetically identical.

The data obtained by gel electrophoresis also show that, as claimed by the balance model of population structure, there is a great amount of genetic variation in natural populations, and thus that there is ample opportunity for evolution to occur. Moreover, the amount of genetic variation in natural populations is probably even greater than estimates of gel electrophoresis indicate. This is so for two reasons. First, because of the redundancy of the genetic code, two or more triplets may code for the same amino acid, and thus not all differences in the DNA result in protein differences. Second, it is known that not all differences between proteins are detected by gel electrophoresis, although it is not yet known what proportion remains undetected.

CLOSER LOOK 3.3 Allele Frequencies in Electrophoretic Gels

The data directly obtained when looking at electrophoretic gels are genotype frequencies. These can be transformed into allelic frequencies simply by counting how many times each allele is found. A homozygous individual (such as the first individual in Figure 3.8, which has the genotype $Pgm^{100/100}$) has two copies of the same allele (Pgm^{100}), so the allele is counted twice. A heterozygous individual (such as the second fly in Figure 3.8, with the genotype $Pgm^{100/108}$) has two different alleles, so each allele (Pgm^{100} and Pgm^{108}) is counted once. Thus the allelic frequencies for the 12 flies shown in Figure 3.8 are as follows: allele *100* appears 15 times (three times in the three heterozygotes, and 12 times in the six homozygotes, twice in each one) and therefore has a frequency of $^{15}/_{24} = 0.625$ (it is divided by 24 because there are 12 individuals each with two alleles); allele *108* appears nine times (six times in the three homozygotes, and three times in the three heterozygotes), and therefore has a frequency of $^{9}/_{24} = 0.375$.

The symbols used to represent gene loci in Table 3.2 use some key letters in the names of the enzymes coded by the gene: *Acp-1* is a gene locus coding for an acid phosphatase enzyme, *Ao* codes for an aldehyde oxidase enzyme, and so on. Sometimes the symbol for a gene locus is followed by a number, because there are several enzymes of the same kind each coded by a different gene locus. For example, there are two aldolase enzymes in *Euphausia superba,* one coded by the locus *Ald-1,* the other coded by *Ald-2.*

The numbers used to represent the alleles are somewhat arbitrary, although they make reference to how far the encoded protein migrates in the gels. In Table 3.2, as it is often done in electrophoretic studies, the most common allele is named *100;* enzymes represented by numbers greater than 100 migrate farther than the enzyme coded by allele *100,* while those with lower numbers do not migrate as far as the enzyme coded by allele *100.*

We see in Table 3.2 that some gene loci, such as *Acp-1,* have only two alleles, while several have three alleles and one locus, *Pgi,* has four alleles. The proportion of heterozygotes at each locus ranges from zero at the 15 invariant loci, and 0.008 at *Acp-1* and other loci, to 0.449 at the *Got* locus. The average frequency of heterozygotes is, for all 36 gene loci studied, 0.057 (5.7 percent).

The Origin of Hereditary Variation

What are the sources of the genetic variation found in natural populations and of the genetic differences among species? The origin of life occurred three-and-a-half to four billion years ago; the primordial living beings were very simple and very small (Chapter 11). Yet all living species have evolved from these lowly beginnings; there live at present more than two million species, most diverse in size, shape, and ways of life, and also in the amount and kind of DNA sequences that contain their genetic information. Because this is so, there must be means in evolution by which existing DNA sequences are changed and new sequences are incorporated into the genome of organisms.

Heredity is a conservative process, but not perfectly so—otherwise evolution could not have occurred. The information encoded in the nucleotide sequence of DNA is, as a rule, faithfully reproduced during replication, so that each replication results in two DNA molecules identical to each other and to the parental one. Occasionally, however, "mistakes" in the process of replication lead to different nucleotide sequences in parental and daughter DNA molecules, or to different amounts of DNA in parental and daughter cells. These changes in the hereditary materials are known as *mutations*. They can be classified in the following categories.

Gene mutations (or point mutations), which affect only one or a few nucleotides within a gene.

Chromosomal mutations (or chromosomal aberrations), which affect the number of chromosomes, or the number or the arrangement of genes in a chromosome. Chromosomal mutations can be subdivided as follows:

1. Changes in the *location of genes* on the chromosomes (Fig. 3.11).

A. *Inversion*, when the location of a block of genes is inverted within a chromosome. If the rotated segment includes the centromere (spindle attachment), the inversion is called *pericentric;* otherwise, the inversion is *paracentric.*

B. *Translocation*, when the location of a block of genes is changed in the chromosomes. The most common forms of translocations are *reciprocal*, involving an exchange of blocks of genes between two nonhomologous chromosomes. A chromosomal segment may also move to a new location within the same chromosome or in a different chromosome without reciprocal exchange; these kinds of translocations are sometimes called *transpositions.*

2. Changes in the *number of genes* in chromosomes (Fig. 3.12).

A. *Deletion* (or deficiency), when a segment of DNA containing one or several genes is lost from a chromosome.

B. *Duplication*, when a segment of DNA containing one or more genes is present more than once in a set of chromosomes. Often duplications occur in tandem, i.e., the two duplicated segments lie next to each other in the same chromosome.

3. Changes in the *number of chromosomes*. These are of four kinds; the first two do not affect the total amount of hereditary material, but the other two do (Figs. 3.13 and 3.14).

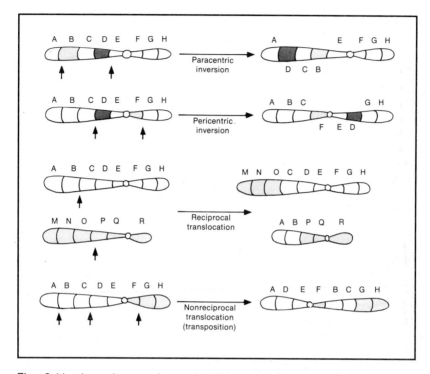

Fig. 3.11. Inversions and translocations are chromosomal mutations that change the location of genes in chromosomes. An inversion occurs when a block of genes rotates 180 degrees within a chromosome. Paracentric inversions are those not involving the centromere. Pericentric inversions include the centromere. A reciprocal translocation occurs when two blocks of genes are exchanged between two nonhomologous chromosomes. Transpositions are a form of translocation shifting the location of a block of genes from one to another position in the same chromosome.

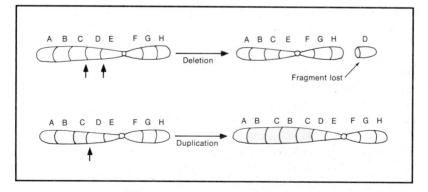

Fig. 3.12. A deletion occurs when part of a chromosome is lost, a duplication when part of a chromosome is doubled.

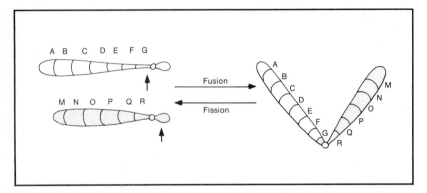

Fig. 3.13. A fusion takes place when two chromosomes fuse into one. A fission occurs when a chromosome splits into two; a new centromere must be produced, because otherwise the chromosomal segment without centromere becomes lost during cell division.

A. *Fusion,* when two nonhomologous chromosomes fuse into one. This involves the loss of one centromere.

B. *Fission,* when one chromosome splits into two. An additional centromere must be produced because otherwise the chromosome without a centromere would be lost when the cell divides.

C. *Aneuploidy,* when one or more chromosomes of a normal set are lacking or present in excess. In diploid organisms, the terms *nullosomic, monosomic, trisomic, tetrasomic,* and so on, refer to the occurrence of a given chromosome zero times, once, three times, four times, and so on.

D. *Haploidy* and *polyploidy,* when the number of *sets* of chromosomes is other than two. Most organisms are *diploid,* i.e., they have two sets of chromosomes in their somatic (body) cells, but only one set in their gametic cells. Some organisms are normally *haploid,* i.e., they have only one set of chromosomes. Haploid and diploid organisms both exist in certain social insects, such as the honeybee, where the males are haploid and develop from unfertilized eggs, while the females are diploid and develop from fertilized eggs. Polyploid organisms have more than two sets of chromosomes; the organism is said to be *triploid* if it contains three sets of chromosomes, *tetraploid* if it contains four sets of chromosomes, and so on. The more common forms of polyploidy are those involving sets of chromosomes in multiples of two, i.e., tetraploids, hexaploids, and octoploids, which respectively have four, six, and eight sets of chromosomes. Polyploidy is very common in some groups of plants but is much rarer in animals. The origin of new species by polyploidy is discussed in Chapter 6.

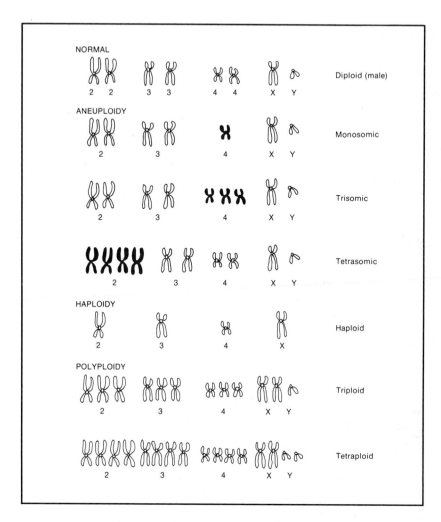

Fig. 3.14. Aneuploidy, haploidy, and polyploidy. Aneuploidy exists when one or more (but not all) chromosomes of a normal set are lacking or present in excess. A haploid is an organism consisting of only one full chromosome set. Polyploids contain more than two full sets of chromosomes.

Gene Mutations

Gene or point mutations occur when the DNA sequence of a gene is altered and the new nucleotide sequence is passed to the offspring. There are two kinds of gene mutations: *base-pair substitutions* are due to the substitution of one or a few nucleotide pairs for others; *frameshift mutations* are due to the addition or deletion of one or a few nucleotides.

Base-pair substitutions in the nucleotide sequence of a structural gene often result in a change in the amino acid sequence of the protein encoded by the gene, but this is not always the case owing to the redundancy of the genetic code. Consider the genetic code shown in Figure 2.21. In the third square on the first line we find the messenger RNA triplet UAU coding for the amino acid tyrosine. A mutation leading to the triplet UAC still codes for tyrosine, while a mutation leading to the triplet UCU will code for serine and will thus result in an amino acid substitution in the encoded protein (Fig. 3.15). Substitutions in the second nucleotide of a triplet always result in an amino acid substitution (or in a terminating signal); changes in the first nucleotide nearly always do (the exceptions are the changes from UUA or UUG to CUA and CUG, or vice versa, all of which code for leucine, and the changes from AGA or AGG to CGA and CGG, or vice versa, all of which code for arginine). However, substitutions in the third nucleotide of a triplet often do not lead to amino acid substitutions in the encoded protein because most of the redundancy of the genetic code affects the third position of triplets.

Some base-pair substitutions may change a triplet coding for an amino acid into a terminating triplet, or vice versa. For example, a mutation changing the messenger RNA triplet UAU, coding for tyrosine, into the triplet UAA, which is a terminating signal (Fig. 3.15). This type of base substitution will result in proteins with changed lengths, since the nucleotide sequence is not translated beyond a terminating signal.

Frameshift mutations often result in a very altered sequence of amino acids in the translated protein. The addition or deletion of one or more (other than exact multiples of three) nucleotide pairs shifts the "reading frame" of the nucleotide sequence from

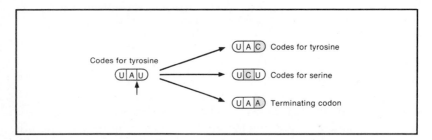

Fig. 3.15. Mutations replacing a nucleotide by another are called base-pair substitutions. Sometimes a base-pair substitution may change a triplet coding for an amino acid into a terminating triplet; the reverse change may also occur.

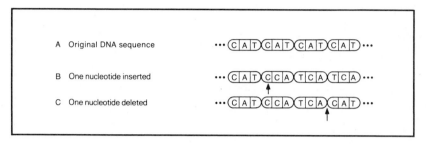

A   Original DNA sequence

B   One nucleotide inserted

C   One nucleotide deleted

Fig. 3.16. Frameshift mutations. *A.* A segment of the original sequence in one of the DNA strands is represented, with the triplets as they would be "read" separated by hyphens. *B.* A nucleotide has been inserted in the position indicated by the arrow: the "reading frame" changes from that position on. *C.* A nucleotide has been deleted from *B* in the position indicated by the arrow: the original reading frame *A* is restored from that position until the end of the molecule.

the point of an insertion or deletion to the end of the molecule (Fig. 3.16). If one nucleotide pair is inserted at some point and another is deleted at some other point, the original reading frame and the corresponding amino acid sequence are restored after the second mutational change.

Gene mutations may occur spontaneously, i.e., without intentional causation by humans. Mutations may also be induced by ultraviolet light, X rays, and other high-frequency radiations, as well as by exposing organisms to certain chemicals, such as mustard gas and many others, called *mutagens*.

Gene mutations may have effects ranging from negligible to lethal. A base-pair substitution that results in no amino acid changes in the encoded protein may have little or no effect on the ability of an organism to survive and reproduce. Mutations changing one or even several amino acids may also have small or no detectable effect on the organism if the essential biological function of the coded protein is not affected. For example, *Drosophila* flies homozygous for either one of the two forms of the phosphoglucomutase enzyme shown in Figure 3.8 are quite able to survive and reproduce. However, the consequences of an amino acid substitution may be severe when the substitution affects the active site of an enzyme or modifies in some other way an essential function of a protein (Fig. 3.17).

The deleterious effects of mutations often depend on particular environmental conditions. For example, in *Drosophila* there is a class of mutants known as "temperature-sensitive." At stan-

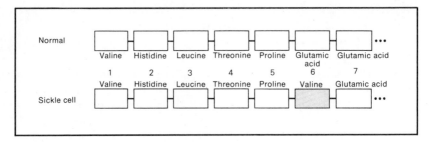

Fig. 3.17. The beta chain of human hemoglobin consists of 146 amino acids. The first seven amino acids are represented. A substitution of valine for glutamic acid in the sixth position is responsible for the severe condition known as sickle-cell anemia. Glutamic acid is coded by the triplets GAA or GAG in messenger RNA. A mutation changing the second nucleotide from A to U results in the triplets GUA or GUG, which code for valine.

dard temperatures of 20°C to 25°C, flies homozygous for these mutants live and reproduce more or less normally. But at temperatures about 28°C these flies become paralyzed or die, although wild-type flies can still function normally. In humans, phenylketonuria (PKU) is a severe disease caused by homozygosis for a recessive allele. However, individuals homozygous for the PKU allele may be effectively normal if they maintain a diet free of phenylalanine, because their problems arise from an inability to metabolize this amino acid.

Newly arisen mutations are more likely to be deleterious than beneficial to their carriers because mutations are random events with respect to adaptation. In other words, they occur independently of whether they have beneficial or harmful consequences. However, the allelic variants already present in a population have been subject to natural selection. If they occur in substantial frequencies in a population it is because they improve the adaptation of their carriers relative to alternative alleles that have been eliminated or kept at low frequencies by natural selection. A newly arisen mutant is likely to have arisen by mutation in the previous history of a population; if it does not exist in substantial frequencies this is because it is not beneficial to the organisms.

The argument advanced in the previous paragraph may be illustrated with an analogy. Assume that we have an English sentence whose words have been chosen because together they express a certain idea. If single letters or words are replaced with others at random, most changes are unlikely to improve the meaning of the sentence; very likely they will destroy it. The nucleotide sequence of a gene has been "edited" by natural selection, because it makes "sense"; if the sequence is changed at random, the "meaning" will rarely be improved and often will be hampered or destroyed.

Occasionally, however, a newly arisen mutation may increase adaptation. The probability of such an event is greater when organisms colonize a new territory, or when environmental changes confront a population with new challenges. In these cases, the adaptation of a population is less than optimal and there is greater opportunity for new mutations to be adaptive. This again illustrates the point that the effects of mutations depend on the environment. A mutation increasing the density of hair may be adaptive in, say, a population of mice that have colonized Alaska, but it is likely to be selected against in a population living in the tropics. Increased melanin pigmentation may be beneficial to people living in tropical Africa, where dark skin protects from the sun's ultraviolet radiation, but not in Scandinavia where the intensity of sunlight is low and light skin facilitates the synthesis of vitamin D. Mutations to drug resistance in microorganisms are additional examples: a mutant making bacteria resistant to streptomycin may be beneficial to the bacteria in the presence of the drug but not in its absence.

Rates of Mutation and Evolution

Mutation rates have been measured in a great variety of organisms, mostly for mutants having conspicuous effects. Mutation rates are generally lower in bacteria and other microorganisms than in multicellular organisms. In the common bacterium of the human intestine, *Escherichia coli,* most mutations appear at rates around one mutant per 100 million (1×10^{-8}) or per billion (1×10^{-9}) cells. In humans and other multicellular organisms, mutants typically appear at about one per 100,000 (1×10^{-5}) or one per million (1×10^{-6}) gametes. There is, however, considerable variation from gene to gene as well as from organism to organism, as shown in Table 3.4.

Important for evolution are mutations that do not severely harm their carriers, but rather modify slightly their ability to survive and reproduce. Allelic variants detected by electrophoresis are of this kind. In *Drosophila,* mutations to new electrophoretic variants occur with frequencies of about four per locus per million gametes. Mutations with small effects on viability have also been measured in *Drosophila* by examining not single genes but whole chromosomes. Second chromosomes of *D. melanogaster* acquire

Table 3.4
MUTATION RATES OF SOME GENES IN VARIOUS ORGANISMS

In bacteria and other microorganisms mutation rates are measured as their frequency per cell; in multicellular organisms they are measured as their frequency per gamete.

Organism and mutation	Rate of mutation
Escherichia coli (bacterium)	
Streptomycin resistance	4×10^{-10}
Sensitivity to phage *T1*	2×10^{-8}
Lactose fermentation	2×10^{-7}
Chlamydomonas reinhardi (alga)	
Streptomycin resistance	1×10^{-6}
Neurospora crassa (fungus)	
Adenine independence	4×10^{-8}
Zea mays (corn)	
Shrunken seeds	1×10^{-6}
Purple seeds	1×10^{-5}
Drosophila melanogaster (vinegar fly)	
Electrophoretic variants	4×10^{-6}
White eye	4×10^{-5}
Mus musculus (mouse)	
Brown coat	8×10^{-6}
Homo sapiens (man)	
Huntington's chorea (degenerative disease of nervous system)	1×10^{-6}
Retinoblastoma (eye tumor)	1×10^{-5}
Neurofibromatosis (tumor of nervous tissue)	2×10^{-4}

such mutations at rates no less than 0.1411 per generation. We do not know how many gene loci exist in the second chromosome of *D. melanogaster,* and consequently we do not know the rate of these mutations per gene. But if we assume that there are 4,000 gene loci in the second chromosome, viability mutations would occur in *D. melanogaster* at a rate no less than $0.1411/4,000 = 3.5 \times 10^{-5}$ per locus, which is not very different from the mutation rate to alleles with visible effects.

The mutation rates given in Table 3.4 are only approximate. Mutation rates are difficult to measure because they are rare events. Moreover, they are under genetic control and thus may be different when measured in different populations. For example, lethal mutation rates observed in second chromosomes of *Drosophila willistoni* were about three times larger in a population from southern Brazil (0.0171 per chromosome per generation) than in a population near the mouth of the Amazon River (0.0057 per chromosome per generation).

Although mutation rates are low, new mutants appear continuously in nature. This is because there are many individuals in any species and many gene loci in each individual. For example, a typical insect species may consist of about 100 million (10^8) individuals. If we assume that the average mutation rate per locus is one per 100,000 gametes (1×10^{-5}), the average number of mutations newly appearing in an insect species would be $2 \times 10^8 \times 10^{-5} = 2,000$ per locus. (The mutation rate is multiplied by the number of individuals, and then by two because each individual results from the union of two gametes.)

Therefore the process of mutation provides species with plenty of new genetic variation every generation. It is thus not surprising to see species becoming adapted to new environmental challenges. For example, many insect species have developed resistance to DDT in different parts of the world where spraying has been intense. Although the insects had never before encountered this synthetic compound, they became adapted to it rapidly by means of mutations that allow them to survive in the presence of DDT. Similarly, *industrial melanism*, that is, darkening of the wings, has evolved in many species of moths and butterflies in industrialized regions (see Chapter 4). If the genetic variants required to face a given environmental challenge are not already present in a population, they are likely to arise soon by mutation.

The probability that a given individual will have a new mutation at a certain locus is low. This probability is simply the mutation rate multiplied by two because each diploid individual arises from two gametes. However, the probability that a given individual will have some new mutation *anywhere in the genome* is not low. Consider, for example, *Drosophila melanogaster*. If we assume that it has 10,000 (10^4) gene loci and that the average mutation rate per locus per gamete is 10^{-5}, the probability that a fly will carry a new mutation is $2 \times 10^4 \times 10^{-5} = 0.2$. Humans may have as many as 100,000 (10^5) gene loci. If we assume the same mutation rate as for *Drosophila*, the probability that each human being will have an allelic variant not present in its parents is $2 \times 10^5 \times 10^{-5} = 2$. That is, on the average each human being carries about two new mutations.

The potential of the mutation process to generate new hereditary variation is indeed enormous. Yet the variation arising in each generation by mutation is only a small fraction of the total amount of genetic variation present in natural populations. Gel electrophoresis indicates that individuals of sexually reproducing species are usually heterozygous at 5 to 20 percent of their loci (Table 3.3). Assume that, on the average, an individual is heterozygous at 10 percent of its loci, and that it has 10,000 loci; it will then have two different alleles at each of 1,000 gene loci. As we have calculated above, the same individual has a 0.2 probability of acquiring a new allelic variant by mutation. That is, the amount of genetic variation present in a population is about 5,000 times greater than that acquired each generation by mutation.

Although these calculations are very rough, they show that newly arisen mutations represent only a small fraction of the genetic variation present in populations at any one time. It follows that changes in mutation rates are not likely to have any immediate effects on the rate of evolution of a population. Even if mutation rates were to increase by a factor of ten, mutations newly induced in any one generation would still represent a very small fraction of the variation already present in populations of sexually reproducing organisms.

Inversions and Translocations

Inversions and translocations are chromosomal mutations that do not change either the number of chromosomes or the number of genes in the chromosomes; they modify the arrangement of genes in the chromosomes. Inversions are 180-degree rotations of chro-

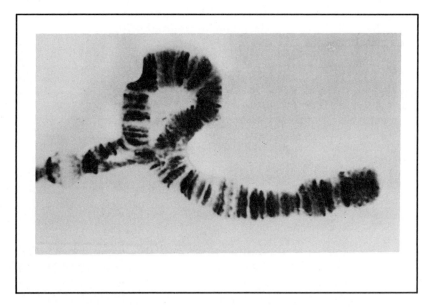

Fig. 3.18. A segment of the third chromosome of a *Drosophila pseudoobscura* fly heterozygous for two chromosomes differing by an inversion. The chromosomes shown are "polytene" or "giant" chromosomes present in the salivary glands, the gut, and other body parts of dipterans (flies and mosquitoes). Polytene chromosomes result from the pairing of the two homologous chromosomes, each replicated many times.

mosomal segments; if the gene sequence of a chromosome is represented as ABCDEF, inversion of the segment BCD will result in a chromosome with the sequence ADCBEF.

In individuals heterozygous for an inverted and a normal chromosome sequence, pairing of the chromosomes at meiosis requires the formation of a loop involving the inverted segment (Fig. 3.18). Genetic recombination is suppressed in the progenies of such inversion heterozygotes, because when recombination occurs the only gametes that can produce viable individuals are those containing the noncrossover chromosomes (Figs. 3.19 and 3.20). Individuals heterozygous for chromosomal inversions may, therefore, have reduced fertility; although sometimes, as in *Drosophila* or the midge *Chironomus,* this is not the case because gametes containing abnormal chromosomes fail to develop or to function.

Reciprocal translocations involve the interchange of blocks of genes between nonhomologous chromosomes. In individuals heterozygous for a translocation, pairing during meiosis results in a cross-shaped configuration (Fig. 3.21). As shown in the figure, segregation at meiosis may occur in a variety of ways. Of the six types of gametes that can be formed, only the two on the left contain all the chromosomal segments once and only once. All other gametes have some chromosome segments duplicated and some missing, and therefore cannot result in normal progeny. Two consequences follow. One is that translocation heterozygotes are usually semisterile, since some gametes cannot produce viable individuals. Another is that all genes in the two translocated chromosomes behave as if they were linked. This is because the only normal gametes are those having either the two nontranslocated chromosomes or the two translocated chromosomes.

Deletions, Duplications, and Evolution of Genome Size

Additions and subtractions of the amount of hereditary material may occur by means of chromosomal duplications and deletions. Chromosomal deletions are often lethal in homozygous condition because genes essential to the organism are missing. Deletions may not be lethal in homozygotes if the deleted genes have been previously duplicated and their function can still be carried out by the remaining genes or when the deleted genes are no longer essential to the organism. There is evidence that chromosomal deletions occur in evolution reducing the overall amount of DNA of an organism. In heterozygous condition, small deletions may not seriously impair organisms if the functions of the missing genes can be carried out by those in the homologous chromosome.

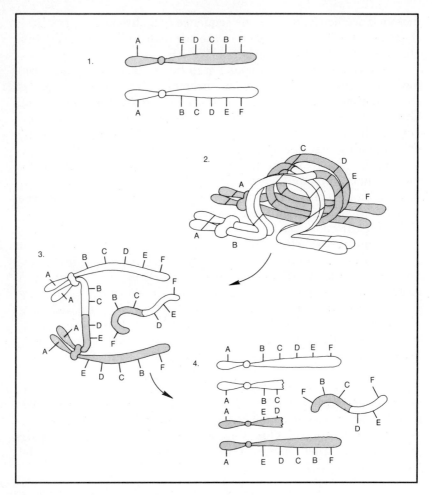

Fig. 3.19. Crossing over in a heterozygote for a paracentric inversion. *1.* Two homologous chromosomes in the body cells of the individual. The chromosomal segment from B to E is inverted in one chromosome relative to the other. Since the centromere is not included in the inverted segment, the inversion is *paracentric*. *2.* Pairing of the homologous chromosomes in inversion heterozygotes requires the formation of a loop so that homologous genes will pair with each other. Crossing over is occurring between two of the duplicated chromosomes. *3.* Separation of the chromosomes at early anaphase of the first meiotic division. One chromosomal segment has two centromeres and will eventually break up as the centromeres move away from each other; another chromosomal segment has no centromere and will be lost. *4.* The resulting chromosomes. Only two chromosomes have complete sets of genes. These are noncrossover chromosomes with the same gene sequences as the two original chromosomes.

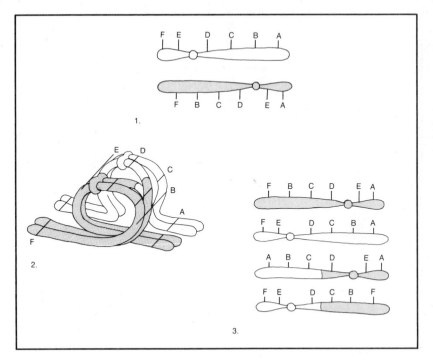

Fig. 3.20.   Crossing over in a heterozygote for a pericentric inversion. *1.* The two chromosomes. The inverted segment from B to E includes the centromere, and therefore the inversion is *pericentric.* *2.* Crossing over between two of the duplicated chromosomes. *3.* The four resulting chromosomes. Only the two top chromosomes have complete sets of genes. They are noncrossover chromosomes with the same gene sequences as the two original chromosomes.

Duplications of genetic material, followed by divergence of the duplicated DNA toward fulfilling different functions, have played a major role in evolution. An organism ancestral to all DNA-containing living organisms probably had a short DNA double helix consisting of very few genes. Today, many organisms contain thousands of gene loci descended from that ancestral DNA segment through multiple duplications and gene mutations. The amount of DNA in organisms has evolved by means of chromosomal duplications and deletions and by polyploidy (see p. 107).

Around 1950 it was discovered that considerable variation in the amount of DNA per cell exists between organisms of different species. Organisms can be classified into four broad classes according to the amount of DNA they carry in each cell (Fig. 3.22). The lowest amounts of DNA are found in some viruses, with about 10^4

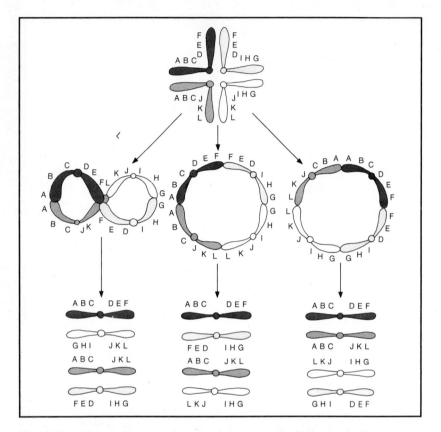

Fig. 3.21. Meiosis in a translocation heterozygote. At the top is the cross-shaped configuration formed when the chromosomes pair at the beginning of meiosis. The second row shows the three configurations that may occur at metaphase—a twisted ring and two open rings. The two lower rows show the six types of gametes formed. Only the two types on the left contain a complete set of genes; the other four contain some duplicated, and some missing, chromosomal segments. (For convenience, the two chromatids of each chromosome are not shown.)

nucleotide pairs per virus. Bacteria have, on the average, about 4×10^6 nucleotide pairs per cell, and fungi about ten times as much, or 4×10^7 nucleotide pairs. Most animals and many plants have considerably more DNA, about 2×10^9 nucleotide pairs per cell on the average. The most advanced plants, gymnosperms and angiosperms, often have 10^{10} and even more nucleotide pairs per cell. The animals with larger amounts of DNA are salamanders and some primitive fishes, with 10^{10} and even more nucleotide pairs per cell.

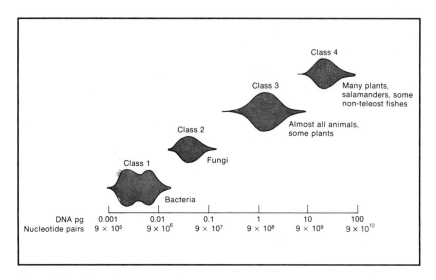

Fig. 3.22. Organisms classified according to their amounts of DNA (viruses are not included). The amount of DNA is given by weight (one picogram, pg = 10^{-12} gram) or by the number of nucleotide pairs. Most organisms within each group have amounts of DNA within one order of magnitude, i.e., differing by a factor from one to ten. The amount of DNA for all organisms from bacteria to plants and animals varies by more than five orders of magnitude, i.e., by a factor of more than 100,000. (After Hinegardner, 1976.)

As a result of evolution there has been considerable change in the amount of DNA per cell. A substantial increase occurs from bacteria to fungi, and then to animals and plants. More complex organisms need more DNA than a bacterium or a mold, but there seems to be no consistent relationship between the amount of DNA of an organism and its complexity of organization. For example, salamanders and flowering plants are not ten times more complex than mammals or birds, although some have ten times more DNA. However, in some groups of organisms, such as the invertebrates, there seems to be an indication of some relationship between complexity of organization and amount of DNA. Sponges and coelenterates have less DNA per cell than echinoderms, annelids, crustaceans, and mollusks. Among mollusks, limpets, snails, and chitons have less DNA per cell than the more advanced squids. It seems that there is a minimum amount of DNA required to achieve a certain degree of complexity of organization. Organisms, however, often have much more DNA than the minimum required to achieve their degree of complexity. Figure 3.23 shows that the minimum amount of DNA in each group of organisms tends to be greater as the complexity of organization of the organisms increases.

How does the amount of DNA in the nucleus increase during evolution? Polyploidy is one process by which the amount of DNA can increase. When the number of chromosomes per cell is doubled, the amount of DNA is also doubled. Some organisms with very large amounts of DNA, such as some primitive vascular plants (*Psilopsida*), are polyploid. Polyploidy is a common phenomenon in the evolution of most groups of plants but is rare in animals.

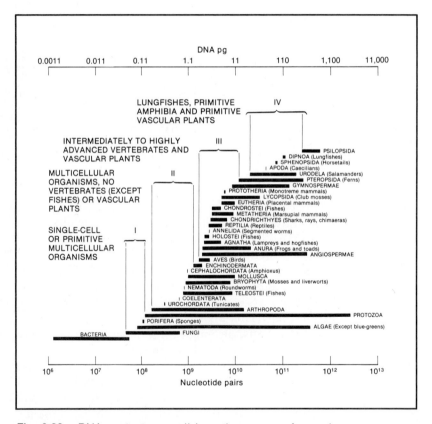

Fig. 3.23. DNA content per cell in major groups of organisms arranged so that the *minimum* amount within a group increases from the lower to the upper part. The minimum amount of DNA per cell tends to be greater in more complex organisms, although there are exceptions; a notable exception is that lungfishes, amphibians, and primitive vascular plants have the largest amounts of DNA although they are not more complex than, say, mammals and birds. (After Sparrow, Price, and Underbrink, 1972.)

Deletions and duplications of relatively small segments of DNA are the most general processes by which evolutionary changes in the amount of DNA have taken place. When the genome sizes (i.e., DNA content per cell) of many fish, frog, and mammal species are arranged in a frequency diagram, they are seen to vary around an intermediate mode (Fig. 3.24). This indicates that evolutionary changes in the genome size of animals are numerous and individually small, as would be the case with duplications and deletions. If changes in the amount of DNA would have occurred mostly by polyploidy, organisms would differ in the

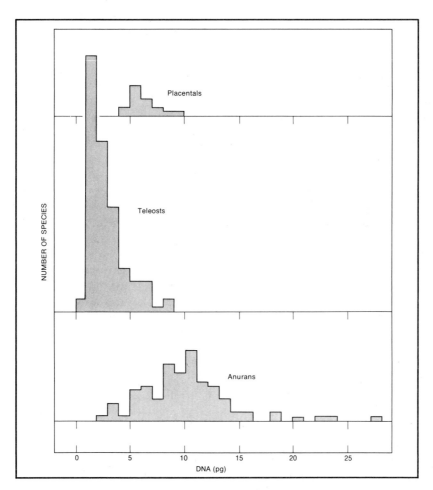

Fig. 3.24. Distribution of DNA per cell in mammals, fish, and amphibians. The distributions vary around an intermediate mode. This suggests that evolutionary changes have been numerous and small. (After Bachmann, Goin, and Goin, 1972.)

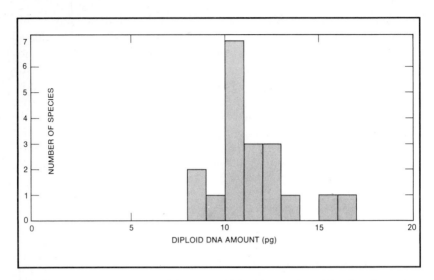

Fig. 3.25. Distribution of DNA per cell in 19 species of toads of the genus *Bufo*. The shape of the distribution suggests that changes in DNA amounts within this genus have occurred by small increments and decrements rather than by polyploidy, which would have produced a distribution where amounts of DNA would differ by exact multiples. (After Bachmann, Goin, and Goin, 1972.)

amount of DNA by exact multiples (double, quadruple, and so on).

The gradual change in DNA amount per cell can also be observed within a single genus, such as the toad *Bufo*. The DNA content has been determined in 19 of the 250 known species and ranges from about 7×10^9 to 15×10^9 nucleotide pairs, with a mode around 10×10^9 (Fig. 3.25). Thus, the transition from about 7×10^9 nucleotide pairs to about twice as much has occurred not by polyploidy, but by cumulative addition of small amounts of DNA, resulting in a fairly smooth and continuous distribution.

Evolution of Duplicated DNA Sequences

Duplications of chromosomal segments often involve only one or a few genes. In recent years it has been discovered that many DNA sequences have originated by duplication, followed in some cases, but not in others, by evolutionary divergence of the duplicated sequences. Of course, if the duplicated DNA sequences have diverged substantially, it may no longer be possible to identify them as originally identical. As stated earlier, all genes must have originated by the duplication of a single or a few original ones. DNA sequences recognizable as duplicated may be classified into these general classes:

1. Duplications of single structural gene loci followed by divergent evolution of the duplicated genes so as to fulfill different functions.

2. Genes that exist in several copies within each genome but which in many cases remain essentially identical to each other in DNA sequence and in function. The presence of several copies of a single gene allows the organism to obtain large amounts of gene product in a short time.

3. Short sequences of DNA in eukaryotes (organisms whose cells have a nucleus) that are repeated many times, from a thousand to more than a million, although not all copies may be identical. These highly repetitive sequences of DNA may be involved in gene regulation.

We shall now discuss these three classes of duplications in turn.

Myoglobins and hemoglobins are proteins involved in respiration (oxygen transport), myoglobins in muscle and hemoglobins in blood. Myoglobins consist of a single polypeptide chain, arranged in a complex three-dimensional structure, with a molecular weight of about 17,000. Hemoglobin molecules generally consist of four subunits—two polypeptide chains of one kind and two of another—and have a molecular weight of about 67,000. In human adults there are two types of hemoglobin, A and A_2. Hemoglobin A, the most common, consists of two α and two β polypeptide chains, which can be represented as $\alpha_2\beta_2$ Hemoglobin A_2, which makes up only about 2 percent of adult hemoglobin, consists of two α and two δ chains ($\alpha_2\delta_2$). A third kind of hemoglobin, found in human embryos, is called fetal hemoglobin and consists of two α and two γ chains ($\alpha_2\gamma_2$). The α, β, δ, and γ polypeptide chains are each coded by a different gene.

The genes coding for myoglobin and the hemoglobins in vertebrates have evolved by a series of gene duplications followed by gradual divergence toward different but related functions (Fig. 3.26). During the evolution of the duplicated globin genes, nucleotide substitutions, as well as nucleotide additions and deletions, have taken place. Human myoglobin consists of 153 amino acids; the α chain has 141 amino acids, and the β, γ, and δ chains consist of 146 amino acids each. Each globin gene has also changed in different lines of descent through evolution. The α chains of humans and chimpanzees are identical, but the α chains of humans and

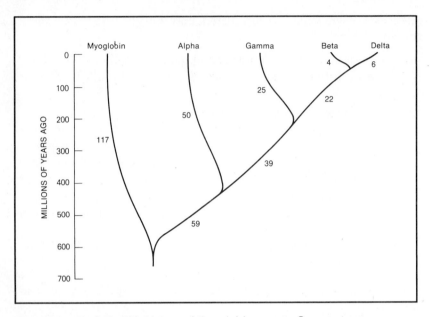

Fig. 3.26. Evolutionary history of the globin genes. Oxygen transport is mediated by myoglobin in muscle and by hemoglobin in blood. The black dots indicate where the ancestral genes were duplicated giving rise to a new gene line. The minimum number of nucleotide replacements required to account for the amino acid differences between the proteins are indicated along the branches. The first gene duplication occurred about 600 million years ago; one gene coding for myoglobin and the other being the ancestor of the various hemoglobin genes. Around 400 million years ago, the hemoglobin gene became duplicated into one leading to the modern alpha gene and another which would duplicate again around 200 million years ago into the gamma and the beta genes. The beta gene duplicated again some 40 million years ago in the ancestral lineage of the higher primates, giving rise to one new gene coding for the delta hemoglobin chain.

horses differ in 18 amino acids, while those of humans and carps differ in 68 amino acids.

Many other instances of gene duplication are known, such as the genes coding for trypsin and chymotrypsin, two enzymes essential in the intestine of humans and other mammals for the digestion of protein. It seems, however, that most structural genes exist in each genome in single copies. This suggests that although gene duplications may not be rare events on the evolutionary time scale, they do not become continuously established. Enough time usually passes between successive duplications of a given gene to allow for evolutionary divergence of the duplicated genes through point mutations.

The second class of duplicated DNA sequences consists of genes repeated from a few to several hundred times in each genome, all copies being identical or nearly so in nucleotide sequence, and thus transcribed in identical RNA sequences. This class includes the genes coding for ribosomal RNA and transfer RNA, both of which are transcribed but not translated, and the genes coding for histone proteins and for antibodies.

Ribosomes are organelles involved in protein synthesis. They are made up of three kinds of ribosomal RNA (rRNA) and nearly 100 different proteins. In eukaryotes, the three kinds of rRNA are designated 5S, 18S, and 28S, larger numbers indicating larger molecules. Ribosomes consist of two subunits, a small one containing an 18S rRNA molecule and a larger subunit containing one 5S and one 28S rRNA molecule. The three kinds of rRNA are encoded by two different genes. One gene codes for the 18S and 28S rRNA; the gene product is transcribed as a unit which is then split into the two types, 18S and 28S. The second gene codes for the 5S rRNA. It is often located in a separate chromosome from the other gene. Both genes are duplicated many times, the number varying from organism to organism: the number of copies of the gene coding for the 18S and 28S rRNA is 130 in *Drosophila melanogaster,* but 400 in the African toad *Xenopus laevis.* The repeated sequences of the gene are arranged in tandem, although they are separated by short interspersed (*spacer*) sequences of DNA. Natural selection may have favored the multiplication of the rRNA genes because cells require large numbers of ribosomes for protein synthesis; the presence of multiple genes makes possible producing large quantities of rRNA in a short time.

Transfer RNA (tRNA) molecules are the carriers of the amino acids to the ribosomes where protein synthesis takes place following the instructions encoded in messenger RNA (mRNA) molecules. The active tRNA molecules consist of about 70 to 80 nucleotides; three of the nucleotides constitute an anticodon that recognizes a corresponding codon in mRNA. Each tRNA carries an amino acid. As tRNA molecules are consecutively associated with the ribosome-mRNA complex they leave their amino acids there and the polypeptide chain gradually grows. In eukaryotes there are about 61 different kinds of tRNA molecules (one for each of the 61 triplets coding for an amino acid), each determined by a different gene. Each of the tRNA genes, however, is repeated many times, about six times in yeast, and about 13 times in *Drosophila melanogaster.* The existence of multiple copies of each gene allows for the efficient production of these molecules needed by cells in large numbers.

The third class of duplicated DNA sequences consists of relatively few sequences that in eukaryotes are each replicated many times. Some of these sequences are very short, about 300 nucleotides in length, and may be repeated each many thousands of times. Little is known at present about the function of these highly repeated sequences, but some authors have suggested that they may be involved in gene regulation. Conceivably, these sequences could either code for molecules sent as signals in order to activate (or deactivate) genes, or they could act as receptors for such signals.

Changes in the Number of Chromosomes

Fusions and fissions are changes in the numbers of chromosomes that do not increase or reduce the amount of hereditary material. Chromosomal *fusion* occurs when two nonhomologous chromosomes fuse into one, thus reducing the number of chromosomes in the genome. Chromosomal *fission* takes place when a chromosome splits into two, thereby increasing the number of chromosomes in the genome.

The gametic chromosome numbers in most animals lie between six and 20, but the range extends from one (the nematode worm *Parascaris equorum univalens*) to about 220 (the butterfly *Lysandra atlantica*). In plants, the most common gametic numbers are seven, eight, nine, 11, 12, and 13, but it can be much greater; the highest gametic number known is 631, in the fern *Ophioglossum reticulatum,* which is almost certainly a polyploid. Chromosomal fusions as well as fissions have occurred in evolution; even species of a single genus may differ in the gametic number of chromosomes, which in *Drosophila,* for example, ranges from three to six (Fig. 3.27).

Aneuploidy occurs when one or more chromosomes of the normal set of a species are lacking or present in excess. Aneuploidy often results in gross abnormalities or failure to develop, although aneuploids are occasionally found in populations of some plants, such as the Jimson weed (*Datura stramonium*), tobacco (*Nicotiana tabacum*), and bread wheat (*Triticum vulgare*). In humans, the severe condition known as Down's syndrome (or, misleadingly, as mongolism) is caused by the presence of an extra chromosome (i.e., by the presence of three copies of the chromosome known as 21); other abnormalities are caused by abnormal numbers of the sex chromosomes.

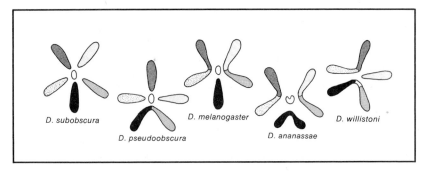

Fig. 3.27.   The chromosome complements of five *Drosophila* species; only one chromosome from each pair is shown. Chromosome
arms that are homologous in different species are identified in the
figure by the shading. The ancestral condition for the genus seems
to be five pairs of rod-shaped chromosomes and one pair of dot-
like chromosomes, as in *D. subobscura*. The other conditions can
be derived from the primitive one through various chromosome fusions. The X chromosomes (black) of *D. melanogaster* and *D. ananassae* differ by a pericentric inversion that has changed the
position of the centromere from the end to the middle of the chromosome in *D. ananassae*.

Polyploidy occurs when the number of *sets* of chromosomes is
greater than two. Polyploidy is a relatively rare phenomenon in animals because it often disrupts the balance between the sex chromosomes and the other chromosomes, which determines the sex in
animals. Natural polyploid species occur in animals among her-
maphrodites (those having male and female organs), such as earthworms and planarians, or in forms with parthenogenetic females
(females producing viable progeny without fertilization), such as
some beetles, moths, sow bugs, shrimps, goldfish, and salamanders.

Polyploid species occur in all major groups of plants. About 47
percent of all flowering plants (angiosperms) are polyploids. Polyploidy is also common among ferns, but rare among gymnosperms
(although the redwood, *Sequoia sempervirens,* is a polyploid). Tetraploids are the most common form of polyploidy; increasingly
higher numbers of ploidy become progressively less common. Polyploids with odd-number multiples of the basic chromosome number have irregular meiosis and are often sterile; in plants they may
subsist through asexual reproduction. Some of the most important
cultivated plants are polyploids (Table 3.5).

Autopolyploids are due to the multiplication of sets of chromosomes of one single species; *allopolyploids* result from the hybridization of two different species. Polyploids may arise because

Table 3.5
SOME EXAMPLES OF POLYPLOIDY IN CULTIVATED PLANTS

Plant	Ploidy	Somatic chromosome number
Banana	Triploid	27 (3 × 9)
Potato	Tetraploid	48 (4 × 12)
Bread wheat	Hexaploid	42 (6 × 7)
Boysenberry	Heptaploid	49 (7 × 7)
Strawberry	Octoploid	56 (8 × 7)

irregularities at meiosis may yield unreduced gametes, i.e., gametes with diploid rather than haploid chromosome complements. Individuals produced by such gametes will have extra chromosome sets. Polyploids may also arise owing to irregular mitosis, resulting in the doubling of chromosomes in somatic cells. This may give rise to tetraploid shoots that yield diploid gametes in their flowers. Self-fertilization will then result in a tetraploid; crossing with a normal diploid plant will yield a triploid. If the chromosome doubling occurs in a hybrid plant, the resulting tissue and eventual progeny will be an allotetraploid. Allotetraploid plant species are far more common than autotetraploids.

QUESTIONS FOR DISCUSSION

1. Many economically important cultivated plants have been selected during many generations with respect to a few characters. As a consequence, they have lost much genetic variability. What are the possible consequences of this loss of variability?

2. Why are most new mutations nonadaptive? Can you conceive of situations that would increase the proportion of new mutations that are adaptive?

3. Explain the various roles of gene duplications in evolution.

4. What proportion of all genes in present living organisms have arisen by gene duplication at some time or other since the origin of life?

5. Given the known amounts of DNA in various organisms, is it likely that some DNA sequences have no genetic function?

Recommended for Additional Reading

Ayala, F. J. ed. 1976. *Molecular evolution.* Sunderland, Mass.: Sinauer Associates.

A multiauthored book that reviews many of the topics presented in this chapter. Moderately advanced.

Lewontin, R. C. 1974. *The genetic basis of evolutionary change.* New York: Columbia University Press.

An excellent review of electrophoretic studies and of the controversies concerning genetic variation in natural populations. Advanced.

Ohno, S. 1970. *Evolution by gene duplication.* New York: Springer Publishing Co.

A moderately advanced and often quite speculative book discussing the role of gene duplication in evolution.

Stebbins, G. L. 1971. *Chromosomal evolution in higher plants.* London: E. Arnold.

White, M. J. D. 1973. *Animal cytology and evolution.* 3rd ed. Cambridge and London: Cambridge University Press.

Two texts on the role of chromosomal evolution, the first in plants and the second in animals. White's book is considerably more extensive and advanced than that of Stebbins.

4

Processes of Evolutionary Change

Evolutionary change affects all aspects of living things—their morphology, physiology, reproductive behavior, ecology, even their evolutionary potential. Underlying all these changes, there are genetic changes, that is, changes in the DNA that—interacting with the environment—determines what the organisms are.

Biological Evolution, a Two-Step Process

Mutations are the ultimate source of all hereditary variability. This variability, however, is sorted out in new ways every generation by the sexual process. The genetic variants are not all equally transmitted from one generation to another. Rather, some variants increase in frequency at the expense of others. Thus, evolution may be seen as a two-step process: we have, first, mutation and recombination, the processes by which variation arises; and, second, we have the processes by which genetic variants are differentially transmitted from generation to generation.

Genetic recombination was discussed in Chapter 2; the process of mutation was discussed in Chapter 3. The present chapter introduces the processes by which genetic variants change in frequency through the generations. These processes are *gene flow* (or migration), *natural selection,* and *genetic drift.* The effects of mutation on changing gene frequencies will also be briefly considered. Before studying these processes of change, however, we shall demonstrate that the process of heredity does not change gene frequencies, a principle known as the Hardy-Weinberg law.

The Hardy-Weinberg Law

The Hardy-Weinberg law says that, by itself, the process of heredity does not change either gene frequencies or (in a random mating population) the genotypic frequencies at a given gene locus.

The genotypic frequencies are given by the square of the sum of the gene frequencies. If there are only two alleles, A and a, with frequencies p and q, the frequencies of the three possible genotypes are

$$(p + q)^2 = p^2 + 2pq + q^2$$
$$A \quad a \qquad AA \quad Aa \quad aa$$

The alleles and genotypes to which the various frequencies correspond are written in the second line of the equation.

If there are only three alleles—say A_1, A_2, and A_3—with frequencies p, q, and r, the genotypic frequencies are

$$(p + q + r)^2 = p^2 + q^2 + r^2 + 2pq + 2pr + 2qr$$
$$A_1 \quad A_2 \quad A_3 \qquad A_1A_1 \quad A_2A_2 \quad A_3A_3 \quad A_1A_2 \quad A_1A_3 \quad A_2A_3$$

The Hardy-Weinberg law was formulated in 1908 independently by the mathemetician G. H. Hardy in England and the biologist W. Weinberg in Germany. A simple way to demonstrate the law is as follows. Assume that a given locus there are two alleles, A and a. Whatever the genotypic frequencies may be, let us assume that the allele frequencies, in males as well as in females, are p for A and q for a. Assume now that males and females mate at random (i.e., have equal probability of mating with any member of the other sex). This is equivalent to saying that the male and female *gametes* meet at random in the formation of the zygotes. The genotypic frequencies among the zygotes of the new generation will be as given:

		Gametic frequencies among males	
		p(A)	q(a)
Gametic frequencies among females	p(A)	p^2 (AA)	pq (Aa)
	q(a)	pq (Aa)	q^2 (aa)

These frequencies result because the probability of an individual with the genotype AA is simply the probability that it will receive an A gamete from the father multiplied by the probability that it will receive an A gamete from the mother, or $p \times p = p^2$; and similarly with the other genotypes. The genotype Aa can arise in two ways (A from the mother and a from the father, or a from the mother and A from the father) and thus will have a total frequency of $pq + pq = 2pq$.

The allelic frequencies among the new zygotes are p and q as in the previous generation. This can be shown as follows. The AA individuals have only allele $A;$ the Aa individuals have alleles A and a in equal frequency. The average frequency of allele A is therefore the frequency of AA individuals (p^2) plus half the frequency of Aa individuals ($2pq$), or

$$\text{Frequency of } A = p^2 + \frac{2pg}{2} = p^2 + pq = p(p + q) = p$$

Note that $p + q = 1$ (since there are only two alleles, their frequencies add to 1), and thus $p(p + q) = p \times 1 = p$.

The genotypic frequencies given will remain the same generation after generation—so long as matings are random and there are no processes changing gene frequencies, such as mutation, migration, selection, or drift. Because the new allelic frequencies are p and q as in the previous generation, using the same procedure as above we will obtain $p^2(AA)$, $2pq$ (Aa), and $q^2(aa)$ as the genotypic frequencies for the following generation; and so on and on.

The Hardy-Weinberg law can be demonstrated for any number of alleles. For example, in the case of three alleles, A_1, A_2, and A_3, with frequencies p, q, and r, the genotypic frequencies in the following generation will be:

		Gametic frequencies among males		
		$p(A_1)$	$q(A_2)$	$r(A_3)$
Gametic frequencies among females	$p(A_1)$	$p^2(A_1A_1)$	$pq(A_1A_2)$	$pr(A_1A_3)$
	$q(A_2)$	$pq(A_1A_2)$	$q^2(A_2A_2)$	$qr(A_2A_3)$
	$r(A_3)$	$pr(A_1A_3)$	$qr(A_2A_3)$	$r^2(A_3A_3)$

Adding up the frequencies of identical genotypes, we obtain

$$p^2(A_1A_1), \; q^2(A_2A_2), \; r^2(A_3A_3), \; 2pq(A_1A_2), \; 2pr(A_1A_3),$$
$$\text{and } 2qr \; (A_2A_3)$$

The Hardy-Weinberg law permits computing the gene and genotypic frequencies in cases where not all genotypes can be distinguished because of dominance. For example, albinism in humans is determined by a recessive allele in homozygous condition. If the allele for normal pigmentation is represented as A, and the allele for albinism as a, albinos have aa genotypes, but we cannot distinguish the other two genotypes, AA and Aa, from each other because both result in normally pigmented individuals. The frequency of albinos is about one in 10,000. According to the Hardy-Weinberg law, the frequency of the homozygotes, aa, will be q^2; thus $q^2 = 0.0001$ and, therefore, $q = \sqrt{0.0001} = 0.01$. It follows that the frequency of the normal allele is $p = 1 - q = 1 - 0.01 = 0.99$, and the frequencies of the other two genotypes are $p^2 = 0.99^2 = 0.98$ for AA, and $2pq = 2 \times 0.99 \times 0.01 = 0.02$ for Aa.

Mutation

The process of mutation generates the allelic variations that make evolution possible, but new mutations change gene frequencies very slowly since mutation rates are low. Assume that the gene allele A mutates to allele a at a rate u per generation, and that at a given time the frequency of A is p_o. In the next generation, a fraction u of all A alleles become a alleles. The frequency of A in the next generation will be the initial frequency (p_o) minus the frequency of mutated alleles $(p_o u)$, or

$$p_1 = p_o - p_o u = p_o \, (1 - u)$$

In the following generation, a fraction u of the remaining A alleles (p_1) will mutate to a, and thus the frequency of A will become

$$p_2 = p_1 - p_1 u = p_1 \, (1 - u)$$

And replacing the value of p_1 as obtained in the first equation in this section we have

$$p_2 = p_1 \, (1-u) = p_o \, (1-u) \, (1-u) = p_o \, (1-u)^2$$

After t generations, the frequency of A will be

$$p_n = p_o \, (1-u)^t$$

The frequency of A alleles decreases gradually because a fraction of them change every generation to a. If the process continues indefinitely, the frequency of A will eventually decrease to zero. The rate of change is nevertheless quite slow. Assume that the mutation rate is 10^{-5} (1:100,000) per gene per generation, and that the initial frequency of A is 1.00. In this case, it will take about 1,000 generations to lower the frequency of A to 0.99. The smaller the initial frequency of A, the longer the time required for a given amount of change. Thus, a change of 0.01 in gene frequency will require about 2,000 generations if the initial frequency of A is 0.50 (change from 0.50 to 0.49) but about 10,000 generations if the initial frequency of A is 0.10 (change from 0.10 to 0.09).

However, mutations are often reversible: the allele a may also mutate to A. Assume that A mutates to a at a rate u as before, and that a mutates to A at a rate v per generation. If at a given time the frequencies of A and a are p and q, after one generation the frequency of A will be

$$p_1 = p - pu + qv$$

This is because pu A genes change to a, but qv a genes change to A. There will be an equilibrium when the number of A alleles changing to a is the same as the number of a alleles changing to A, that is when

$$pu = qv$$

Since $p + q = 1$, and therefore $q = 1 - p$, the equilibrium will occur when

$$pu = (1 - p)v,$$
$$pu = v - pv,$$
$$pu + pv = v,$$
$$p(u + v) = v,$$
$$p = \frac{v}{u + v}$$

Assume that the mutation rates are $u = 10^{-6}$ and $v = 10^{-5}$. Then at equilibrium $p = 10^{-6}/(10^{-6} + 10^{-5}) = 1/(1 + 10) = 0.09$, and $q = 0.91$.

Changes in gene frequencies occur at slower rates when there is forward and backward mutation than when mutation occurs in only one direction, because backward mutation partially counteracts the effects of forward mutation. However, allelic frequencies usually are not in mutational equilibrium, because some alleles may be favored over others by selection. The equilibrium frequencies are then decided by the interaction between mutation and selection, as we shall see.

Gene Flow

Gene flow or genetic migration occurs when individuals migrate from one to another population and interbreed with its members. Gene flow may not change gene frequencies for the whole species, but it can change them locally if different populations have different allele frequencies.

Assume that in a population a proportion, m, of all reproducing individuals are migrants, and that the frequency of allele A is p_0 in the population but P among the migrants. In the next generation the frequency of A in the local population will be

$$p_1 = p_0 (1 - m) + mP = p_0 - mp_0 + mP = p_0 - m(p_0 - P)$$

That is, the new gene frequency will be the original gene frequency (p_0) multiplied by the proportion of reproducing individuals from the population ($1 - m$), plus the proportion of reproducing migrant individuals (m) multiplied by their gene frequency (P). Or, after reorganizing the terms in the expression, the new gene frequency will be the original gene frequency (p_0) minus the proportion of migrant individuals (m) multiplied by the difference in gene frequency between the resident and the migrant individuals ($p_0 - P$). It can be shown that after t generations of migration the frequency of A will be

$$p_n = (1 - m)^t (p_0 - P) + P$$

Random Genetic Drift

Populations of organisms consist of limited numbers of individuals, although the numbers may be extremely large in some cases. Because populations are finite in numbers, gene frequencies may change due to a pure chance process known as *random genetic drift, genetic drift,* or simply *drift.* Assume that in a certain population two alleles, A and a, exist in frequencies 0.40 and 0.60. The frequency of A in the following generation may be smaller (or greater) than 0.40 simply because by chance allele A is present less (or more) often among the gametes that form the zygotes of that generation.

Genetic drift is a particular case of the general phenomenon known as *sampling errors*. The magnitude of the errors due to sampling is inversely related to the size of the sample—the smaller the sample, the larger the effects. With respect to organisms, the smaller the number of individuals in a population, the larger are likely to be the allele frequency changes due to genetic drift.

It is simple to see why there should be an inverse relation between sample size and sampling errors. Assume that we have a coin, say a penny, and that the probability of getting heads in a throw is 0.5. If we throw the coin only once, we can get heads or tails, but not both. Although the probability of getting heads is 0.5, we get heads either once or not at all, but not half the time. If instead we throw the coin ten times we are likely to get several heads and several tails; we would be surprised (and suspicious of the coin) if we got only heads, but not if we got, say, six heads and four tails—the frequency of heads would be in such case 0.6 rather than the "expected" 0.5, but we would attribute such deviation from expectation to chance. Assume now that we throw the coin 1,000 times. We would be extremely suspicious of the coin if we got only heads, or even if we were to get as many as 600 heads and only 400 tails, although the frequency of heads in such case would be 0.6, the frequency observed without surprise when we threw the coin ten times. If the coin was thrown 1,000 times, we however would not be surprised if we got 504 heads and 496 tails, a frequency of heads of 0.504, although the expected frequency is 0.500.

The point of the coin examples is that the larger the sample the more nearly will be the agreement between the expected frequency of heads (0.5) and the observed frequency (1, 0.6, and 0.504 for one, ten, and 1,000 throws in the examples). With populations we also expect that the larger the number of individuals producing the next generation, the closer the agreement between the expected allelic frequency (which is that in the parental generation) and the observed frequency (which is the allelic frequency among the progeny); the number of parents contributing to the following generation plays a similar role to the number of throws of the coins.

Consider the following example. Assume that we have a large number of pea plants, *Pisum sativum,* like those used by Mendel, and that the frequency of the allele responsible for yellow peas, Y, is 0.5, the same as the frequency of the y allele that produces green peas in homozygous condition. Assume also that the three genotypes occur in the expected frequencies of ¼ YY, ½ Yy, and ¼ yy. Assume now that we pick up one pea, without looking at the phenotype, and obtain a plant from it. What is the frequency of the Y allele among the seeds produced by this plant? Clearly, there are three possibilities; the frequency of Y will be one, one-half, or zero depending on the genotype of the pea used to produce the plant. The probability is ¼ that the pea was YY, and also ¼ that it was yy; thus the frequency of Y is likely to change to either one or zero (from its frequency of 0.5 in the parental population) with a probability of one-half. Assume instead that we collect 1,000 peas from the original population and obtain 1,000 plants from them. The frequency of the Y allele among the peas produced by these plants is likely to be very nearly 0.5.

Whenever we know, as in the previous examples, the number of parents used to produce the following generation, and the allelic frequencies, it is possible to calculate the probability of obtaining a given allelic frequency in the following generation. In order to do this we need to know the *variance* of the allelic frequencies in the following generation, which is a measure of the amount of variation that would be found among different samples. If there are two alleles with frequencies p and q, and the number of parents is n (so that the number of genes in the sample used to produce the next generation is $2n$), the *variance* (s^2) of the allelic frequency in the following generation is

$$s^2 = \frac{pq}{2n}$$

This equation shows an inverse relationship between sample size, n, and the expected variance in gene frequencies.

Founder Effect and Bottlenecks

Unless a population is very small, changes in gene frequencies due to genetic drift will be small from one generation to another. The effects over many generations, however, may be large. If no other processes (mutation, migration, selection) affect allelic frequencies at a gene locus, evolution will ultimately result in the fixation of one allele and the elimination of all others. When only drift is operating, the probability that a given allele will be ultimately fixed is precisely its frequency. For example, if a certain allele has a frequency of 0.2 at a given time, it has a probability of 0.2 of ultimately being the only allele in the population. But this requires a very long time; generally the number of generations required for fixation is of the same order of magnitude as the number of parents per generation.

Assume that a new allele arises by mutation in a population with effective size n. Since there are $2n$ alleles in the population, the frequency of the new mutant is $1/2n$, which is also the probability that the population will become fixed for that mutant. With drift alone, the population will eventually be made up of genes all descended from the new mutant (or of one of the other $2n$ alleles in the population), but this will require approximately $4n$ generations. If the effective size of the population is one million individuals, the process will require four million generations.

It is unlikely that random drift alone will affect allelic frequencies at any locus during long periods of time because mutation, migration, and selection are likely to take place at one time or another. These three processes are the *deterministic* processes of evolutionary change because they favor one allele, rather than affecting all alleles equally as is the case with genetic drift. A simple rule can be applied to ascertain whether drift will have important effects relative to the deterministic processes. Using x to represent either the mutation rate (u), or the migration rate (m), or the selection coefficient (s, to be defined later in this chapter), then gene frequency changes will be primarily governed by random genetic drift if, and only if, the product nx is much smaller than one, or in symbols

$$nx \ll 1$$

If the product nx is about one or greater, then gene frequency changes will be determined for the most part by the deterministic processes.

Assume that the mutation rate from allele A to allele a is $u = 10^{-5}$. This will have little effect in gene frequencies relative to the effects of drift in a population of 100 individuals because $nu = 10^2 \times 10^{-5} = 10^{-3} \ll 1$. However, it will be the main determining factor in a population of one million breeding individuals because, then, $nu = 10^{+6} \times 10^{-5} = 10 > 1$. If the migration rate is 0.02 (or two individuals for every 100) per generation, gene frequencies will change toward the frequencies in the population from which the migrants come, even in a small population with only 100 individuals, because in such case $nm = 100 \times 0.02 = 2 > 1$.

Extreme cases of random genetic drift occur when a new population is established by only very few individuals. This has been called the *founder effect* by the eminent evolutionist Ernst Mayr. Populations of many species living on oceanic islands, although they may now consist of millions of individuals, are descendants of one or very few colonizers that arrived long ago by accidental dispersal. The situation is similar in lakes or other isolated water bodies, in isolated forests, and in other ecological isolates. Because of sampling errors, gene frequencies at many loci are likely to be different in the few colonizers than in the population from which they came, which may have lasting effects on the evolution of such isolated populations.

An experimental demonstration of the founder effect is shown in Figure 4.1. Laboratory populations of *Drosophila pseudoobscura* were begun with samples from a population in which a certain genetic constitution, represented as *PP*, had 0.50 frequency. There were two types of populations, some ("large") were started with 5,000 individuals each, the other ("small") with 20 individuals each. After one-and-a-half years, or about 18 generations, the mean frequency of *PP* was about 0.30 in the large as well as in the small populations, but the range of frequencies was considerably greater in the small populations. Starting the small populations with few founders resulted in considerable variation among populations in the frequency of *PP*.

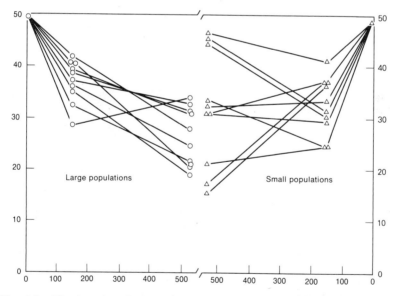

Fig. 4.1. The founder effect in laboratory populations of *Drosophila pseudoobscura*. The graphs show the changes in the frequency of a certain genetic variant known as *PP*. Note that time proceeds from left to right for the large populations, but from right to left for the small populations, so as to facilitate comparison of the final results. (After Dobzhansky and Parlovsky, 1957.)

Chance variations in allelic frequencies similar to those due to the founder effect occur when populations go through *bottlenecks*. When climatic or other conditions are unfavorable, populations may be drastically reduced in numbers and approach the risk of extinction. Such populations may later recover their typical size, but random drift may alter considerably their allelic frequencies during the bottleneck. In early human society, whole tribes were decimated owing to various calamities, although some recovered from a few survivors or migrants from other tribes. Differences between human populations in the frequency of the ABO blood-group alleles may, at least in part, have resulted from population bottlenecks (Fig. 4.2).

The Concept of Natural Selection

We have so far considered in this chapter three of the four processes that change gene frequencies—mutation, gene flow, and drift. Now we introduce the fourth and most important one—natural selection. If the appropriate parameters are known (mutation

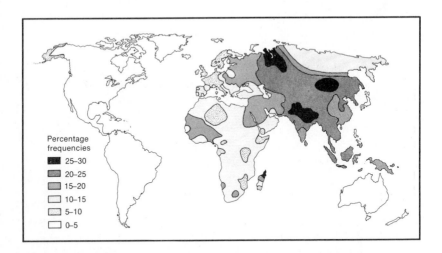

Fig. 4.2. Frequency of the blood group gene I^B in aboriginal populations of the world. The ABO blood groups are determined by three alleles, I^A, I^B, and i, with varying frequencies in different human populations. The I^B allele is absent among American Indians and aboriginal Australians unmixed with Europeans, but it is present in all Old World populations. The frequency of I^B is highest in Mongolia, Central Asia, northern India, and in some aboriginal peoples from Siberia.

or migration rate, and allele frequencies) we may predict the *direction* and *rate* of change in allelic frequencies due to mutation or to migration. With respect to drift, knowledge of the relevant parameters (allele frequencies and population size) allows calculating the expected magnitude of allelic frequency change—i.e., the expected *rate* of change—but not the direction of change.

There is, however, an important attribute that mutation, gene flow, and drift share in common: none of them is oriented with respect to adaptation. These processes change gene frequencies independently of whether or not such changes increase or decrease the adaptation of organisms to their environments. Because these processes are random with respect to adaptation, by themselves alone they would destroy the organization and adaptation that characterize living creatures. Natural selection is the process that promotes adaptation and keeps in check the disorganizing effects of the other processes. In this sense, natural selection is the most critical evolutionary process, because only natural selection accounts for the adaptive and highly organized nature of living beings. Natural selection explains also the diversity of organisms because it promotes their adaptation to different ways of life.

The idea of natural selection as the fundamental process of evolutionary change was independently reached by Charles Darwin and Alfred Russel Wallace. In 1858 they made a joint presentation of their discovery to the Linnean Society of London. The significance of this communication went apparently unnoticed: the annual report of the Society states that "nothing of significance happened in 1858." This apathy disappeared the following year when Darwin, who had been working on the subject for 20 years, published *The Origin of Species*. The book immediately attracted considerable attention from scientists as well as from laymen.

The central argument of the theory of evolution by natural selection was eloquently summarized by Darwin in *The Origin of Species* as follows:

> Can it, then, be thought improbable, seeing that variations useful to man have undoubtedly occurred, that other variations useful in some way to each being in the great and complex battle of life, should sometimes occur in the course of thousands of generations? If such do occur, can we doubt (remembering that more individuals are born than can possibly survive) that individuals having any advantage, however slight, over others, would have the best chance of surviving and of procreating their kind? On the other hand, we may feel

sure that any variation in the least degree injurious
would be rigidly destroyed. This preservation of favorable variations and the rejection of injurious variations,
I call Natural Selection.

This argument is simple, yet powerful. It starts from the existence of hereditary variation. Some variations are "useful in some
way," "favorable," or "advantageous" in comparison with others;
adaptive variations are likely to increase the chances of survival
and procreation and thus will gradually increase at the expense of
the less adaptive alternatives. This differential reproduction is the
process of natural selection. As a result of the process organisms
are well adapted to their environments.

Natural selection may be simply defined as *the differential reproduction of alternative genetic variants,* determined by the fact that
some variants increase the chances of survival and reproduction of
their carriers relative to the carriers of other variants. As Darwin
saw it, natural selection may be due to either differential survival
or to differential fertility or to both. Differential rate of development, differential mating success, and differences in other components of the life cycle may contribute to natural selection, but
these may be subsumed under survival and fertility. In fact, all
components of the process of natural selection may be simply incorporated under differential reproduction; differential survival,
mating success, and so on, result in natural selection only if they
become translated into differential reproduction.

According to Darwin, competition for limited resources results in natural selection of the most effective competitors. In the
paragraph quoted, he reminds us "that more individuals are born
than can possibly survive." Elsewhere he writes that because this
is so, ". . . there must in every case be a struggle for existence, either one individual with another of the same species, or with the
individuals of distinct species."

Calling attention to competition for limited resources is appropriate because in the presence of relevant hereditary variations, competition leads necessarily to natural selection. But
natural selection may occur without competition. Inclement
weather and other aspects of "the physical conditions of life," to
use Darwin's expression, result in differential mortality even
without competition for limited resources. Populations of all sorts
of organisms are often depleted during unfavorable seasons. Moreover, natural selection may occur even if no organism dies before
completing its reproductive period, simply because some organisms produce more progeny than others.

Darwinian Fitness

The parameters used to measure mutation, gene flow, and random drift are, respectively, the mutation rate, the migration rate, and the variance of allelic frequencies. The parameter used to measure natural selection is Darwinian fitness, or simply *fitness* (also called *selective value* and *adaptive value*). Fitness is a measure of the reproductive efficiency of a genotype *relative* to other genotypes.

Assume that at a certain locus there are three genotypes, and that on the average for every one progeny produced by *AA* homozygotes, the heterozygotes *Aa* produce also one progeny, but the *aa* homozygotes produce only 0.8 progeny. By definition, the first two genotypes have identical fitnesses, while the fitness of the *aa* genotype is 80 percent of the fitness of the other two. For mathematical convenience, the genotype with the highest reproductive efficiency is assigned 1 as its fitness value; the fitnesses of the other genotypes are then calculated relative to that of the genotype with the highest fitness. In our example, the fitnesses of the three genotypes are, then, 1, 1, and 0.8.

Fitness is often symbolized by the letter w. A related measure is the *selective coefficient* (not to be confused with *selective value*, which is the same as fitness), usually represented as s, and defined as $s = 1 - w$ (therefore, $w = 1 - s$), and which measures the reduction in fitness. In the example given in the previous paragraph, the selective coefficient is zero for genotypes *AA* and *Aa*, but 0.2 ($= 1 - 0.8$) for *aa*. Table 4.1 shows how fitnesses can be computed when the number of progeny produced by each genotype is known. If we know the genotypic fitnesses, we may predict the rate of change in the frequency of the genotypes. The converse is also true, and population geneticists often compute fitnesses based on the changes in genotypic frequencies. As a simple example, assume that in a haploid organism, such as a bacterium, the frequency of the two genotypes *A* and *a* in a large population is 50 percent at a given time, but the frequencies change to 0.667 *A* and 0.333 *a* in one generation; we infer that the fitnesses of *A* and *a* are 1 and 0.50, respectively.

The fitnesses of genotypes are measured *relative* to each other rather than as absolute values, because natural selection is *differential* reproduction. The course of selection, i.e., the change in gene frequencies, depends on the relative reproductive efficiency of the genotypes. Therefore, fitness values do not tell us how well the population is doing as a whole, that is, whether the population is increasing or decreasing in numbers. For example, assume that in Table 4.1 the numbers of progeny produced are 40, 40, and 5.

Table 4.1
CALCULATION OF FITNESSES OF GENOTYPES
When the number of progeny produced by each genotype is known, there are two steps in the calculation. First, we calculate the average number of progeny per individual produced in the next generation by each genotype. Second, we divide the average number of progeny of each genotype by that of the best genotype.

| | Genotypes | | | |
	AA	Aa	aa	Total
a. Number of zygotes in one generation	40	50	10	100
b. Zygotes produced	80	80	10	170
Step 1: average number of progeny(b/a)	80/40 = 2	80/50 = 1.6	10/10 = 1	
Step 2: fitness(relative reproductive efficiency)	2/2 = 1	1.6/2 = 0.8	1/2 = 0.5	

The relative fitnesses would be the same as in the table although the total number of individuals in the population would have decreased from 100 to 85, rather than increased from 100 to 170.

All aspects of the life of an individual that may affect its reproductive success contribute to natural selection and therefore to the fitness of organisms. These various aspects—survival, rate of development, mating success, fertility—are called *fitness components.* Fitness differences may be due to one, or several, fitness components. Phenylketonuria (PKU) is a human disease caused by the inability to convert the amino acid phenylalanine into tyrosine. This results in an accumulation of phenylalanine, causing severe mental defects and usually an early death. The fitness of phenylketonurics is thus zero due to their inability to survive to reproductive age. Achondroplastic dwarfs reproduce, on the average, only 20 percent as efficiently as normal individuals. The low fitness of PKU patients is due to poor viability; that of achondroplastics is due to reduced fertility.

The effects of natural selection on gene frequencies depend on the fitnesses of the genotypes involved. The ultimate outcome may be elimination of one or other allele (although mutation pressure may keep deleterious alleles at low frequencies—see the section on normalizing selection), or a stable polymorphism with two or more alleles. The effects of natural selection can be simply treated when there are only two alleles, and therefore three genotypes, at a single gene locus.

The cases considered in the appendix assume that genotypic fitnesses are constant, i.e., independent of the frequency of the alleles themselves, or of the allele frequencies at other loci, of the density of the population, or of any other factors. Selection against a recessive allele (Case 1), or against a dominant allele (Case 2), or against an allele without dominance (Case 3), all lead to the eventual elimination of the disfavored allele. However, when the heterozygotes have higher fitness than either homozygote (Case 4), selection leads to stable polymorphism with both alleles present at frequencies determined by the selection coefficients against the two homozygotes. Examples of these various modes of selection are considered in this and the following chapter.

Normalizing Selection

Natural selection, the differential reproduction of alternative genotypes, does not always result in change. If we look at the effects of natural selection on the phenotypes of a population, we may distinguish three types of selection: *normalizing* (or stabilizing) selection, *directional* selection, and *diversifying* (or disruptive) selection. These three modes of selection are diagrammatically represented in Figure 4.3. Although real situations may be much more complex, it is assumed in the figure that the phenotypic trait under consideration is of such a nature that the individuals can be arranged along a linear scale. Traits of this kind are, for example, height, weight, number of progeny, and longevity; individuals can be ordered from shorter to taller, or from lighter to heavier, and so on. It is also assumed that the distribution of phenotypes is bell-shaped (the so-called "normal" distribution), so that the number of individuals is greater at the intermediate values, and gradually decreases towards the extremes (Fig. 4.4).

Normalizing selection occurs when the norm—i.e., individuals with intermediate phenotypes—is favored and extreme phenotypes are selected against. The phenotypic distribution will remain approximately the same from one generation to another, and it is for this reason also called stabilizing selection. If the selection in favor of intermediate phenotypes is very strong, the distribution may keep the same mean but have less variance, i.e., the distribution of phenotypes may become somewhat narrower from one generation to another. If selection is less strong, the distribution may increase in variance from generation to generation (Fig. 4.5).

Natural selection is often normalizing; those individuals are favored that have intermediate values. In humans, for example, the mortality among newborn infants is highest when they are either very small or very large; infants of intermediate size have a better chance of surviving.

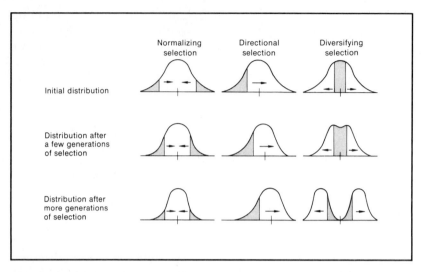

Fig. 4.3. An idealized representation of three modes of selection and their effects on the phenotypic distribution of populations. *Normalizing selection:* natural selection favors individuals with intermediate phenotypes; it will tend to maintain the mean value of the phenotypic distribution but the variance of the distribution may be reduced more or less depending on the intensity of the selection. *Directional selection:* natural selection favors individuals with phenotypes at one end of the distribution; the phenotypic distribution will gradually change in the direction of selection. *Diversifying selection:* natural selection favors individuals with phenotypes at either end of the distribution; if the selection is sufficiently strong, the end result may be two nonoverlapping phenotypic distributions. The shaded regions of the distributions indicate the phenotypes being selected against. The arrows point in the direction of the phenotypes most favored by selection. The mean value of each distribution is indicated by a notch on the abscissa.

The effects of normalizing selection are often experienced after artificial selection. Breeders want, for example, chickens that produce larger eggs, and they succeed by artificial selection in obtaining flocks where the eggs are large. If the artificial selection is stopped, natural selection gradually reverts the effects of selection, and the eggs decrease in size until they reach some intermediate optimum size.

An example of normalizing natural selection with respect to a behavioral trait is shown in Figure 4.6. Theodosius Dobzhansky and his collaborator Boris Spassky established two laboratory populations of *Drosophila pseudoobscura.* One was selected for positive phototactic behavior, the other for negative phototactic behavior.

Fig. 4.4. The distribution of height among a group of individuals recruited for the army around the turn of the century. The distribution is bell-shaped: the number of individuals of intermediate height is great and decreases towards the extremes. Bell-shaped ("normal") distributions are common for traits that can be arranged linearly along a scale, such as height, weight, and number of progeny. (Courtesy of the *Journal of Heredity*.)

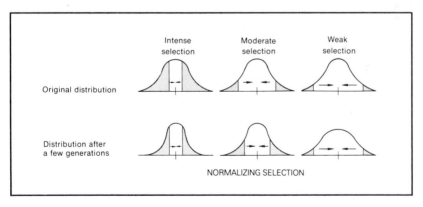

Fig. 4.5. Effects of normalizing selection of different intensities. *Intense selection:* the variance of the distribution may decrease over the generations, i.e., the distribution may become gradually narrower. *Moderate selection:* the distribution may remain approximately the same from generation to generation. *Weak selection:* the distribution may increase its variance—i.e., become more widely spread—over the generations if selection in favor of the intermediate phenotypes is only very weak.

When *Drosophila* flies are placed undisturbed in a maze where they can choose to go toward the light or toward the dark, they move as often in one direction as in the other. If the most photopositive flies are chosen to breed the following generation, the population becomes gradually more photopositive. A similar result is obtained when selection is practiced in the opposite direction. After several generations of selection, the flies go more often toward the dark than toward the light. However, if the artificial selection is stopped, natural selection favors the individuals with neutral behavior toward light, and both populations gradually revert toward an intermediate phototactic score.

As a result of normalizing selection, populations often maintain a steady genetic constitution with respect to many traits. This attribute of populations is called *genetic homeostasis.* When the genetic constitution is temporarily changed, as occurs through artificial selection, natural selection may favor returning to the original genetic constitution.

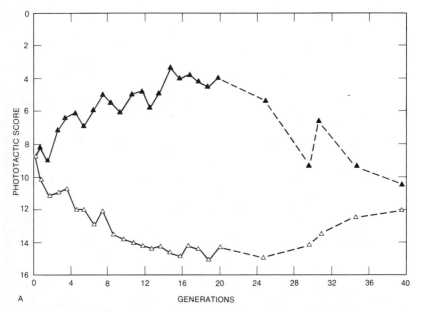

Fig. 4.6. Artificial selection for positive and negative phototaxis followed by normalizing natural selection in laboratory populations of *Drosophila pseudoobscura. A. Mean phototactic score.* Artificial selection for 20 generations is successful for both negative (dark triangles) and positive (light triangles) phototaxis. When the artificial selection is stopped after generation 20, natural selection fa-

The effects of normalizing selection are manifested in the case of deleterious mutations. Most mutations are deleterious when their carriers live in their usual environments. Deleterious mutations may reduce only the fitness of the homozygotes (recessive mutations), or that of the heterozygotes as well (dominant mutations). The accumulation of deleterious mutations in populations is impeded by normalizing selection; the greater the deleterious effect, the stronger the selection against the mutants.

Consider, first, the case of dominant mutations. Assume that the fitness of the homozygotes, *aa,* is 1 and that of the heterozygotes, *Aa,* is $1 - s$; and that mutations from *A* to *a* occur at a rate *u.* (Back mutation from *A* to *a* may be ignored because the frequency of *A* is very low.) Let the frequency of *A* be *p,* and the frequency of *a* be *q,* so that $p + q = 1$. The frequency of *A* will increase every generation due to mutation by a fraction *qu,* because a fraction *u* of all *a* genes mutate to *A.* But the frequency of *A* will decrease owing to natural selection by *ps,* because the individuals

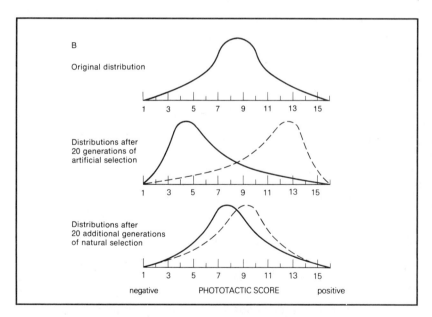

vors individuals with intermediate phototactic behavior and the populations gradually return towards the original phototactic score. *B. Distribution of phototactic behavior in the populations.* The original population has a normal distribution, which is gradually shifted towards the left or towards the right in response to artificial selection. (After Dobzhansky and Spassky, 1969.)

having the A gene reproduce less efficiently by a factor s. There will be an equilibrium when the increase in A genes due to mutation is equal to the decrease in A genes due to selection, that is, when

$$qu = ps$$

If selection is strong, the frequency of a will be nearly 1, and therefore, approximately

$$u = ps, \text{ or } p = \frac{u}{s}$$

The equilibrium frequency of the deleterious dominant allele is simply the mutation rate divided by the selection coefficient. Note that the equilibrium frequency is higher when the mutation rate, u, becomes higher, or when the selection coefficient, s, becomes smaller.

In the case of lethal dominant mutations, $s = 1$, and therefore $p = u$; the equilibrium frequency is simply the mutation rate. This is as expected, because when a dominant mutation is lethal, every time the mutation occurs its carrier dies and thus there is no accumulation of deleterious alleles from one generation to another; the only mutants present in a population are those that have arisen in the current generation. Down's syndrome, a severe deleterious condition in humans, is caused by a chromosomal mutation—the presence of an extra chromosome. Since Down's syndrome patients rarely reproduce, $s = 1$, and therefore $p \simeq u$. In other words, the frequency of Down's syndrome in human populations is simply the frequency with which it arises.

We now consider the counteracting effects of mutation and selection in the case when the deleterious allele is recessive. Assume that the mutation rate from A to a is u (back mutation is ignored, as in the case of dominance, and for the same reason, namely that the frequency of the deleterious allele is low). Assume also that the fitness of the recessive homozygotes, aa, is $1 - s$. As before, let the frequencies of A and a be p and q respectively. The frequency of a will increase every generation owing to mutation by pu. According to the Hardy-Weinberg law, the frequency of the recessive homozygotes is q^2. Since the recessive homozygotes reproduce s times less efficiently than the other genotypes, the frequency of a will decrease by q^2s owing to selection. There will be an equilibrium when

$$pu = q^2s$$

Since p will be close to one, we will have approximately

$$u = q^2 s, \text{ or } q^2 = \frac{u}{s}, \text{ and } q = \sqrt{\frac{u}{s}}$$

The equilibrium frequency is the square root of the mutation rate divided by the selection coefficient. As in the case of dominant mutations, the frequency of the deleterious allele is higher when the mutation frequency becomes higher, or when the selection coefficient becomes lower. But for any given values of u and s, the equilibrium frequency is greater for a deleterious recessive allele than for a deleterious dominant. For example, if $u = 0.00001$ and $s = 0.1$, then $u/s = 0.0001$; the equilibrium frequency for a recessive allele will be $q = \sqrt{0.0001} = 0.01$, while for a dominant allele with the same mutation rate and selective coefficient is $p = 0.0001$.

Directional Selection

Environments change over long periods of time. Changes may occur in the physical conditions, for example, climate, or in the biotic conditions, that is, in other species that may be predators, prey, parasites, or competitors sharing the same resources. Environmental changes often promote genetic changes. The fitness of variant genotypes may be shifted so that the alleles favored are different from before; directional selection then comes into operation changing the genetic constitution of the population. Directional selection also operates when organisms colonize a new territory where the conditions are different from those in the native habitat. The appearance of a new favorable allele or a new genetic combination may also start directional selection: the new genetic constitution gradually replacing the preexisting one.

Like other modes of natural selection, directional selection is only possible if there is genetic variation. We have seen in Chapter 3 that natural populations contain large reserves of genetic variation, and that new variants continuously arise by mutation. That this variation may provide the materials for directional selection is evidenced by the experience of artificial selection and by the rapid response of populations to new environmental challenges. It was pointed out in Chapter 3 (pp. 68–70) that artificial selection has been successfully practiced in a great variety of organisms and for a great variety of traits. In a previous section of this chapter (Fig. 4.6) we saw one instance of successful artificial selection, this time for a behavioral trait; strains phototactically positive as well as negative were successfully obtained. Artificial selection has been successful virtually every time it has been attempted.

Mankind has transformed, intentionally or not, the environments of many organisms. These organisms have rapidly responded through directional selection in order to meet the new environmental challenge. The resistance of insect species to pesticides is a case in point. The story is usually the same: when a new insecticide is applied to control a certain pest species, the results are encouraging at first—a small amount of the insecticide is sufficient to control the pest organism. However, the amount required to achieve a certain level of control needs to be repeatedly increased until finally it becomes totally ineffective or economically impractical. This occurs because the organisms become resistant to the pesticide through directional selection. The first known case of resistance to a pesticide was reported in 1947—a population of the housefly, *Musca domestica,* had become resistant to DDT. Today resistance to one or more pesticides has been recorded in more than 200 species of insects and other arthropods. The efficacy of directional selection is particularly noticeable because the pesticides usually are synthetic substances never previously present in the natural environments of organisms.

Industrial melanism is another example of directional selection provoked by artificial changes to the environment. The phenomenon has been best studied in the moth *Biston betularia* in England. Until the middle of the nineteenth century, the moths were uniformly peppered light gray. Darkly pigmented variants started to appear at that time in industrial regions where the vegetation was blackened by soot and other pollution. In some localities the dark varieties have now almost completely replaced the lightly pigmented forms, while in unpolluted areas the light ones are still the most common. Directional selection is due to predation by birds: on pale, lichen-covered tree trunks, the light-gray moths are well camouflaged, while the dark ones are conspicuously visible. The opposite is the case on trees darkened by pollution (Fig. 4.7).

H. B. D. Kettlewell of Oxford University released hundreds of marked light and dark moths in two areas, one near Birmingham, a heavily industrialized area, and the other in an unpolluted region of Dorset. A greater proportion of dark moths than of light ones were recaptured in the polluted area, while a greater proportion of light moths than of dark ones were recovered in the unpolluted region (Table 4.2). Direct observation confirmed that birds more easily captured light moths in the polluted region and dark moths in the unpolluted area. The phenomenon of industrial melanism has been discovered in about one hundred different species of moths. An interesting development has, however, taken

Fig. 4.7. Industrial melanism in the peppered moth, *Biston betularia.* A typical, light-gray moth and a darkly pigmented variant are resting on two oak trunks. *Left:* On a soot-covered trunk near the industrial city of Birmingham, England, the light-gray moth is more conspicuous than the darkly pigmented form. *Right:* On a lichened tree trunk in the unpolluted countryside, the typical form is much less conspicuous than the melanic. (Courtesy of H.B.D. Kettlewell.)

place as a consequence of strict pollution control near several industrial cities in England and elsewhere: the soot covering the tree trunks has largely disappeared and been replaced by lichens. In such places the melanic forms are gradually being replaced by light-gray moths, which are once again favored by natural selection.

Table 4.2
BISTON BETULARIA MOTHS RECAPTURED IN TWO AREAS

Birmingham is a polluted area with soot-covered tree trunks, while Dorset is unpolluted. (After H. B. D. Kettlewell)

Locality	Moths released		Moths recaptured	
	light	dark	light	dark
Birmingham	64	154	16(25%)	82(53%)
Dorset	393	406	54(13.7%)	19(4.7%)

Directional selection often operates over long periods of time, although it may not act at every moment or always with similar intensity. Changes more or less continuous over long periods of time are known as evolutionary trends. An example is the increase in cranial capacity of the human lineage. *Homo habilis,* our ancestor of two million years ago, had a small brain, about 700 cc (cubic centimeters) in volume. *Homo erectus,* who lived up to a few hundred thousand years ago, had a brain 50 percent larger, while modern humans, *Homo sapiens,* have a brain about twice as large, about 1,400 cc on the average.

A well-studied example of directional selection occurred in the horse family. The earliest known member of the family is eohippus (*Hyracotherium*), which lived in the early Eocene, about 50 million years ago. Eohippus stood about 50 cm high. It had four toes on the front feet and three on the hind, each toe ending in a small hoof. The small head had large eyes set near the middle, not far back as in modern horses, and lacked the heavy muzzle of today's horses. The evolution of the horse is complex, but several trends can be recognized. Between about 30 and 5 million years ago there was an on-and-off gradual increase in size, until the size of the modern horse was reached. Changes in the feet occurred after the Eocene, starting about 40 million years ago, and led rapidly in different lineages to three different kinds of feet, of which one type alone has survived—that of the modern horse, with only one functional toe ending in a big hoof on each leg (Figs. 4.8 and 4.9).

Diversifying Selection

Natural environments are not homogeneous. Rather, they are mosaics made of more or less similar subenvironments. Environments may be heterogeneous with respect to various aspects: climate, food resources, living space, and so on. Moreover, the heterogeneity may be temporal or spatial. Temporal environmental heterogeneity exists when the same organisms experience different environments at different times of their lives; we may speak of *stable* or *unstable* environments, depending on whether they vary little or much through time. Spatial environmental heterogeneity occurs when at any one time different organisms experience different environments; environments may be called *uniform* or *patchy,* depending on whether they are spatially homogeneous or heterogeneous. Spatial heterogeneticity does not necessarily involve temporal heterogeneity or vice versa. A patchy environment may be highly stable, while a spatially uniform environment may experience temporal instability.

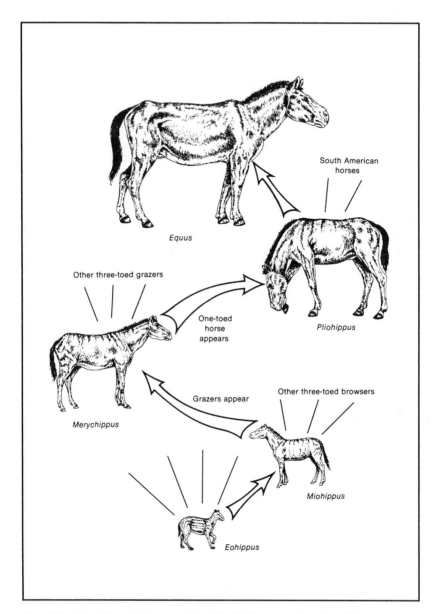

Fig. 4.8. Evolution of the horse family. Eohippus, a browser living in the Old World about 50 million years ago, evolved into several forms. One of these (*Miohippus*) was a three-toed browser that evolved into several other browsers as well as into one form (*Merychippus*) that became a grazer. *Merychippus* was still three-toed and evolved into other three-toed grazers as well as into a one-toed horse (*Pliohippus*), which eventually gave rise to the modern horse (*Equus*) as well as to the South American horses.

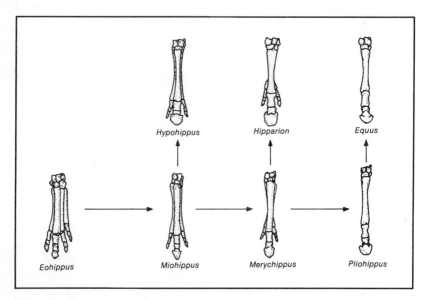

Fig. 4.9. Evolution of the forefoot in the horse family. Eohippus (*Hyracotherium*) had four toes in the forefoot. Several of his successors had three toes but one evolved a single toe, characteristic of today's horses. (After Simpson, 1951.)

How is a species to cope with the variety of environments it meets both in space and time? Two genetic strategies are possible. One strategy is to have a genotype whose carriers possess well-developed homeostasis and thus are well-adapted in various sub-environments. Another strategy is polymorphism, a diversified gene pool with different genotypes adapted to different sub-environments. What strategy is more often followed in nature is not a settled question. Some authors claim that the most efficient strategy is genetic monomorphism to confront temporal heterogeneity (unstable environments), but polymorphism to confront spatial heterogeneity (patchy environments). The rationale is as follows. If the environment is unstable (relative to the length of life of the organisms), each individual has to face the diverse environments that appear successively. If there are a variety of genotypes, each well-adapted to the conditions that prevail at one time but not at others, the population will not fare very well; the organisms survive well at one period of their lives but not at others. A single genotype that survives well in all the successive environments is a better strategy.

The situation may be different with respect to spatial heterogeneity. While a single genotype well-adapted to the various subenvironments is a possible strategy, a variety of genotypes with some individuals optimally adapted to each environmental "patch" might be still better, since it would maximize the ability of the population to exploit the environmental patchiness. The strategy of monomorphism could be favored when the environment is patchy but also unstable, for the reasons advanced in the previous paragraph. If the environment is spatially patchy but temporally stable, the polymorphic strategy might be best.

Natural selection favoring different genotypes in different subenvironments is *diversifying selection* (also called *disruptive selection*). Various authors have shown mathematically that environmental patchiness may lead to stable polymorphism. It has also been shown experimentally that increased patchiness results in higher levels of genetic polymorphism. For example, J. F. McDonald and F. J. Ayala studied eighteen experimental populations of *Drosophila melanogaster,* all started with progenies from a single set of flies. Some populations had only one or another culture medium, or one or another kind of yeast which grows in the culture medium and is also used as food by the flies. Other populations had both kinds of culture medium, and some had both kinds of yeast. After many generations, the populations with greater patchiness were significantly more polymorphic (Table 4.3).

Diversifying selection may produce populations genetically differentiated within short distances. In Britain, A. D. Bradshaw and D. Jowett have studied populations of bentgrass growing on soils contaminated with heavy metals, such as lead and copper. On heaps of mine spoils, the soil is so contaminated with the metals that it is toxic to most plants, including bentgrasses growing in the surrounding, uncontaminated soils. Yet, dense bentgrass stands grow over the contaminated spoil heaps. These bentgrass stands

Table 4.3
GENETIC POLYMORPHISM WITH DIFFERENT DEGREES OF ENVIRON-MENTAL PATCHINESS

The value given is the average heterozygosity at 20 gene loci coding for enzymes. Some populations had one or another kind of food (*a, b*), some had both (*a* and *b*). (After McDonald and Ayala, 1974.)

Factor	Kinds of food present		
	a	*b*	*a* and *b*
Culture medium	0.181	0.173	0.192
Yeast	0.176	0.161	0.194

have genes that make them resistant to high concentrations of lead and copper. The efficiency of diversifying selection is manifest. Resistant bentgrasses are surrounded by nonresistant plants growing a few meters from the contaminated soils. Since bentgrass is normally cross-pollinated, resistant bentgrass receive wind-borne pollen from nonresistant plants and vice versa. The genetic differentiation is maintained because nonresistant seedlings are unable to grow on contaminated soils, while they outgrow the resistant ones on noncontaminated soils. Diversifying selection has produced the resistant strains in a relatively short period of time: some mines are less than 400 years old.

Diversifying selection associated with mimicry occurs in the African swallowtail butterfly, *Papilio dardanus,* studied by C. A. Clarke and P. M. Sheppard (Fig. 4.10). This species lives in tropical and southern Africa. Everywhere the males have yellow and black wings with typical "tails" characteristic of swallowtail butterflies. In Madagascar and parts of Ethiopia, where mimicry has not evolved, the females are similar to the males. In most of tropical Africa, however, the females are conspicuously different from the males in that their wings lack "tails" and have different color patterns. Moreover, this appearance varies from place to place. *P. dardanus* are edible to predator birds; in many places they coexist with other butterfly species that are noxious to birds and thus carefully avoided as food. In these places *P. dardanus* females mimic the appearance of these noxious species. *P. dardanus* females have evolved phenotypes that are similar in external appearance to the noxious species, since the mimics are then confused with their models and thus not preyed upon by birds. In some localities there exist two or even three different female forms, each mimicking the appearance of a different noxious species: diversifying selection has produced different phenotypes of *P. dardanus* because natural selection favors the mimicking of whichever model species happen to be present.

Polymorphism and Balancing Selection

Normalizing, directional, and diversifying selection are distinguished by the effects that natural selection has on the phenotypes of the population. Diversifying selection is a mode of selection promoting the maintenance of genetic polymorphism. Normalizing and directional selection may or may not favor polymorphism at the genetic level. Normalizing selection operates against deleterious mutations and favors "normal" alleles, leading to a mutation-selection equilibrium with the deleterious alleles

Fig. 4.10. Mimicry in the African swallowtail butterfly, *Papilio dardanus. Top row.* Nonmimetic female (left) and male (right) forms of *P. dardanus* from Madagascar. Second to fifth rows show noxious model species on the left and the corresponding mimetic forms of *P. dardanus* on the right. The model species on the left are *Bematistes poggei* (second row), *Danaus chrysippus* (third row), *Amauris albimaculata* (fourth row), and *Amauris niavius dominicanus* (bottom row).

kept at very low frequencies. But the normal alleles need not be only one at each locus; the normal phenotype may be produced by any one of several alternative genotypes. Similarly, directional selection may keep a population genetically polymorphic at the loci involved in the selection.

The experience with artificial selection suggests that such possibilities are common. When natural populations that have long been under normalizing selection are subject to artificial selection, the response is usually rapid. This indicates that alleles affecting the character under selection exist in the populations with frequencies that are not very low. And whenever artificial selection is stopped, populations usually have a homeostatic response that changes them back toward the original phenotypic characteristics. This indicates that the directional selection has preserved alternative alleles.

One mechanism maintaining genetic polymorphism is *heterosis,* or selection in favor of the heterozygotes, described as Case 4 in the appendix. A classical example of heterosis in humans involves sickle-cell anemia, a disease fairly common in some African and Asian populations. The anemia is due to homozygosis for a gene, *a,* which produces an abnormal hemoglobin instead of the normal hemoglobin produced by allele *A.* The homozygotes *aa* usually die before reaching sexual maturity, so that their fitness is only slightly greater than zero. In spite of this, the *a* allele has fairly high frequencies in certain parts of the world, precisely in

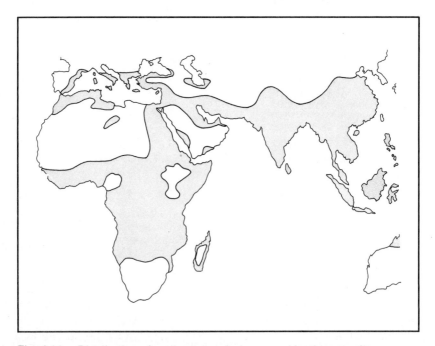

Fig. 4.11. Distribution of malignant malaria caused by the parasite *Plasmodium falciparum* in the Old World.

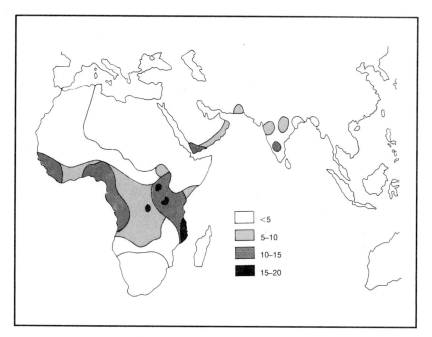

Fig. 4.12. Distribution of the allele *a* which in homozygous condition is responsible for sickle-cell anemia. The frequency of *a* is high in parts of the world where falciparum malaria is endemic, because individuals heterozygous for the *a* allele are highly resistant to malarial infection.

regions where a certain form of malaria (caused by the parasite *Plasmodium falciparum*) is common (Figs. 4.11 and 4.12).

The reason for the high frequency of the *a* allele in malarial regions is that heterozygotes *Aa* are effectively immune against malarial infections, while homozygotes *AA* are not. In some malarial regions, the fitness of the *aa* homozygotes is approximately 0.1 (the selective coefficient is, therefore, $t = 1 - 0.1 = 0.9$). The fitness of the *AA* homozygotes is, however, about 0.8 ($s = 0.2$), while the fitness of the *Aa* heterozygotes is 1. According to the formula given in the appendix in the case of heterosis the expected equilibrium frequency of the *A* allele is

$$p = \frac{t}{s + t} = \frac{0.9}{0.9 + 0.2} = \frac{0.9}{1.1} = 0.82$$

The equilibrium frequency of the *a* allele is, therefore $q = 1 - 0.82 = 0.18$, which is the frequency it has in some malarial regions.

In such regions, $q^2 = 0.18^2 = 0.032$, that is 3.2 percent of the infants born will be homozygotes *aa*, most of whom will die before adulthood. The frequency of homozygotes AA is $p^2 = 0.82^2 = 0.672$, or 67.2 percent of the individuals, all of whom have a fitness of only 0.8 due to susceptibility to malaria. The frequency of the heterozygotes is $2pq = 2 \times 0.82 \times 0.18 = 0.295$, or 29.5 percent of the individuals, who are resistant to malaria and do not suffer from sickle-cell anemia. If members of such a population migrate to a malaria-free country, or if malaria is eradicated, the fitness of the AA homozygotes will be 1, the same as the fitness of the heterozygotes Aa. Normalizing natural selection will, then, be of the type called Case 1 in the appendix, and will reduce the frequency of a, and eventually eliminate it entirely. A gradual reduction of the a allele has occurred in U.S. Blacks whose ancestors migrated from malarial regions to the U.S., where falciparum malaria is effectively absent.

Other forms of selection besides heterozygote advantage may lead to balanced genetic polymorphisms. One is *frequency-dependent selection,* probably common in nature. Selection is frequency dependent when the genotypic fitnesses vary with their frequency. In the model cases discussed in the appendix, it is assumed that fitnesses are constant, no matter what the frequencies of the genotypes are. This simplifies the mathematical treatment of selection but is often unrealistic. Assume that the fitnesses of two genotypes AA and aa are inversely related to their frequencies, high when a genotype is rare and low when it becomes frequent. If a genotype is rare at a given time, natural selection will enhance its frequency; but as its frequency grows, its fitness diminishes while the fitness of the alternative genotype increases. If there is a frequency at which the two genotypes, AA and aa, have equal fitness, a stable polymorphic equilibrium will occur even without heterosis.

Frequency dependence is related to diversifying selection. In heterogeneous environments, a genotype may have high fitness when it is rare because the subenvironments in which it is favored are relatively abundant. When a genotype is common, its fitness may be low because its favorable subenvironments are saturated. Frequency-dependent selection has been extensively demonstrated in experimental populations of *Drosophila* and in cultivated plants. For example, in the lima bean, *Phaseolus lunatus,* R. W. Allard and his collaborators have shown that the fitnesses of three genotypes, SS, Ss, and ss, change over the generations as their frequencies change. The fitness of the heterozygotes is equal to that of the homozygotes when the heterozygotes represent about 17 percent of the population, but is nearly three times as high when the heterozygotes are only 2 percent of all individuals. Frequency-dependent *sexual* selection occurs when the probability of mating is affected by the genotypic frequencies. Examples of this phenomenon are given in Chapter 5.

QUESTIONS FOR DISCUSSION

1. What are the assumptions of the Hardy-Weinberg law?

2. Which process is more important with respect to the rate of evolution, mutation or selection? What about with respect to the direction of evolution?

3. Assume that in a population no individual dies between birth and reproductive maturity (or that all individuals die of old age). Could there be natural selection in that population?

4. In view of your answer to the previous question, would you say that "struggle for existence" is an adequate name for natural selection? Explain.

5. Contrast directional and diversifying selection. Can you think of examples of these processes in nature?

Recommended for Additional Reading

Fisher, R. A. 1930. *The genetical theory of natural selection.* Oxford: Clarendon Press.

Haldane, J. B. S. 1932. *The causes of evolution.* London and New York: Harper & Brothers.

Two classical texts that greatly contributed to the development of the neo-Darwinian theory of evolution. Although mathematically advanced, they can be usefully read skipping the mathematics.

Levins, R. 1968. *Evolution in changing environments.* Princeton, N.J.: Princeton University Press.

Original and mathematically sophisticated discussion of the role of spatial and temporal variation in evolution.

Mettler, L. A., and Gregg, T. G. 1969. *Population genetics and evolution.* Englewood Cliffs, N.J.: Prentice-Hall.

Wilson, E. O., and Bossert, W. H. 1971. *A primer of population biology.* Sunderland, Mass.: Sinauer Associates.

Two elementary texts dealing with the fundamental concepts of population genetics introduced in the chapter.

5

Natural Selection in Action

In Chapter 4 we introduced natural selection as the evolutionary process responsible for the adaptation of organisms to the environments in which they live. Fitness is the parameter used to measure selection at the genetic level. The effects of natural selection were first considered in simple cases in which only one gene locus is affected. But natural selection is a process affected by many factors and with diversified effects in different cases. Considering its effects on a given phenotypic trait, natural selection may be normalizing, directional, or diversifying; which of these forms of selection depends largely on the conditions of the physical environment. In this chapter we continue the study of natural selection. We shall consider the effects of various interactions, first between different gene loci, then between different organisms of the same or of different species. In the latter part of the chapter we will consider the genetic and physiological basis of adaptation and we shall see that natural selection is, in some sense, a creative process.

Genetic Coadaptation

Imagine a zygote made of some human genes, some shark genes, and some wheat genes. Such a chimera could not develop into a functional organism. The genes that make up the gene pool of a species interact well with one another so that they produce viable zygotes—this is known as *genetic coadaptation*—but not usually with the genes of other species. The process responsible for the co-adaptation of gene pools is natural selection. Whenever a new gene or chromosomal mutation arises that does not interact well with the rest of the genome, it is eliminated (or kept at low frequency) because the organisms possessing it are unable to survive or to leave progeny. Most conspicuous evidence of genetic coadaptation is provided by the failure of interspecific hybrids. Most living species cannot be intercrossed; interspecific fertilization is sometimes possible between closely related species, but more often than not the hybrid zygotes fail to develop or develop into sterile organisms, such as the mule. Horse genes and donkey genes are not mutually coadapted.

Owing to genetic coadaptation, a certain allele or set of alleles may be favorably selected in one species, but a different allele or set in a different species. One example involves *Drosophila equinoxialis* and *D. tropicalis,* two vinegar fly species abundant in the rain forests of the American tropics. Electrophoretic studies of the gene locus *Mdh,* coding for the enzyme malate dehydrogenase, have shown that one allele, called *94,* has a frequency of about 0.99 in *D. equinoxialis,* while allele 86 also has a frequency of about 0.99 in *D. tropicalis.* Two observations are particularly relevant. The first is that both alleles exist in either species (Table 5.1), but the one that is common in *D. equinoxialis* is very rare in *D. tropicalis,* and vice versa. The second observation is that the situation described occurs everywhere throughout the northern half of South America where the two species exist (see next chapter, Fig. 6.6).

Table 5.1
ALLELIC FREQUENCIES AT THE *MDH* LOCUS IN TWO *DROSOPHILA* SPECIES

The frequencies were observed in a large sample of flies collected in Tame, Colombia. "Other" refers to alleles with very low frequencies in both species. (After Ayala and Anderson, 1973.)

	Alleles		
Species	86	94	Other
D. equinoxialis	0.005	0.992	0.003
D. tropicalis	0.995	0.004	0.001

The two alleles, 86 and 94, are very similar to each other, and flies can function whether they carry one or the other. It might be thought that the two alleles are equivalent, and that which one of the two alleles is common in a given population has been determined by random drift. But this is unlikely, because we would then expect to find local populations of *D. equinoxialis* with allele 86 in high frequency, and populations of *D. tropicalis* with allele 94 in high frequency. Another conceivable explanation is that allele 86 is favored in some environments, but allele 94 in other environments. This is also unlikely, because the two species exist in the same habitats, feed on the same fruits, and have very similar ways of life.

The following experiment corroborates the hypothesis that genetic coadaptation is responsible for the different allelic frequencies at the *Mdh* locus (Fig. 5.1). Laboratory populations of each species were set up with the two alleles, 86 and 94, present in all populations, but the frequency of the rare allele was artificially increased in some populations above its natural frequency. For about eight generations the populations were sampled at regular intervals to ascertain the allelic frequencies. Natural selection acted towards restoring the natural frequencies—in *D. tropicalis* allele 94, which is rare in nature, decreased in frequency, while the same allele was favored by natural selection in *D. equinoxialis,* a species in which allele 94 is common in nature.

The enzymes coded by alleles 86 and 94 differ in their net electric charge (this difference is responsible for their separation in gels after electrophoresis). Alleles with a given electric charge are favored in one species, but selected against in the other. The different directions of the selection process in the two species could not be due to the external environment, because this was the same in all experimental populations. Whether one or another allele is favored by natural selection appears to be determined by the genetic background in which the allele is present, i.e., by which alleles are present at other loci.

Genetic coadaptation is a property of the gene pool of a species as a whole, but also of local populations. Alleles may be favored in one locality but not in others, because they interact well with other alleles in the first but not in the second locality. An illustrative example is provided by the African swallowtail butterfly, *Papilio dardanus,* mentioned in Chapter 4 (p. 139, Fig. 4.10). Recall that various mimetic forms of this species exist, each mimicking a different species of butterfly noxious to bird predators. Several mimetic forms exist in some localities, while only one is found in other localities, depending on what model species happen

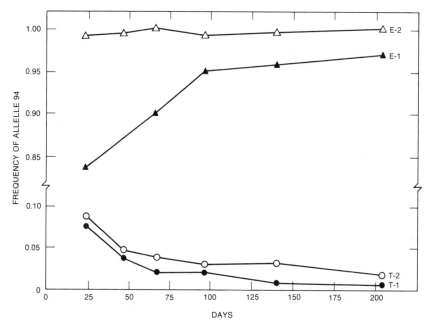

Fig. 5.1. Natural selection in laboratory populations of *Drosophila equinoxialis* (E-1 and E-2) and *Drosophila tropicalis* (T-1 and T-2). Two alleles, *86* and *94,* of *Mdh* (a gene coding for the enzyme malate dehydrogenase) were initially present in all four populations. Allele *94* is very common in natural populations of *D. equinoxialis* but very rare in natural populations of *D. tropicalis*. Natural selection proceeds toward restoring the frequencies occurring in nature: allele *94* increases in frequency in population E-1, where it had an initial frequency lower than it has in nature. The same allele *94* decreases in populations T-1 and T-2 where its initial frequency was above its frequency in nature. The experimental environment is the same in all four populations.

to exist in such localities. Crosses can be made between any two mimetic forms, which we may call A and B. The interesting phenomenon is that crosses between A and B give different results depending on whether or not the two forms come from the same locality. If both parents come from the same locality, only perfect mimics are produced in the F_1, F_2, and backcross generations. However, when the two mimicking forms come from different regions, the F_1 progenies are intermediate between the two parents in appearance, and the F_2 and backcross progenies also show intermediate phenotypes.

The explanation for these results is as follows. The mimetic patterns are determined for the most part by two major gene loci.

At one locus there are two alleles, one determining presence, the other absence, of the "tails" that are typical of swallowtail butterflies. The other locus consists of several alleles, each determining the main color pattern of one mimetic form. There are, moreover, a number of "modifier" gene loci that affect the expression of the major genes. Alleles have been selected at these loci that maximize the mimetic characteristics of the butterflies. This, however, is accomplished by different sets of alleles in different local populations. Because in nature *P. dardanus* butterflies from different regions do not intercross, natural selection has not coadapted the sets of modifier alleles from separate regions. When mimetic forms from different regions are intercrossed, alleles that are not mutually coadapted are joined together and imperfectly mimicking forms arise.

Linkage Disequilibrium

At polymorphic gene loci genetic coadaptation may exist between certain alleles at one locus and certain alleles at another locus. The following represents a simple case. Assume that there are two loci, A and B, and that at each locus there are two alleles, A_1 or A_2 and B_1 or B_2. Assume further that alleles A_1 and B_1 interact well with each other so that they produce well-adapted phenotypes, and that the same is true for A_2 and B_2, but that the combinations A_1B_2 and A_2B_1 yield poorly adapted phenotypes. The adaptation of the population would be increased if the alleles would always (or most often) be transmitted in the combinations A_1B_1 and A_2B_2, while the combinations A_1B_2 and A_2B_1 would never (or rarely) occur.

When alleles at different loci are not associated at random, the loci are in *linkage disequilibrium*; *linkage equilibrium* exists when alleles at different gene loci are associated at random (Closer Look 5.1). Favorably interacting alleles may be transmitted together in disproportionately large frequencies if they are in linkage disequilibrium. Linkage disequilibrium may result from natural selection because individuals with favorable allelic combinations reproduce more efficiently than those with less favorable combinations.

As shown in Closer Look 5.1, linkage disequilibrium is decreased by genetic recombination. Thus, the possibility of maintaining linkage disequilibrium is enhanced by reduction of the frequency of recombination between loci. This may be accomplished through translocations and inversions. Assume that two loci, A and B, are located in different chromosomes; a translocation might bring them together in the same chromosome. Assume, now, that the two loci are separated within the same chromosome by a number of loci that we represent as $FG \ldots MN$, so that the gene sequence along the chromosome is

$$\ldots AFG \ldots MNB \ldots$$

An inversion comprising the segment $FG \ldots MNB$ would bring together A and B; the new gene sequence would be

$$\ldots ABNM \ldots GF \ldots$$

Whenever linkage disequilibrium is favored by natural selection, chromosomal rearrangements increasing linkage between the loci will also be favored by natural selection. The term *supergene* is used to refer to several closely linked gene loci that affect a single trait or a series of interrelated traits.

Another way of maintaining linkage disequilibrium between loci is through inversion polymorphisms. Assume as before that A_2B_1 and A_2B_2 are favorable allelic combinations, while A_1B_2 and A_2B_1 are unfavorable ones. Let us represent the gene sequence in the chromosome as

$$\ldots DEAF \ldots NBOP \ldots$$

Assume that an inversion of the segment from E to O takes place and that the alleles A_1 and B_1 are included in the segment. We would have the following sequence:

$$\ldots DOB_1N \ldots FA_1EP \ldots$$

(Subscripts at loci other than A and B are not added, because we are not now concerned with alleles at these other loci.)

Assume now that an individual heterozygous for the inversion and the original chromosome sequence carries alleles A_2 and B_2 in the original chromosome sequence, i.e., that an individual has the following genetic constitution:

$$\ldots DOB_1N \ldots FA_1EP \ldots$$
$$\ldots DEA_2F \ldots NB_2OP \ldots$$

As explained in Chapter 3 (p. 95; Figs. 3.19 and 3.20), recombination is suppressed in the progenies of inversion heterozygotes. Therefore, such an individual will produce only two kinds of gametes, one containing the alleles A_1 and B_1, and the other containing the alleles A_2 and B_2. Natural selection may, then, favor original chromosome sequences that have alleles A_2 and B_2 and lead to the elimination of alleles A_1 and B_1 from the uninverted chromosome sequence. The population will then consist of only three types of individuals: (1) homozygotes for the chromosomal inversion, and therefore for alleles A_1 and B_1; (2) homozygotes for the original chromosome sequence, and therefore for alleles A_2 and B_2; (3) heterozygotes for the inverted and the original sequence. Only the two gametic combinations A_1B_1 and A_2B_2 exist in such individuals.

CLOSER LOOK 5.1 Linkage Disequilibrium

Assume that there are two loci, A and B, each with two alleles in the following frequencies

$$A_1 = p \quad A_2 = q$$
$$B_1 = r \quad B_2 = s$$

Since only two alleles exist at each locus, we have

$$p + q = 1$$
$$r + s = 1$$

If the alleles at the two loci are associated at random, we expect the two-loci combinations to have frequencies that are the product of the frequencies of the two alleles involved, that is,

$$A_1 B_1 = pr$$
$$A_1 B_2 = ps$$
$$A_2 B_1 = qr$$
$$A_2 B_2 = qs$$

Since those are the only four possible combinations, the frequencies must add to 1; indeed:

$$pr + ps + qr + qs = p(r + s) + q(r + s)$$

and since $r + s = 1$, and $p + q = 1$, we have

$$p(r + s) + q(r + s) = p + q = 1$$

If the alleles at the two loci are associated at random, the product of the expected frequencies of the two gametic combinations, $A_1 B_1$ and $A_2 B_2$, is $(pr)(qs) = pqrs$, which is the same as the product of the other two gametic combinations, $A_1 B_2$ and $A_2 B_1$, namely $(ps)(qr) = pqrs$. However, if the alleles are not randomly associated, then the two products will be different, and the degree of nonrandom association can be measured by the difference, d, between the products of the two pairs of gametic combinations:

$$d = (\text{freq. of } A_1 B_1)(\text{freq. of } A_2 B_2) -$$
$$(\text{freq. of } A_1 B_2)(\text{freq. of } A_2 B_1)$$

d is, therefore, a measure of *linkage disequilibrium* between alleles at two gene loci. When the alleles at two loci are randomly associated we have $d = 0$, i.e., there is no linkage disequilibrium, or to put it conversely, the two loci are in *linkage equilibrium.*

Linkage disequilibrium is complete when only two gametic combinations occur, either A_1B_1 and A_2B_2, or A_1B_2 and A_2B_1. The maximum absolute value that d may take is 0.25, namely, when linkage disequilibrium is complete and the allelic frequencies are 0.5 at both loci. The two possible cases are

	Gametic frequencies				d
	A_1B_1	A_1B_2	A_2B_1	A_2B_2	
Case 1	0.5	0	0	0.5	$(0.5)(0.5) - (0)(0) = 0.25$
Case 2	0	0.5	0.5	0	$(0)(0) - (0.5)(0.5) = -0.25$

When the allelic frequencies are not all 0.5, the absolute value of d is always less than 0.25. (You might want to check this for various cases; for example when $A_1 = 0.7$, $A_2 = 0.3$, $B_1 = 0.7$, $B_2 = 0.3$, d cannot be greater than 0.21.)

In a random mating population (ignoring natural selection and other processes of genetic change), the value of d will decrease every generation by a proportion that is precisely the frequency of recombination, r, between the two loci. When the two loci are unlinked (equal in different chromosomes), $r = 0.5$, and d will be halved every generation. If $r = 0.1$—i.e., there is 10 percent recombination between the two loci—then d will decrease by 10 percent each generation. In general, linkage disequilibrium, d_1, in any one generation in terms of the disequilibrium in the previous generation, d_0, is

$$d_1 = (1 - r)\, d_0$$

where r is the frequency of recombination as defined earlier.

Therefore, the change in the value of d is

$$\Delta d = d_1 - d_0 = (1 - r)d_0 - d_0 = d_0 - rd_0 - d_0 = -rd_0$$

As stated above the value of d is reduced each generation by a fraction r of its value.

Supergenes

A supergene is responsible for the expression of two flower phenotypes, known as *pin* and *thrum,* found in the primrose and other species of the genus *Primula* (Fig. 5.2). Pin-thrum polymorphisms were made famous by Darwin, who gave a detailed account of them in 1877. The pin phenotype is characterized by a long style above the ovary, which places the stigma at the same level as the mouth of the corolla; the pollen-bearing anthers are halfway down the corolla-tube. The thrum phenotype has a short style, so that the stigma is halfway down the corolla-tube, while the stamens are long, placing the anthers at the mouth of the corolla. Pin and thrum are also phenotypically different in other ways, such as the configuration of the stigma and the size of the pollen grains. Moreover, they differ physiologically—thrum pollen succeeds better in fertilization when deposited on pin stigmas than on thrum stigmas, and vice versa, pin pollen fertilizes thrum flowers more successfully than pin flowers.

The pin-thrum phenomenon is known as *heterostyly* (meaning "different styles"). Heterostyly assures cross-pollination. An insect that visits both pin and thrum flowers will receive pollen

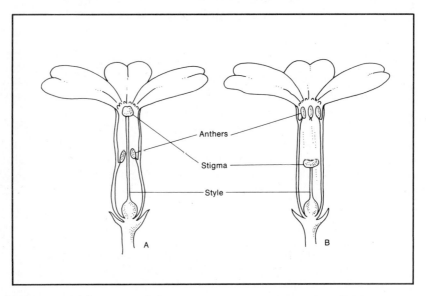

Fig. 5.2.   *A,* the pin and *B,* thrum phenotypes in the primrose *Primula officinalis.* The pin phenotype has a highly placed stigma but low-placed anthers, while the opposite occurs in the thrum phenotype. This complementary arrangement facilitates cross-fertilization between the two phenotypes.

from one type on parts of the body that get close to the stigma of the other type. The physiological differences reinforce the chances of cross-fertilization.

Pin and thrum phenotypes behave, as a rule, as if they were controlled by a single gene locus, with two alleles: S (for thrum) is dominant over s (for pin). Thrum plants, however, are generally heterozygous (Ss); when they are self-pollinated or intercrossed they produce pin and thrum plants in a typical 3:1 Mendelian ratio. Pin plants are homozygous (ss) and produce only pin types when intercrossed or self-pollinated. In nature, most crosses are between thrum (Ss) and pin (ss) plants, producing thrum and pin progenies in about 1:1 ratio. Pin and thrum flowers are found in approximately similar frequencies in natural populations.

The set of traits characteristic of the thrum or the pin phenotypes is not, however, determined by a single gene locus, but by several closely linked loci making up a supergene. The existence of multiple gene loci might be suspected because the phenotypic and physiological differences between the thrum and pin types are multiple. And indeed this has been confirmed by examination of large progenies from thrum x pin crosses; examples of mixtures between components of the two complex phenotypes are occasionally found, due to recombination within the supergene. In nature, mixed phenotypes are occasionally found as well, but these remain rare owing to their low fitness relative to the thrum and pin phenotypes. The supergene has become established in *Primula* because it makes possible the joint transmission of sets of alleles that produce adaptive phenotypes. The supergene control saves *Primula* populations from a high proportion of ill-adapted phenotypes.

A. J. Cain and P. M. Sheppard have found in the snail *Cepaea nemoralis* a supergene controlling the color and banding pattern of the shells. Two main loci are involved. Alleles at one locus determine the background color of the shell, which may range from light yellow through pink to brown, with lighter-color alleles being recessive to darker-color alleles. The other locus determines whether or not darkly pigmented bands will be superimposed upon the background color, the unbanded type being dominant over the banded type. Still other loci control the number of bands. The loci determining the background color and the presence or absence of bands are closely linked, but occasional crossovers have been obtained in experimental studies.

The frequency of the various color and banding patterns in *C. nemoralis* varies considerably from colony to colony, and it is associated with the ecology of the site. The commonest phenotypes are those that are most inconspicuous against the prevailing background, which protects the snails from birds and other predators. Thus, yellow shell colors are most common in green areas, pinks on leaf litter, and browns in beechwoods; also, the unbanded phenotypes are common in relatively uniform backgrounds such as in dense woodlands, while the banded phenotypes prevail in diversified habitats, such as a mixed hedgerow (Fig. 5.3). The supergene makes it possible for favorable allelic combinations to be

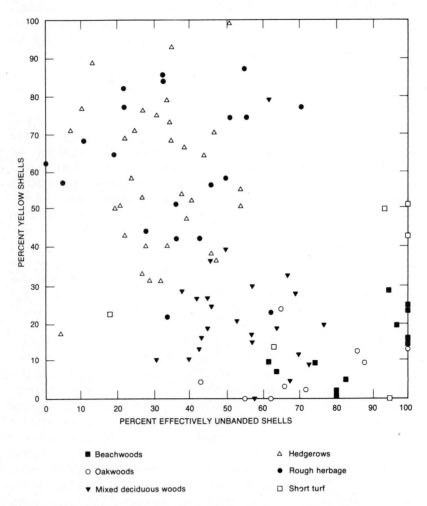

Fig. 5.3. Banding and shell color in the snail *Cepaea nemoralis* in different environments. Yellow and banded shells are common in short turf, rough herbage, and hedgerows, while brown and unbanded shells prevail in mixed woods, oakwoods, and beechwoods. The common phenotypes found in a given site are those that are less conspicuous there.

maintained in a given population without production of unfavorable recombinants.

The formation of supergenes by means of translocations and inversions that bring interacting loci into close linkage is evidenced by the situation found in different species of grouse locusts. As shown by R. K. Nabours, the color patterns are determined by alleles at some 13 gene loci: in one species, *Acridium arenosum,* the genes are spread throughout one single chromosome and recombine fairly freely. In another species, *Apotettix eurycephalus,* the genes are combined into two groups (supergenes) of closely linked genes, the recombination between the groups being only 7 percent. In a third species, the relevant genes are all tightly linked, forming a single supergene. The formation of supergenes has advanced most in this last species.

Inversion Polymorphisms

As a rule inversions and other chromosomal changes produce no visible effects on the external appearance of their carriers. They can, however, be detected by microscopic examination. This is easiest in flies and mosquitoes, which have giant banded chromosomes (called *polytene chromosomes*) in the cells of larval salivary glands. The polytene chromosomes are bundles of chromosomes replicated

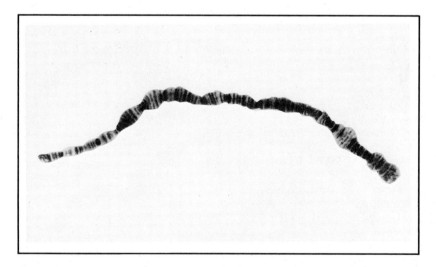

Fig. 5.4. A segment of a polytene chromosome from the salivary gland of *Drosophila.* Polytene chromosomes are bundles of chromosomes replicated hundreds or thousands of times, all of which are perfectly paired lengthwise; their "giant" size makes it possible to study their fine structure under a microscope.

hundreds or thousands of times, all of which are perfectly paired lengthwise (Fig. 5.4). Inversion heterozygotes have two chromosomes of a homologous pair with gene arrangements differing in one or more chromosome segments placed in inverted orders—*ABCDE* and *ADCBE,* for example. The pairing of such chromosomes results in a loop, easily visible under the microscope in the salivary glands (Fig. 3.18).

Inversion polymorphisms have been studied in many species of *Drosophila.* Some species have polymorphisms in all chromosomes—for example, the European species *D. subobscura* and the American tropical species *D. willistoni*—while others have inverted segments concentrated mostly in one chromosome—for example, the North American *D. pseudoobscura,* which exhibits extensive polymorphism in only one of the five chromosomes, the third (Table 5.2).

As shown in Table 5.2, the frequencies of the various chromosomal arrangements in *Drosophila pseudoobscura* vary from one

Table 5.2
VARIATION IN THE FREQUENCIES OF CHROMOSOMAL INVERSIONS

Relative frequencies of third chromosomes with different gene arrangements in populations of *Drosophila pseudoobscura* in various localities. Each gene arrangement is designated by two capital letters.

	ST	AR	CH	PP	TL	SC	OL	EP	CU
Methow, Washington	70.4%	27.3%	0.3%	—	2.0%	—	—	—	—
Mather, California	35.4	35.5	11.3	5.7%	10.7	0.9%	0.5%	0.1%	—
San Jacinto, California	41.5	25.6	29.2	—	3.4	0.3	—	—	—
Fort Collins, Colorado	4.3	39.9	0.2	32.9	12.3	—	2.1	7.2	—
Mesa Verde, Colorado	0.8	97.6	—	0.5	—	—	—	0.2	—
Chiricahua, Arizona	0.7	87.6	7.8	3.1	0.6	—	—	—	—
Central Texas	0.1	19.3	—	70.7	7.7	—	2.4	—	—
Chihuahua, Mexico	—	4.6	68.5	20.4	1.0	3.1	0.7	—	—
Durango, Mexico	—	—	74.0	9.2	3.1	13.1	—	—	—
Hidalgo, Mexico	—	—	—	0.9	31.4	1.7	13.5	1.7	48.3%
Tehuacan, Mexico	—	—	10.3	—	7.9	—	0.9	1.6	71.4
Oaxaca, Mexico	—	—	—	—	20.2	1.1	—	3.2	74.5

locality to another. Moreover, the frequencies may change from month to month throughout the year (Fig. 5.5). These changes are cyclic with the seasons of the year, and thus are repeated in successive years. This suggests that the chromosomal arrangements differ in the sets of alleles they carry, and that these differences are adaptive: one arrangement is adaptively superior to the other during some time of the year, but inferior during some other time. This hypothesis was tested with laboratory populations started with known frequencies of the chromosome arrangements and allowed to breed freely within the laboratory cage. Typical results are shown in Figure 5.6. The frequencies of the inversions change rapidly in the early generations, more slowly later on, and eventually reach an equilibrium with both chromosome arrangements present. From the rate of change and the equilibrium frequencies, the fitnesses of the three genotypes can be estimated; these are 1 for the heterozygote (ST/CH), 0.89 for one homozygote (ST/ST), and 0.41 for the other homozygote (CH/CH).

These results show that heterosis (heterozygote superiority) contributes to maintain the chromosomal polymorphism (see Case 4 in the appendix). Other laboratory experiments have shown that the fitnesses of the two homozygous genotypes depend on the temperature and on the density of the population, which may account for the seasonal oscillations observed in nature.

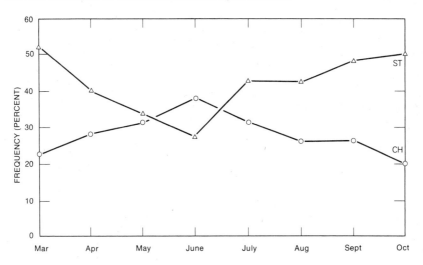

Fig. 5.5. Frequencies of the *ST* and *CH* chromosomal arrangements of *Drosophila pseudoobscura* in San Jacinto, California. The frequencies of the two chromosomal arrangements change throughout the year. *CH* reaches its highest frequency at the beginning of summer, when the frequency of *ST* is lowest. (After Dobzhansky, 1970.)

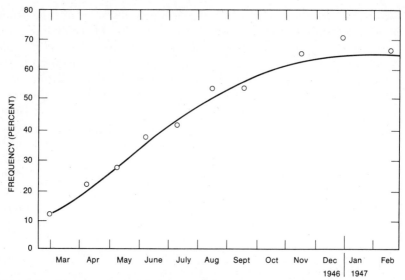

Fig. 5.6. Changes in the frequency of the *ST* chromosomal arrangement of *Drosophila pseudoobscura* in a laboratory population. Two chromosomal arrangements, *ST* and *CH,* are present in the population. The frequency of *ST* gradually increases from its initial frequency of 12 percent until reaching an equilibrium frequency around 70 percent. Correspondingly, *CH* decreases from its initial frequency of about 88 percent to an equilibrium frequency around 30 percent. (After Dobzhansky, 1970.)

Direct evidence that chromosomal inversions differ in their allelic content has been obtained in *D. pseudoobscura* by S. Prakash and R. C. Lewontin. Two gene loci, *Pt-10* and *α-Amy,* coding for two proteins, were examined by gel electrophoresis. It was discovered that the allelic frequencies were quite different in different chromosomal arrangements (Table 5.3).

Inversion polymorphisms have been observed in natural populations of mosquitoes, black flies, midges, and other dipterans. It is uncertain how widespread such polymorphisms are in other organisms since inversions are difficult to detect in the absence of polytene chromosomes. Nevertheless, inversion heterozygotes are known in many animals, such as grasshoppers, and in some plants.

Population Growth

Organisms are able to multiply in numbers, without intrinsically imposed bounds. Bacteria, such as *Escherichia coli,* multiply by fission, so that each bacterium may divide into two every hour or so. Assume that this multiplication process goes on for 1,000 generations. Then for every bacterium in the original population there

Table 5.3
ALLELIC FREQUENCIES AT TWO GENE LOCI IN DIFFERENT CHRO-
MOSOMAL ARRANGEMENTS OF *DROSOPHILA PSEUDOOBSCURA*

The two gene loci are *Pt-10*, coding for a larval protein, and *α-Amy*, coding
for the enzyme α-amylase. The Pikes Peak chromosomal arrangement is
evolutionarily closely related to Standard, while Santa Cruz is closely related to Tree Line. The first two chromosomal arrangements have alleles
104 of *Pt-10* and *100* of *α-Amy* in high frequencies, but alleles *106* of *Pt-10*
and *84* of *α-Amy* in low frequencies; the opposite is true for the Santa
Cruz and Pikes Peak chromosomal arrangements. (After Prakash and Lewontin, 1968.)

Chromosomal	*Pt-10*		*α-Amy*	
arrangement	*104*	*106*	*84*	*100*
Standard	1.00	0.00	0.15	0.85
Pikes Peak	1.00	0.00	0.00	1.00
Santa Cruz	0.00	1.00	1.00	0.00
Tree Line	<0.01	>0.99	>0.90	0.05

will be $2^{1000} = 10^{301}$ bacteria at the end. It is clear that this cannot
happen: 10^{301} (one followed by 301 zeros) is a number larger than
the number of atoms in the universe.

Sexually reproducing organisms also have a capacity to multiply which is potentially unlimited. A pair of individuals may produce during their lifetime from a few to several million progeny.
Assume that each two individuals leave four surviving progeny in
the following generation. For every two original individuals we
will have four after one generation, eight after two generations,
sixteen after three generations, and so on. After 1,000 generations
(which represent only a brief instant on the evolutionary time
scale), we would have, as before, $2^{1000} = 10^{301}$ individuals for every
organism in the original population.

The kind of increase in numbers described in the previous
two paragraphs is known as "exponential" growth, "logarithmic"
growth, or "geometric" growth (Fig. 5.7). It is obvious that organisms cannot grow exponentially forever, since soon they would exhaust the resources they need for survival. This is what Thomas
Robert Malthus, the English clergyman and economist, pointed out
in his famous *Essay on the Principle of Population* (1798). Since
mankind multiplies geometrically, humans would in relatively few
generations exhaust all possible means of subsistence. Thus he
concluded that either mankind must intentionally restrict its growth,
or famine, disease, and other catastrophies will eventually do it for
us. Darwin said that reading Malthus' essay helped him to discover the principle of natural selection: because not all progeny
produced by organisms can possibly survive and reproduce, the op-

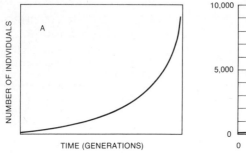

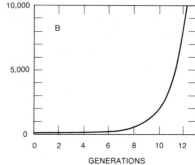

Fig. 5.7. Exponential population growth. Exponential or log-arithmic population growth occurs when the number of individuals in the population is multiplied by a fixed number (the *rate* of population growth) every generation. *A:* General representation. *B:* Exponential population growth when the rate of growth is two per generation; starting from two individuals there would be 2,048 individuals after 10 generations, more than two million after 20, and more than two billion after 30 generations.

portunity exists for differential multiplication, those carrying useful variations surviving and reproducing better than others.

What are the factors limiting the multiplication of organisms? For convenience, we may classify them into two categories: density-dependent and density-independent factors. *Density-dependent* factors have effects that depend on the density of the population, such as food and living space. As plants, animals, or any organisms increase in numbers within a given habitat, the amount of resources available *per individual* decreases. A point may be reached when there is not sufficient food for all of them, and some will die or fail to reproduce owing to starvation. Or it may be that there is not any suitable soil for the plants to grow on, or not enough places for the animals to rest, or not enough territories where the animals can establish themselves for feeding and reproduction. As the organisms increase in number, their rate of growth may gradually decrease until an equilibrium density is established, at which the environmental resources are saturated (Fig. 5.8).

Density-independent limiting factors affect population growth in a way that is largely independent of population density. Density-independent limiting factors are, for the most part, those encompassed by the term "weather," including the daily and seasonal cycles of light and dark, temperature and rainfall variations, as well as more irregular changes such as storms, earthquakes, and others. Organisms may multiply when the weather is favorable but fail to do so and be drastically reduced in numbers when the weather is unfavorable or natural catastrophes take place (Fig. 5.9).

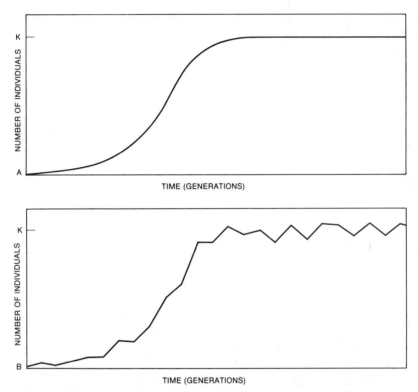

Fig. 5.8. Effects of density-dependent factors on population growth. *A:* The model known as *logistic* growth assumes that the initial *rate* of growth (*r*) decreases by a constant amount (*c*) for each individual added to the population. Population growth will cease when (*c*) multiplied by the number of individuals is equal to *r,* since at that point the rate of growth will be zero. The number of individuals at the saturation density is called the carrying capacity of the environment (*K*). Some plant and animal populations grow following approximately the logistic model. No population does so perfectly, since perturbations occur due to various factors. *B:* A representation of how logistic growth might be approximated in an actual population.

The fitness of genotypes in a population may be a function of the density of the population. Consider, for example, a population of starfish in the deep-sea, such as *Nearchaster aciculosus* that lives at more than 1,000 meters off the California coast. Temperature and other environmental variables are much more constant for organisms living in the deap-sea than virtually anywhere else. There is also a constant supply of food resources, mostly in the form of organic debris that falls through the water column down to the bottom of the sea; but the amount of food is limited. (Fig. 5.10).

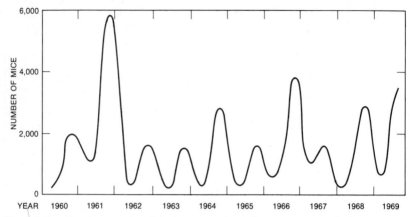

Fig. 5.9. Numbers of wild house mice (*Mus musculus*) in Skokholm, a small (100 hectares) island of the southwest tip of Wales. The numbers of mice increase particularly during the summer and decrease during winter. The population reaches "plague" numbers following a mild winter (for example, 1961, and to a lesser extent, 1966). (After Berry, 1978.)

Fig. 5.10. *Thrissacanthias,* a starfish found off the California coast that, like *Nearcaster aciculosus,* is bathymetrically restricted in range. (Courtesy California Academy of Sciences.)

Natural selection favors genotypes that produce organisms able to compete effectively for the limited food supply. Populations living in nearly saturated environments (i.e., having population numbers near the equilibrium saturation density, K, as shown in Figure 5.8) are said to be K-selected, because the favored genotypes are those that perform well at near-saturation densities.

Consider now a population of organisms that has been decimated during winter, such as a species of cricket living in the Central Valley of California. In the spring, the conditions for
reproduction are favorable and food resources are abundant. The
genotypes favored by natural selection are those responsible for
high fecundity rather than those that increase competitive ability
for limited resources. That is, simply, because organisms with high
fecundity will produce more progeny and most of the progeny encounter enough food and other resources necessary to reach maturity. Populations that are expanding are said to be subject to r
selection, because genotypes causing higher rates of reproduction
under unsaturated conditions are the favored ones. The mice in
Figure 5.9 are subject to r selection every summer as the population increases dramatically in numbers. (r is the symbol used to
represent the initial rate of population growth in the logistic
model, Figure 5.8.)

No population of organisms is exclusively r-selected or K-
selected. Even populations generally living under environmental
saturation levels occasionally experience more or less drastic re-
ductions in numbers through natural catastrophes, such as storms
or diseases, or through the invasion of a predator, or some other
way. Most populations experience times when they are rapidly expanding and other times when competition for density-dependent
resources is keen. If some genotypes are favored by selection when
a population is at low densities, while other genotypes are favored
at high densities, then oscillations in population density will be accompanied by changes in genotypic frequencies, and may contribute to maintaining genetic polymorphisms.

We mentioned earlier that the frequencies of chromosomal arrangements oscillate cyclically in natural populations of *Drosophila
pseudoobscura*. On Mount San Jacinto, California, the frequency of
the *CH* chromosomal arrangement increases in the spring but decreases in the summer relative to the *ST* arrangement. The populations remain polymorphic because the heterozygotes, *CH/ST*,
have higher fitness than either one of the two homozygotes. The
oscillations in frequency require, nevertheless, that the *ST/ST* homozygotes have lower fitness than the *CH/CH* homozygotes in the
spring, but higher in the summer. Theodosius Dobzhansky originally thought that temperature variations might be responsible for
the fitness differences. Indeed, it was shown that at 25°C the
ST/ST homozygotes had higher fitness than the *CH/CH* homozygotes (Fig. 5.6). However, tests at lower temperatures, similar to
the temperatures that occur in the natural population during the
spring, failed to show any advantage of the *CH/CH* homozygotes. It

was later experimentally shown that population density affects the fitness of the homozygotes: at low densities the CH/CH homozygotes are superior to the ST/ST homozygotes. In nature, food is abundant in the spring when density is low and the populations are experiencing r selection. In the summer, population density and competition for food are high, so that the populations are subject to K selection.

Interspecific Competition and Character Displacement

Competition for resources occurs not only between organisms of the same species but also between organisms of different species. If a limiting resource, such as food or living space, is better exploited by members of one species than another species, the second species may be eliminated from the habitat. However, often one species exploits one resource better (say, one kind of food) and a different species another (a different kind of food), or one species exploits the resource better in some subdivision of the habitat, but a different species in some other subdivision. Coexistence of the two species may then result, and in fact closely related species of plants or animals, exploiting similar resources, commonly coexist in the same habitat.

An interesting example is provided by two species of barnacles, *Balanus balanoides* and *Chthamalus stellatus*. Along the rocky seacoast of Scotland these species of barnacles compete for space in the rocks on which they grow. *Chthamalus* occupies the upper intertidal zone, while *Balanus* occupies the lower intertidal zone; but when *Balanus* is absent, *Chthamalus* can occupy both zones. J. H. Connell demonstrated that in the lower intertidal zone *Balanus* develops more rapidly and crowds out *Chthamalus* individuals; but in the upper zone, where desiccation occurs when the tide goes out, *Balanus* individuals grow slowly or die out while *Chthamalus* individuals grow better because they tolerate desiccation better.

One consequence of competition may be what has been called character displacement by W. L. Brown and E. O. Wilson. *Character displacement* is the situation in which, when two species overlap geographically, the differences between them are accentuated in the zone of coexistence but weakened or entirely lost outside this zone. Character displacement occurs when two species coexist because natural selection may favor genotypes that minimize competition for resources. The more different in their ecological requirements two species become, the less intensely they will compete. Natural selection may directly promote character dis-

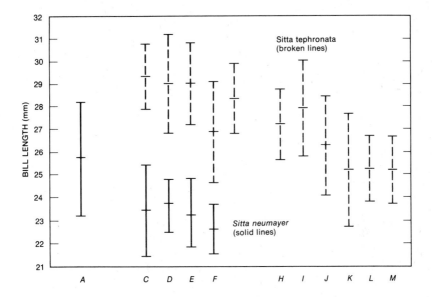

Fig. 5.11. Character displacement in beak length in two species of nuthatches. *A* to *M* represent localities in Asia arranged from east to west. *Sitta neumayer* occurs in localities *A* through *F*, *Sitta tephronata* in localities *C* through *M*. Where the two species do not occur together, they often have similar beak sizes (compare *A* with *K, L,* and *M*), but where the two coexist, the bill lengths are quite different.

placement where the species coexist, but not in habitats where only one or the other species is present. The displacement may affect physiological, behavioral, and/or morphological characters.

Character displacement is apparent in many kinds of organisms, such as in the nuthatches, *Sitta tephronota* and *S. neumayer* (Fig. 5.11). These species occur in southeastern Europe and southwestern Asia. In localities where only one or the other species lives, their beak sizes are very similar; where they coexist, *S. tephronota* has longer beaks than *S. neumayer*. Different beak sizes or shapes reflect specialization for different kinds of food, as exemplified by the famous "Darwin's finches" (Fig. 5.12).

Predation, Parasitism, and Coevolution

The ultimate source of energy for all life on earth is the sun. Plants obtain energy from the sun's radiation through photosynthesis and use that energy to synthesize organic matter and to carry out the living processes. All other organisms depend on the

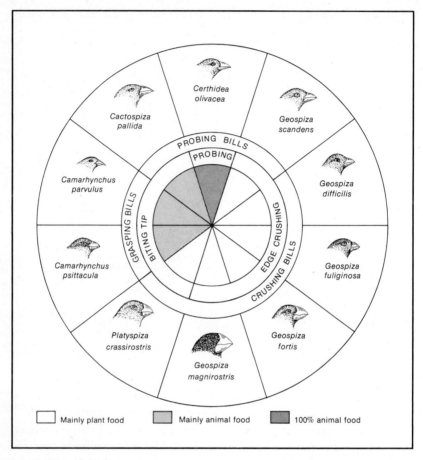

Fig. 5.12. Darwin's finches: ten species from Indefatigable Island in the Galapagos. The beaks have evolved as adaptations to the kinds of foods eaten by the finches.

energy captured by plants. Animals obtain the energy needed to synthesize their constituents and to live from plants, or from other animals that themselves obtained it from plants. Fungi and bacteria also draw the energy they need either directly or indirectly from plants.

A sequence of species through which materials and energy pass is a *food chain* (Fig. 5.13). Food chains start with *producers,* the photosynthetic plants that use energy from the sun in order to synthesize organic materials out of inorganic substances. *Consumers* are the animals that depend on the organic chemicals synthesized by plants. Consumers exist at various levels. The first level con-

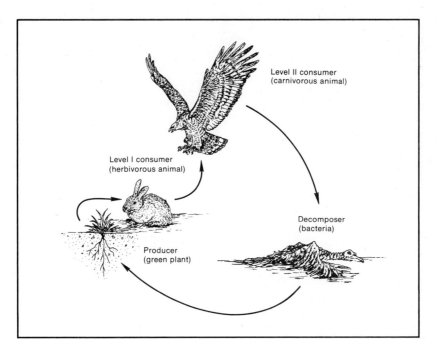

Level II consumer
(carnivorous animal)

Level I consumer
(herbivorous animal)

Decomposer
(bacteria)

Producer
(green plant)

Fig. 5.13.   A simple food chain.

sists of the herbivores—such as insects, rabbits, and deer—that feed directly on plants. First-level consumers serve as food for second-level consumers—such as spiders, frogs, fish, and birds that feed on insects, or snakes and coyotes that feed on rabbits. Then there are third-level consumers that feed on animals which themselves feed on other animals. Predatory birds eating other birds, fish, frogs, and spiders are an example of third-level consumers. Food chains end in *decomposers* (also called reducers)—including fungi and especially bacteria—that reduce organic components into inorganic matter. The source of energy for decomposers may be the producers or the consumers at any level. Actually the flow of materials and energy in a natural community is much more complicated than suggested by the term "food chain" and might be more appropriately described as a *food web* (Fig. 5.14).

Competition, whether intraspecific or interspecific, is an interaction between organisms at the same level of a food chain. Interactions between organisms at different levels include plant-herbivore, prey-predator, and host-parasite. As with interspecific competition so it is with these interactions—the evolution of one population affects the evolution of other populations that interact with it. This process is known as *coevolution*.

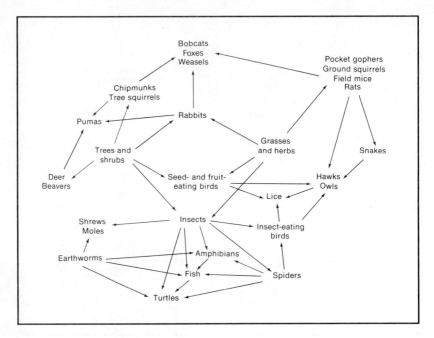

Fig. 5.14.   A simplified food web.

Predation occurs when an animal (predator) eats another animal (prey). The abundance of prey limits the possible abundance of the predators, and at the same time the abundance and efficiency of the predators reduce the abundance of the prey. Predation differs from the plant-herbivore relationship in that the herbivore usually does not kill the plant on which it feeds. In this respect the plant-herbivore relationship is similar to that between host and parasite; in most cases, parasites do not kill their hosts directly, although they usually weaken them.

Coevolution may be seen as a real-world evolutionary game where moves alternate, and the nature of a move determines what the appropriate countermove might be. Consider, first, predation. Among the prey, animals better able to escape predators will be more likely to survive and reproduce. Hence natural selection favors the development of adaptations that allow prey organisms to avoid being found, caught, and eaten. On the other hand, the better a predator is at capturing the prey, the fitter it will be. Therefore natural selection will tend to favor higher-preying efficiency in the predator. Within the limits imposed by the makeup of the organisms, and by the availability of genetic variation, prey species evolve towards avoidance of predation, but predators evolve toward maintaining predation and making it more efficient. Long-term evolution has consequently produced rather complex adaptations, such as, among predators, the long sticky tongues and accurate aim of some insectivorous lizards, the elaborate webs of spiders, and the social hunting behavior of wolves and lions. Examples of prey adaptations to escape predation are cryptic coloration and shape (which avoid detection by making the prey hardly distinguishable from the background, Figure 5.15), fleetness (Fig. 5.16), noxious substances and toxins (for example, some frogs and

Fig. 5.15.   A stick insect, *Carausius morosus,* resting with front pair of legs directed forward.

Fig. 5.16.   The bighorn or Rocky Mountain sheep, *Ovis canadensis,* found in mountains from Canada to Mexico, uses speed and agility to escape from its predators, such as this mountain lion.

toads secrete foul-tasting or poisonous substances; South American Indians use a preparation of the skins of frogs of the family Dendrobatidae to poison their arrows), and behaviors such as alarm calls and the posting of sentinels.

An interesting defense behavior exists in starlings. Starlings usually fly in loose flocks, but when a peregrine falcon is sighted starlings quickly assume a very tight formation. Peregrine falcons display magnificent adaptations for preying on birds. They have very keen vision, and fly high until a potential prey is discovered, when they dive at fantastic speeds that may reach 300 kilometers per hour. To make their detection more difficult, they often dive directly from the direction of the sun. Most prey are instantly killed by the sudden jolt of the peregrine's talons. A peregrine falcon, however, will not usually attack a tight flock of starlings because the falcon might itself be injured were it to dive into the flock (Fig. 5.17).

Fig. 5.17.   Peregrine falcon, *Falco peregrinus,* shown in a nineteenth century engraving swooping down on a swallow. Falcons prey almost entirely on birds, although small mammals and insects may also be taken.

Chemical "warfare" is the most common form of defense of plants against herbivores, although thorns and other morphological defenses are not rare. Many plants store alkaloids and other toxic chemicals in their tissues that make them unpalatable or poisonous to potential eaters. The existence of poisonous mushrooms is known to would-be gourmets, but toxins exist in innumerable plant species. A plant with an effective toxin protecting it from insect attack will be at a selective advantage; the adaptation may rapidly spread throughout the species. Often, however, insects evolve the ability to resist the toxicity of a plant poison. In fact, an additional evolutionary twist has taken place repeatedly: insects are not only able to resist the toxic effect of the poison, but they incorporate it into their own bodies and thus become unpalatable or poisonous to their predators. A typical example is the monarch butterfly, *Danaus plexippus,* which incorporates cardiac glycosides from juices of milkweeds, *Asclepias,* the plants on which it feeds. Bluejays, *Cyanocitta cristata,* will avoid monarch butterflies after tasting one and experiencing the severe vomiting provoked by the glycosides.

The coevolution of host-parasite combinations is illustrated by a well-known Australian example. Twenty-four pairs of the European rabbit, *Oryctolagus cuniculus,* were introduced into Australia in 1859. They rapidly multiplied, eventually growing to several hundred million and becoming a pest harmful to agriculture. In 1950, a virus that causes a disease known as myxomatosis was released as a control agent. The immediate effects were drastic. During the first year the disease killed about 98 percent of the rabbits throughout Australia. However, mortality was reduced to about 90 percent during the second year, and to about 50 percent in the third year. Laboratory tests showed that both the rabbits and the virus had evolved. The rabbits had become more resistant to the myxomatosis virus, which is hardly surprising since more resistant rabbits were more likely to resist the epidemic, and selection was extremely intense. The virus had also evolved, but by becoming *less* rather than more virulent. This may seem at first surprising but should not be unexpected: rabbits infected by less potent viruses take longer to die and thus are available for a longer time to mosquitoes which transmit the disease as they feed on the rabbits. The more virulent viruses kill their hosts rapidly and thus are selected against because such viruses are less likely to be passed on from rabbit to rabbit.

The Australian rabbit-virus example shows that parasites, herbivores, and predators do not always evolve toward greater efficiency. If they kill their hosts too efficiently, they may reduce their own fitness and perhaps extinguish themselves in the process. Often natural selection favors "prudent" predators, herbivores, and parasites that moderately exploit their hosts allowing many of these to survive and reproduce, so that the predators might be provided with a continuing food supply.

Sexual Selection

In many animal species, males and females are morphologically quite similar except for the sexual character and perhaps other minor differences. However, in some species, particularly among birds and mammals, the sexes are strikingly dimorphic. The males are often larger, stronger, more brightly colored, or possessing various "adornments" (Fig. 5.18). Conspicuous ornamentations may appear as handicaps to their carriers; the bright plumage of some male birds makes them more visible to predators. How could these apparently disadvantageous traits have evolved? Darwin considered this problem at length and concluded that such traits arise by *sexual selection,* which "depends not on a struggle for existence in relation to other organic beings or to external conditions, but on a struggle between the individuals of one sex, generally the males, for the possession of the other sex."

Fig. 5.18. Two examples of sexual dimorphism. *Left,* male and female pronghorn antelope, *Antilocapra americana,* found in the western United States and northern Mexico, and, *right,* peacock and peahen, *Pavo cristatus,* native to east Asia.

Sexual selection is a special case of natural selection. Natural selection is due to differential reproduction; organisms more proficient in securing mates have higher fitness if they produce more progeny than organisms less successful in obtaining mates. Sexual selection illustrates a general truth, namely that the outcome of natural selection depends on the summation of advantages and disadvantages. Fitness components are not always correlated. For example, a trait may decrease the probability of survival, but increase the probability of siring offspring. Conspicuous coloration or overgrown antlers may be nuisances to their possessors, but they will be promoted by selection if they increase the reproductive success of their possessors by helping them to secure mates.

Sexual selection sometimes occurs because a given trait makes the organism more appealing in some way to the opposite sex, something similar to the "sex appeal" recognized in people. Mate preferences have been experimentally demonstrated in such different organisms as *Drosophila* flies, pigeons, mice, dogs, and rhesus monkeys. For example, when yellow-body and normally pigmented females and males of *Drosophila* are placed together, both yellow and normal females prefer as mates normal males over yellow males.

Sexual selection also occurs when a trait, the antlers of deer for example, increases competitive prowess relative to members of the same sex. Antlers are used in contests of strength between males; a dominant male wins more female mates. Increased size and aggressiveness are common results of sexual selection. Males are often larger and more aggressive than females, though the opposite sometimes occurs. Male baboons are two to three times as large as their females; and visitors to zoological parks notice the contrast between the behavior of the docile females and the aggressive males. Mature males of the California sea lion, *Zalophus californianus,* weigh nearly 500 kg, while the females weigh only about 100 kg. The sexual dimorphism is not as great in the northern sea lion, *Eumetopias jubata,* where males weigh about 1000 kg, about twice as much as females, but the effects of sexual selection are quite clear. Battle-scarred winners occupy the rocky islets where most copulations occur, and may have harems of ten to twenty females, while losing males fail to reproduce at all (Fig. 5.19).

Experimental studies have shown that sexual selection may often be combined with frequency dependent selection: the extent to which a type of mate is preferred depends on its frequency. Often the mates preferred are those that happen to be rare, a phe-

Fig. 5.19. Fur seal bull, *Callorhinus ursinus,* and harem of cows. Males average 1.8 meters in length and weigh about 270 kilograms; females are much smaller, averaging 1.2 meters in length and weighing only about 45 kilograms.

nomenon not surprising perhaps to people who have experienced the exotic appeal of blonds in Mediterranean countries or of brunettes in Scandinavia. This phenomenon known as the *rare mate advantage* has been thoroughly studied in *Drosophila,* where it commonly affects the males. The results of an experiment conducted by C. Petit and L. Ehrman are shown in Table 5.4. *Drosophila pseudoobscura* males and females from California (C) and from Texas (T) were placed together in variable proportions. When flies from the C and T localities occur in equal frequencies (12C:12T), they mate with about equal frequencies (55:49 for males, 50:54 for females). But when the two localities are unequally represented, the less common males mate disproportionately more often than the more common males. For example, when the C and T flies exist in the ratio 23:2 the proportion of matings among males is 77:24 (3.2:1), each T male mating nearly four times as often as each C male. When the proportions of flies are reversed (2C:23T), the now rare C males mate more efficiently than the common T males.

Frequency-dependent fitness in favor of rare genotypes is a mechanism contributing to the maintenance of genetic polymorphism, since the fitness of a genotype increases as it becomes rare. Frequency-dependent selection may be particularly impor-

Table 5.4
FREQUENCY DEPENDENT SEXUAL SELECTION
Numbers of matings of two strains of *Drosophila pseudoobscura* when the
proportions of the two kinds of flies are varied. C, California; T, Texas.
Each line summarizes the results of several replicate experiments con-
ducted in observation chambers. (After Petit and Ehrman, 1969.)

Number of flies in each chamber	Males mated			Females mated		
	C	T	C:T	C	T	C:T
23C:2T	77	24	3.2:1	93	8	11.6:1
20C:5T	70	39	1.8:1	84	25	3.4:1
12C:12T	55	49	1.1:1	50	54	1:1.1
5C:20T	39	65	1:1.7	30	74	1:2.5
2C:23T	30	70	1:2.3	12	88	1:7.3

tant in cases of migration. Immigrating individuals may have a
mating advantage, thus making it more likely that their genes will
become established in the population where they have arrived.

Kin and Group Selection

Opponents of Darwin quoted the common occurrence of altruistic
behavior as evidence against the theory of natural selection. *Altruism* may be defined as a behavior that benefits other individuals
at the expense of the altruist. If altruistic behavior is genetically
determined, it would seem that natural selection will favor the development of selfish behavior and eliminate altruism. The fitness
of the altruist is diminished, while that of individuals that refuse
to act altruistically is increased by the altruist's behavior. Altruism, however, is no evidence against the theory of evolution by
natural selection. Indeed, two kinds of selection, kin selection and
group selection, have been proposed to explain altruism and other
related behaviors.

Kin selection is simply a name for natural selection when the
kin or relatives of an individual are taken into consideration. Natural selection favors the multiplication of genes that increase the reproductive success of their carriers. But not all individuals
carrying a given genotype need to have higher reproductive success; it is enough that they reproduce more successfully on the average. Because a parent shares half of its genes with each progeny,
a gene that promotes parental altruism will be favored by selection
if a parent's altruistic death saves the life of more than two of its
progeny. Such a gene will increase in frequency in a population
more effectively than an alternate gene that does not promote al-

truistic behavior so that the life of the parent is saved at the expense of several progeny. Altruistic behavior does not always, of course, entail the death of the altruist.

Parental care is a form of altruism, which is readily explained by kin selection. The parent spends some energy caring for the progeny because this increases the reproductive success of the parent's genes. In general, kin selection will facilitate the development of altruistic behavior when the risk taken, or the energy invested, by an individual is more than compensated by the benefits ensuing to its relatives. The closer the relationship between the altruist and the beneficiaries and the larger the number of beneficiaries, the greater the risks and efforts warranted by natural selection in the altruist. Members of herds or troops, such as baboons and horses, are more or less close relatives, and altruism is displayed among their members.

Altruism may also exist in groups that do not involve relatives when the behavior is reciprocal and the loss to the altruist is only minor but the benefits to the recipient are relatively large. This is known as *reciprocal altruism* and is manifested, for example, in grooming behavior and in the posting of sentinels. Grooming behavior occurs in such primates as chimpanzees; individuals will clean each other of lice and other pests. A sentinel crow sitting in a tree watching for predators while the rest of the flock forages is incurring a small loss by not feeding, but this is more than compensated by receiving sentinel protection from the rest of the flock when it forages itself.

The theory of kin selection has proved particularly valuable in understanding the evolution of social behavior in such insects as ants, bees, wasps, and hornets. Consider, for example, the honeybee. The female workers toil selflessly building the hive, caring for the young, and providing food. Yet, they are sterile; perpetuation of the colony is the charge of the queen bee. Assume that there are genetic variants that induce behaviors that benefit the colony such as that found in workers. It would seem that these genetic variants would rapidly be eliminated from the population since the workers contribute to the reproductive success of the queen's genes, not their own. W. D. Hamilton demonstrated that such inference is mistaken; the situation is indeed most interesting.

Queen bees produce two kinds of eggs. Some are parthenogenetic, i.e., unfertilized and haploid. These are the males or drones, which have a mother but no father, and whose main role in the colony is for one of them to fertilize the queen at the "nuptial flight." The other eggs laid by queen bees are diploid and fertilized. These result in females, most of which are workers, while occasionally one develops into a queen. As a rule most bees in a colony are the progeny of a single queen. Consider, now, the genetic relationships among members of the colony. The queen mates once during her lifetime and stores one male's sperm, using it gradually as she is ready to lay fertilized eggs. Thus all her progeny have the same father. The consequence is that sisters are more related to one another than daughters are to their mothers. Daughters and their mothers share one half of their genes: the daughters receive half their genes from the mother and half from the father. But sisters share three quarters of their genes: half their genes come from the father and are identical in all workers because the father is haploid and all sperm therefore carry the same set of genes; the other half of their genes come from the mother, and on the average half of them will be identical in any two sisters. Thus, the workers' genes are one-and-a-half times more effectively multiplied when they raise a sister than if they would produce a daughter (Fig. 5.20). The existence of a sterile caste of worker females is thus explained by natural selection.

Group selection refers to selection between colonies, or populations, rather than selection between individuals. It has been invoked by V. C. Wynn-Edwards and others to explain some forms of altruism and other behaviors, as well as physiological processes that result in reduced reproductive rates of populations that have grown beyond optimal members. The general argument may be presented as follows. Assume that in a certain organism some colonies possess genes promoting reproduction even beyond the carrying capacity of the environment; these colonies may overexploit the environment, destroy their food source, and thus become extinct. This would not occur in colonies possessing genes that restrict population growth to the numbers that can survive in the environment; these colonies, and therefore their genes, will be perpetuated in the species. The persistence of altruistic genes can be explained by the same argument of selection between groups or colonies, since colonies with altruists are more likely to survive than colonies without them.

The difficulty with group selection is that it will only work if "altruistic" colonies go extinct faster than "selfish" genes arise within altruistic colonies. This is because within a colony, a non-altruistic gene would still be favored by selection since its carriers do not restrict their fertility or assume the risks of being altruistic. Thus, there will be two opposing processes: selfish genes will continuously spread within each colony, but colonies with higher frequencies of selfish genes are more likely to become extinct.

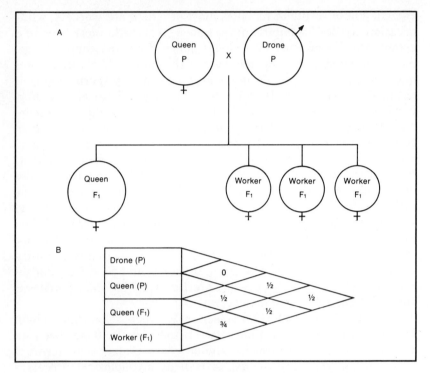

Fig. 5.20. A honeybee pedigree (*A*) and the coefficients of relationship expressing the average proportion of genes in common
between the various individuals in the pedigree (*B*). Female hon-
eybees are diploid; males are haploid. Half of the genes of a progeny female come from the queen and half from the drone; a female
progeny, therefore, shares half of her genes in common with the
mother. Since the genes coming from the drone are the same for all
female progeny, these share three-quarters of their genes with each
other (the genes coming from the drone plus one-half of the genes
coming from the mother). Thus worker bees have more genes in
common with a queen sister than they would have with their own
daughters if they would produce progeny.

Evolutionists are still investigating and debating the extent to
which group selection operates in nature. Theoretical arguments
and empirical evidence suggest that group selection occurs, although the process may not be as extensive as some have claimed.

Adaptation and Fitness

In *The Origin of Species* Darwin supplied ample evidence for biological evolution. Even more important, however, was that he provided a causal explanation of the evolutionary origin of living

things—the theory of natural selection. Before Darwin, the adaptations and the diversity of organisms were accepted as facts without explanation or, more frequently, they were attributed to the omniscient design of the Creator. Organisms are *functionally designed*. In a similar way as a watch is made to tell time, organisms are made to live in certain environments, and their parts are made to serve certain functions, wings for flying, hands for grasping, and gills for breathing in water. Theologians and others argued that in the same way as functionally designed mechanical objects (such as a watch) betray the existence of a designer (a watchmaker), so the functional design of organisms betrays the existence of a deity who designed them.

Darwin accepted that organisms are functionally designed to live in certain ways and their parts to serve certain functions. (In biological language, we would say that organisms are adapted to their environments, and their organs, limbs, and so on, are adaptations to serve certain functions.) But then he provided a natural explanation for functional design: organisms are adapted because variations useful as adaptations to the environment are favored by natural selection. Individuals carrying useful variations are likely to produce more progeny than organisms lacking them. Useful variations will thus gradually become established in populations. Just as Newton had provided a natural explanation of the motion of the planets, Darwin advanced a natural explanation of the origin, diversity, and functional organization of living beings. Biology thereby came into maturity as a scientific discipline because it could now explain not only the workings of organisms but also their origin and configuration.

Natural selection increases the frequency of genotypes having higher fitness. This is simply a matter of definition: genotypes that reproduce more effectively are said to have higher fitness. Such genotypes will, of course, increase in frequency through the generations. But the synthetic theory of evolution claims, in addition, that fitness and adaptation are correlated, i.e., that as a rule genotypes with higher fitness also provide useful adaptations to their carriers. If fitness and adaptation are not correlated, natural selection would not be a satisfactory explanation of the adaptation of organisms.

An adaptation is something that is useful, that serves some function, such as wings for flying, eyes for seeing, and kidneys for regulating the composition of the blood. Fitness (in the sense the word is used in genetic and evolutionary studies as a measure of reproductive efficiency, not in the vernacular) is not, therefore, the same as adaptation. If this were not sufficiently clear from the definition of the terms, it can be confirmed by showing that fitness and adaptation do not always go together. We shall consider two examples.

The first example is provided by the "sex-ratio" chromosome in *Drosophila*. A male carrying this genetic variant in its X chromosome produces only daughters, and no (or few) sons, when crossed to any female. Thus a "sex-ratio" male transmits his X chromosome to his entire progeny, while a normal male transmits his X chromosome only to half of his progeny; the "sex-ratio" chromosome has, therefore, higher fitness than the normal X. As the "sex-ratio" chromosome spreads, the proportion of females increases in the population. But in a species unable to reproduce parthenogenetically, a considerable increase in the proportion of females may mean a decrease in the adaptation of the population, which may become extinct if it reaches the point where it consists only of females and no males.

The second example comes from P. Dawson's studies of competition between the grain beetles, *Tribolium confusum* and *T. castaneum*. Under certain experimental conditions, *T. confusum* is usually eliminated from the population in a few generations. However, shortly after the beginning of one experiment an eye color mutant appeared in *T. castaneum*. The mutant increased gradually in frequency while the numbers of *T. castaneum* gradually decreased until eventually this species was eliminated from the competition (Fig. 5.21). The eye color mutant had higher fitness than its wild-type allele (and thus it increased in frequency) but the competitive ability of the species (i.e., its adaptation to the experimental environment) decreased as the mutant increased in frequency. The mutant had higher fitness than the normal allele, but it decreased the adaptation of the population.

Other examples could be provided. It is clear, for instance, that genotypes increasing the efficiency of a predator at capturing their prey will have higher fitness than those less efficient. Yet if the predator becomes extremely efficient it may drive the prey species to extinction and thus become extinct itself. Genotypes that lead a species to commit suicide can hardly be considered as useful adaptations.

Most species that lived in the past have become extinct (see Chapter 10). It is likely that many were driven to extinction by natural selection. Nevertheless, fitness and adaptation must often go together, otherwise life on earth would have become extinct long ago. Moreover, the correlation between fitness and adaptation stands to reason. As Darwin argued, organisms carrying useful adaptations are likely, as a rule, to produce more progeny than organisms lacking them. Adaptations are often the result of complex

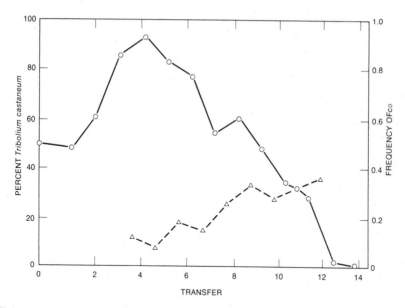

Fig. 5.21. Frequency of *Tribolium castaneum* in competitions with *T. confusum*, and of the gene c^D in *T. castaneum*. Under the experimental conditions, *T. castaneum* gradually eliminates *T. confusum*; this was happening, as expected, during the first five generations of the experiment. In generation 4, a mutant appears that has high fitness and gradually increases in frequency. The mutant, however, decreases the adaptation of *T. castaneum* to the environmental conditions, and as the mutant increases in frequency, *T. castaneum* is gradually eliminated by *T. confusum*.

genetic combinations, making it difficult to show whether genetic variants favored by selection do also serve as adaptations. We shall, however, consider in the next section two simpler examples, one of which illustrates that regulatory gene variants (as well as structural gene variants) are involved in adaptive evolution.

The Genetic Basis of Adaptation

Sickle-cell anemia is a genetically determined human disease fairly common in some African and Asian regions where malaria is rife. The disease is expressed in individuals homozygous for a certain allele. It was pointed out in Chapter 4 (pp. 140–142) that the high frequency of the allele in malarial regions is due to heterosis: heterozygotes for the "normal" and the sickle-cell alleles have higher fitness than homozygotes for the "normal" allele because they are largely resistant to malarial infection. We shall now examine how the genotype at this locus affects adaptation.

Hemoglobins are molecules involved in respiration. They transport oxygen (O_2) from the lungs to the rest of the body and carbon dioxide (CO_2) from the body cells to the lungs, where CO_2 is breathed out and exchanged for O_2. Hemoglobins are one of the most common proteins in the body. Adult humans have about one kilogram of hemoglobin, most of which (about 98 percent) is of the kind known as hemoglobin A. This is a tetramer, i.e., four unit, molecule consisting of two α and two β polypeptide chains, the α and β chains being each coded by a different gene locus. The gene locus coding for the β chain is the one responsible for sickle-cell anemia.

The β chain of hemoglobin consists of 146 amino acids. Figure 3.17 (p. 90) shows the first seven amino acids in the β chain of normal individuals and of sickle-cell patients. The only difference between the two is that normal β has glutamic acid in the sixth position while sickle β has valine in that position. If we look at the genetic code given in Figure 2.21, we see that glutamic acid (Glu) can be coded by either one of two codons, GAA or GAG, while valine (Val) can be coded by any one of four codons, GUU, GUC, GUA, or GUG. Thus, a mutation changing the second A into a U in the triplet coding for glutamic acid will result in a triplet coding for valine, thereby being responsible for sickle-cell anemia. There are $146 \times 3 = 438$ nucleotide pairs coding for 146 amino acids in the gene for β hemoglobin; the difference between the normal and the sickle-cell allele is a mutation changing only one of the 438 pairs. This seemingly trivial difference has serious health consequences—about 100,000 people die every year in the world because they are homozygotes for the sickle-cell allele.

The anemic condition of sickle-cell patients is due to the different chemical properties of valine and glutamic acid. Proteins have convoluted configurations with some amino acids lying in the inside of the molecule and others towards the outside (Fig. 5.22); the sixth position in the β chain is represented by an "outside" amino acid. Glutamic acid is a hydrophilic amino acid ("hydrophilic" means "water liking," i.e., tending to associate readily with water molecules); but valine is a hydrophobic amino acid ("hydrophobic" means "water disliking," i.e., not associating with water). When valine rather than glutamic acid is present in the sixth position of the β chain, the solubility of hemoglobin is considerably decreased, at least under conditions of low oxygen tension. In the narrow blood capillaries sickle-cell hemoglobin tends to crystalize and the red blood cells tend to break open. Hemoglobin molecules fail, then, to perform as efficient oxygen carriers, and

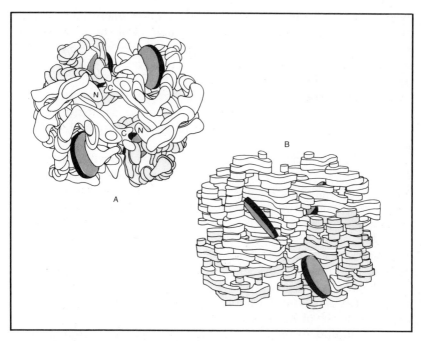

Fig. 5.22. Schematic three-dimensional representation of the configuration of hemoglobin, seen from the top (*A*) or from the side (*B*). The molecule is made up of two alpha and two beta chains, as well as four iron-containing heme groups (dark disks) that bind oxygen to the molecule.

severe anemia ensues. Individuals homozygous for the sickle-cell allele usually die before adulthood.

Let us represent the allele coding for the "normal" β chain as A and that coding for the sickle-cell β chain as a. In a population polymorphic for both alleles, there will be three genotypes: AA, the normal homozygotes; aa, the sickle-cell patients; and Aa, the heterozygotes. Where malaria is endemic, the heterozygotes have higher fitness than the "normal" homozygotes. What is the adaptive basis for this?

The late J. B. S. Haldane noticed that the frequency of the a allele was high in regions infected with malaria (Figs. 4.11 and 4.12), and suggested in 1949 that heterozygotes might be more resistant to malaria than individuals with normal hemoglobin. Experiments to test this suggestion were carried out using volunteers. Fifteen Aa heterozygous individuals and 15 AA homozygotes were inoculated with *Plasmodium falciparum,* the protozoan causing malaria. The results were striking: 14 of the 15 *AA* homozygotes, but only two of the 15 *Aa* heterozygotes, were later found infected with the *Plasmodium* parasite. (The experiments were not dangerous, because malaria can be readily cured.) These experimental results are moreover confirmed by statistical data. The incidence of severe malarial infection is disproportionately low—virtually nil—in *Aa* heterozygotes compared with *AA* homozygotes.

The heterozygotes' resistance to malaria derives from the properties of sickle-cell hemoglobin. Heterozygotes produce both forms of hemoglobin, normal and sickle-cell. The normal hemoglobin allows them to function normally, although heterozygotes tend to show fatigue more readily when exercising strenuously. On the other hand, red blood cells containing sickle-cell hemoglobin tend to break open. Therefore, the malarial parasite, which multiplies in the red cells and feeds on hemoglobin, finds a much less favorable environment in individuals containing sickle-cell hemoglobin than in those with only normal hemoglobin. Hence the lower incidence of malaria among the former.

Our second example of adaptation involves the tolerance of *Drosophila* flies to high levels of alcohol. Populations of *D. melanogaster* often live in environments, such as around breweries and wineries, with high alcohol concentration. Laboratory tests show that flies from such environments tolerate alcohol better than flies from other environments less rich in alcohol. The genetic basis of the adaptation to high alcohol concentrations has been investigated by J. F. McDonald and his colleagues. A laboratory population of *D. melanogaster* flies was divided into two strains. One (*S,* for selected) was subjected every generation to increasing levels of alcohol; the other was used as a control (*C*). In the *S* strain, only the flies that survived the exposure to alcohol were allowed to reproduce each generation. After twenty-eight generations of selection, the *S* flies were definitely more tolerant of alcohol than the *C* flies (Fig. 5.23).

Drosophila melanogaster flies metabolize alcohol mostly by means of the enzyme alcohol dehydrogenase (ADH), which is coded by a gene locus (*Adh*) on the second chromosome. Laboratory tests showed that, indeed, *S* flies had higher ADH activity than *C* flies. However, when the ADH enzyme was studied chemically, no differences were found between the *S* and the *C* flies. On the other hand, it was discovered that the *S* flies had more ADH molecules than the *C* flies. Adaptation of the *S* flies to high alcohol concentrations had taken place not by allelic substitutions at the *Adh*

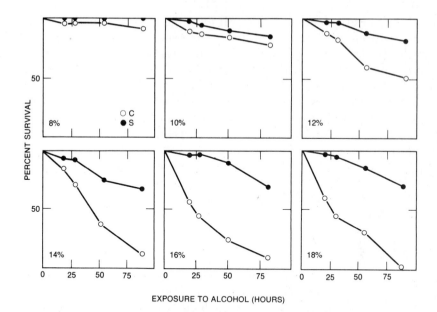

Fig. 5.23. Percent of *Drosophila melanogaster* flies surviving exposure to various alcohol concentrations. Two strains were tested: the Selected (*S*) strain had been selected for alcohol tolerance. The Control (*C*) strain was not selected. The proportion of flies surviving exposure to alcohol for various lengths of time is greater for *S* flies than for *C* flies. The differential response of the two strains becomes more apparent as the concentration of alcohol (indicated in the lower left of each graph) becomes greater. (After McDonald et al, 1977.)

structural locus, but rather by changes in regulatory genes that control the amount of ADH present in the flies, probably by controlling the rate of synthesis of ADH (Table 5.5). Regulatory genes controlling the amount of ADH enzyme were found to be located on the third chromosome, although the *Adh* locus is on the second chromosome.

Table 5.5
SELECTION FOR ALCOHOL TOLERANCE

Relative amounts of ADH protein (in arbitrary units) in a Selected and a Control strain of *Drosophila melanogaster*. The difference between the two is given with its standard error in the last column.

	Selected	Control	Difference (Selected-Control)
Females	18.88	15.25	3.63 ± 0.24
Males	20.25	17.12	3.13 ± 1.26

Adaptation in the sickle-cell example depends on the genetic constitution at a structural locus, the gene coding for the β chain of hemoglobin. On the other hand, adaptation to high levels of environmental alcohol in the *D. melanogaster* example depends on the genotype at regulatory gene loci rather than at the structural locus directly involved in alcohol metabolism. Doubtless, structural genes and regulatory genes are both involved in the evolution of adaptations. Recent studies suggest that changes in gene regulation may play a more important evolutionary role than was formerly believed. In fact, the morphological and physiological differences that are observed between closely related species may be largely due to differences in gene regulation, and only to a lesser extent due to differences in structural genes.

Natural Selection as a Creative Process

About one-and-a-half million living species have been described and named, but the total number is doubtless much larger. Organisms are extremely diversified in their ways of life. Consider for example, a bacterium, a pine tree, a starfish, and a human being. Natural selection is the process responsible for the adaptations of organisms because, as a rule, it promotes the multiplication of genetic variations useful to their carriers. As pointed out earlier, natural selection is the only "adaptively oriented" process of genetic change; mutation, recombination, and drift are random processes with respect to adaptation. We may now raise two questions: (1) Is it justified to consider natural selection as a "creative" process since it accounts for the origin and adaptive configuration of organisms? (2) Why are there so many kinds of organisms?

Natural selection has been compared to a sieve which retains the rarely arising useful mutations, but lets pass the more frequently produced deleterious mutations. Natural selection acts in this way, but it is much more than a purely negative process that simply eliminates what is bad. Natural selection is a process able to generate novelty by increasing the probability of otherwise extremely improbable genetic combinations, and in this sense it can be said to be a creative process.

A dictionary definition of "create" is "to bring into being: cause to exist." Natural selection does not bring into being, or cause to exist, the genetic entities upon which it operates, but it does bring into being, or cause to exist, adaptive genetic combinations that would not have existed otherwise. The creative role of natural selection must not be understood in the sense of the "absolute" creation that Christian theology predicates of the divine act by which the universe was brought into being from nothing. Natural selection is creative rather in a sense similar to the way in which an artist, say a painter, is creative. The canvas and the pigments are not created by the painter, but the painting is. Natural selection does not create the atoms from which an eye, a kidney, or a person is made, but natural selection is responsible for the occurrence of adaptive combinations of such atoms, such as they occur in the eye, the kidney, or in human beings. The probability that a painting, such as Leonardo's *Mona Lisa,* should come about by a random combination of pigments is nearly infinitely small. In the same way, the combination of genes carrying the hereditary information responsible for the formation of the vertebrate eye would never have come about by random mutations and random combinations of such mutations without the organizing process of natural selection; not even if we allow for the four billion years or so during which life has existed on earth. The complicated anatomy of the eye, like the exact functioning of the kidney, is the result of a nonrandom process—natural selection.

How natural selection, a purely material process, can generate novelty in the form of accumulated hereditary information may be illustrated by the following example. To be able to reproduce in a culture medium, some strains of the colon bacterium *Escherichia coli* require that a certain substance, the amino acid histidine, be provided in the medium. When a few bacteria are added to a cubic centimeter of liquid culture medium, they multiply rapidly and produce between two and three billion bacteria in a few hours. Spontaneous mutations to streptomycin resistance occur in normal, i.e., sensitive, bacteria at rates of the order of one in 100 million (1×10^{-8}) cells. In our bacterial culture we expect between 20 and 30 bacteria (2 or 3×10^9 bacteria $\times 10^{-8}$ mutations = 20 or 30) to be resistant to streptomycin due to spontaneous mutation. If a proper concentration of the antibiotic is added to the culture, only the resistant cells survive. The 20 or 30 surviving bacteria will start reproducing, however, and allowing a few hours for the necessary number of cell divisions, several billion more bacteria are produced, all resistant to streptomycin.

Among cells requiring histidine as a growth factor, spontaneous mutants able to reproduce in the absence of histidine arise at rates of about four in 100 million (4×10^{-8}) bacteria. The streptomycin resistant cells may now be transferred to an agar-medium plate with streptomycin but with no histidine. Most of them will not be able to reproduce, but about a hundred (2-3 $\times 10^9$ bacteria $\times 4 \times 10^{-8}$ mutations = 80–120) will start dividing and form colonies until the available medium is saturated. Natural selection will thus have produced in two steps bacterial cells resistant to streptomycin and not requiring histidine for growth. The probability of the two mutational events happening in the same bacterium is of about four in ten million billion ($1 \times 10^{-8} \times 4 \times 10^{-8} = 4 \times 10^{-16}$) cells. An event of such low probability is unlikely to occur even in a large laboratory culture of bacterial cells. With natural selection cells having both properties are the common result.

Critics of evolution have sometimes alleged as evidence against the synthetic theory of evolution examples showing that random processes cannot produce meaningful, organized results. For example, a series of monkeys randomly striking letters on a typewriter would never write *The Origin of Species,* even if we allow for millions of years and many generations of monkeys pounding at typewriters. The criticism would be valid if evolution would depend only on random processes. But natural selection is a nonrandom process that promotes adaptation by selecting combinations that "make sense," i.e., that are useful. The analogy of the monkeys would be more appropriate if a process existed by which, first, meaningful words would be chosen every time they appeared in the typewriter; and then we would also have typewriters with previously selected words rather than just letters in the keys, and again there would be a process to select meaningful sentences every time they appeared. If every time words such as "the," "origin," "species," and so on, appeared in the first kind of typewriter they each became a key in the second kind of typewriter, meaningful sentences would occasionally be produced in these. If such sentences became incorporated into keys of a third type of typewriter where meaningful paragraphs were selected, it is clear that pages and even chapters "making sense" would eventually be produced.

We need not carry the analogy too far, since no analogy is fully satisfactory, but the point is clear. Evolution is not the outcome of purely random processes, but rather there is a "selecting" process, which picks up adaptive combinations because these reproduce more effectively and thus become established in populations. These adaptive combinations constitute, in turn, new levels of organization upon which the mutation (random) plus selection (nonrandom or directional) process again operates.

As illustrated by the bacterial example, natural selection produces combinations of genes that would otherwise be highly improbable because natural selection proceeds stepwise. The vertebrate eye did not appear suddenly in all its present perfection. Its formation requires the appropriate integration of many genetic units, and thus the eye could not have resulted from random processes alone. The ancestors of today's vertebrates had for more than half a billion years some kind of organs sensitive to light. Perception of light, and later vision, were important for these organisms' survival and reproductive success. Accordingly, natural selection favored genes and gene combinations increasing the functional efficiency of the eye. Such genetic units gradually accumulated, eventually leading to the highly complex and efficient vertebrate eye. Natural selection can account for the rise and spread of genetic constitutions, and therefore of types of organisms, that would never have existed under the uncontrolled action of random mutation and recombination of the hereditary materials. In this sense, natural selection is a creative process, although it does not create the raw materials—the genes—upon which it acts.

Natural Selection as an Opportunistic Process

There is an important respect in which an artist makes a poor analogy of natural selection. A painter usually has a preconception of what he wants to paint and will consciously modify the painting so that it represents what he wants. Natural selection has no foresight, nor does it operate according to some preconceived plan. Rather it is a purely natural process resulting from the interacting properties of physicochemical and biological entities. Natural selection is simply a consequence of the differential multiplication of living beings. It has some appearance of purposefulness because it is conditioned by the environment: which organisms reproduce more effectively depends on what variations they possess that are useful in the environment where the organisms live. But natural selection does not anticipate the environments of the future; drastic environmental changes may be insuperable by organisms that were previously thriving.

The team of typing monkeys is also a bad analogy of evolution by natural selection, because it assumes that there is "somebody" who selects letter combinations and word combinations that make sense. In evolution there is no one selecting adaptive combinations. These select themselves, as it were, because they multiply more effectively than less adaptive ones. But there is a sense in which the analogy of the typing monkeys is better than the analogy of the artist, at least if we assume that no particular statement was to be obtained from the monkeys' typing endeavors, but just any statements making sense. As stated in the previous paragraph, natural selection does not strive to produce predetermined kinds of organisms, but only organisms that are adapted to their present environments. What characteristics will be selected depends on what variations happen to be present at a given time in a given place. This in turn depends on the random process of mutation, as well as on the previous history of the organisms (i.e., on what kind of constitution they have as a consequence of their previous evolution). Thus natural selection is an "opportunistic" process. The variables determining in what direction it will go are the environment, the preexisting constitution of the organisms, and the randomly arising mutations and genetic combinations.

Natural selection is thoroughly opportunistic. A new environmental challenge is responded to by appropriate adaptations in the population (or results in its extinction). Adaptation to the same environment may, however, occur in a variety of different ways. An example may be taken from the adaptations of plant life to desert climate. The fundamental adaptation is to the condition of dryness, which involves the danger of desiccation. During a major part of the year, sometimes for several years in succession, there is no rain. Plants have accomplished the urgent necessity of saving water in different ways. Cacti have transformed their leaves into spines, having made their stems into barrels containing a reserve of water; photosynthesis is performed in the surface of the stem instead of in the leaves. Other plants have no leaves during the dry season, but after it rains they burst into leaves and flowers and produce seeds. Ephemeral plants germinate from seeds, grow, flower, and produce seeds—all within the space of the few weeks while rainwater is available; the rest of the year the seeds lie quiescent in the soil.

The opportunistic character of natural selection is also well-evidenced by the phenomenon of adaptive radiation (Chap. 10, p. 327). The evolution of *Drosophila* flies in Hawaii is a relatively recent adaptive radiation. There are about 1,500 *Drosophila* species in the world. Approximately 500 of them have evolved in the Hawaiian archipelago, although this has a small area, about half the size of the Netherlands, or one twenty-fifth the size of California (Fig. 5.24). Moreover, the morphological, ecological, and behavioral diversity of Hawaiian *Drosophila* exceeds that of *Drosophila* in the rest of the world (Fig. 5.25).

Why should have such "explosive" evolution occurred in Hawaii? The overabundance of drosophilids there contrasts with the scarcity, or absence, of other insects. The ancestors of Hawaiian *Drosophila* reached the archipelago before most other groups of insects did, and thus they found a multitude of unexploited opportunities for living. They responded by a rapid adaptive radiation; although they are all probably derived from a single colonizing species, they adapted to the diversity of opportunities available in diverse places or at different times by developing appropriate adaptations, which range broadly from one to another species.

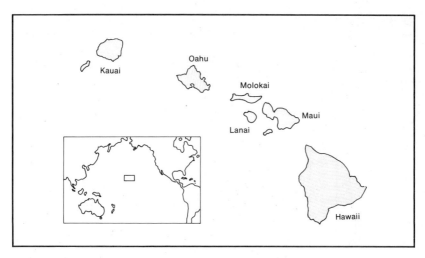

Fig. 5.24. The Hawaiian archipelago has an area of 6,424 square miles. Because of its geographical isolation from other lands, relatively few organisms have colonized the archipelago. However, the plants and animals that got there evolved into multiple species because they found few competitors and many unexploited environments.

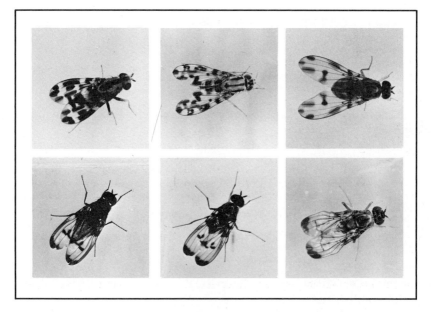

Fig. 5.25. Diversity of Hawaiian *Drosophila* flies. *Top row: D. grimshawi, D. diaglaia, D. engyochracea spieth. Bottom row: D. neopicta, D. silvestrio, D. crytoloma.* (Courtesy H. T. Spieth.)

QUESTIONS FOR DISCUSSION

1. Contrast the action of natural selection in a stable environment and in a changing environment.

2. Where do you think that "founder effects" are likely to be more common, on islands or on continents? In lakes or in the oceans?

3. Do you think that the human population can continue increasing in numbers indefinitely? There now are about four billion people in the world. If the number of people were to double every 35 years (as it would under the present growth rate), how long would it take before there would be 32 billion people in the world? 256 billion? 2,048 billion?

4. Most animal populations consist of approximately equal numbers of females and males. Do you think that this might be the result of natural selection? Why would you expect natural selection to favor an even sex ratio?

5. Do you think that natural selection might lead a species to extinction? Explain.

Recommended for Additional Reading

Emlen, J. M. 1973. *Ecology: an evolutionary approach.* Reading, Mass.: Addison-Wesley.

A well-known book on evolutionary ecology. Moderately advanced.

Ford, E. B. 1971. *Ecological genetics.* 3rd ed. London: Chapman and Hall.

A classic, with extensive discussion of research on butterflies and snails. Moderately advanced.

MacArthur, R. H. 1972. *Geographical ecology.* New York: Harper & Row.

A well-known book. Moderately advanced.

McNeil, W. H. 1976. *Plagues and peoples.* Garden City, N.Y.: Doubleday.

A fascinating account of the adaptation of human populations to infectious diseases and the critical role that these have played in human history.

Williams, G. C. 1966. *Adaptation and natural selection.* Princeton, N.J.: Princeton University Press.

A readable but profound discussion of natural selection as the process that accounts for the adaptations of organisms.

6

The Origin of Species

The title of Darwin's classic, *The Origin of Species,* suggests its contents: an argument showing that the diversity of organisms and their characteristics could be explained as the result of natural processes. Species come about as the result of gradual change due to natural selection. Because the environmental conditions vary from time to time and from place to place, natural selection favors different adaptations in different situations, thus giving rise to different species. However, before proceeding with the question of how species come about we need to ask, "What is a species?"

What Is a Species?

Our everyday experience tells us that there are different *kinds* of organisms and allows us to identify them. We recognize people as belonging to the human species, and as different from cats, which in turn are different from dogs, from redwood trees, and from daisies. We are aware of differences among people, as well as among cats, dogs, redwood trees, and daisies. We saw in Chapter 3 that such population variation is due to genetic differences as well as to interactions between the genotype and the environment. But differences between individuals of different species are considerably larger than differences between individuals of the same species.

Common experience relies on external similarity when identifying individuals as members of the same species or not. But there is more to it than that. A terrier, a Chihuahua, and a Pomeranian have very different appearances, yet we call them all dogs because they can interbreed. People can also interbreed, and so can cats, but people cannot interbreed with dogs or cats, nor these with each other.

These considerations bring forth two points. The first point is that species are usually identified by appearance. The other point is that there is something basic, of great biological significance behind the similarity of appearance—species are made up of individuals able to interbreed, but not of breeding with members of other species. This is expressed in the following definition proposed by Ernst Mayr: *Species are groups of interbreeding natural populations that are reproductively isolated from other such groups.*

A species is, therefore, a natural unit or system, defined by the possibility of interbreeding between its members. The ability to interbreed is of considerable evolutionary import, because it establishes species as discrete and independent *evolutionary units.* Consider an adaptive mutation or some other genetic change originating in a single individual. Over the generations, this may spread by natural selection to all members of the species, but not to individuals of other species. This can be stated differently: individuals of a species share in a common gene pool (Chapter 3, p. 61), which is however not shared by individuals of other species. Owing to reproductive isolation, different species have independently evolving gene pools.

The term "species" is used as a category of classification; other categories are "genus," "family," "order," and so on (see Chapter 7). A group of individuals identified as members of a species is given a Latin name consisting of two words (such as *Homo sapiens* for humans, or *Felis cattus* for domestic cats, or *Sequoia sempervirens* for redwoods). This Latin name represents a *taxon* (plural, *taxa*) within the species category. The genus category is made of taxa identified by a single word (*Homo, Felis, Sequoia*) and often consists of several species (such as *Felis cattus; Felis pardus,* the African leopard; *Felis lynx,* the bobcat, and so on).

The species is sometimes said to be a "natural" category because individuals of the same species share in a common gene pool, which is not true of the genus and other higher categories of classification. The grouping of various species into a genus, or of various genera into a family, and so on, is based on morphological

Fig. 6.1.   Species of cats. A male African lion, *Panthera leo,* a young Indian tiger, *Panthera tigris,* a cheetah, *Acinonyx jubatus,* found in Africa, and a puma, *Felis concolor,* found in the western United States and northern Mexico.

similarity, common ancestry, and other considerations with some natural basis, but which are to a certain extent arbitrary. House cats, leopards, bobcats, tigers, and lions are included in a single genus (*Felis*) by some authors, while others place the lion, tiger, leopard, and jaguar in the genus *Panthers* and leave the domestic cat, the European wildcat, and the bobcat in the genus *Felis* (Fig. 6.1). This arbitrariness, however, does not occur with respect to the species category—all individuals sharing in a common gene pool belong to one single species.

Ambiguity may, however, exist within the species category for two reasons. One concerns the extent of our knowledge: sometimes we do not know for sure whether or not individuals from two different populations belong to the same species because we do not know whether they could naturally interbreed. The other reason is based on the nature of evolution, which is a gradual process. Two groups of populations that at some time were members of the same species may gradually diverge into two different species. Because the process is gradual, there is not a particular point in time at which we can say that the two groups constitute two different species, while in previous generations there was only one species (Fig. 6.2). This kind of ambiguity occurs in all gradual processes of change. It is clear that at noon it is daytime and at midnight it is nighttime, but there is not a precise moment at which the day becomes night, just as there is no precise moment in life at which a person becomes an adult.

One additional issue should be considered here. Species have been defined in terms of contemporary populations. What about organisms living at different times? Clearly, we cannot test whether or not today's humans could interbreed with those that lived many generations in the past. It is, however, a reasonable guess that living people, or living cats, would be able to interbreed with people, or cats, exactly like those living ten or a hundred generations earlier. But what about the ancestors removed by one million, or 100 million generations? Our ancestors of one million years ago (about 50,000 generations) are classified in the species *Homo erectus* (see Chapter 12), while we are classified in the same genus, *Homo,* but a different species, *sapiens.* Was there a precise time at which

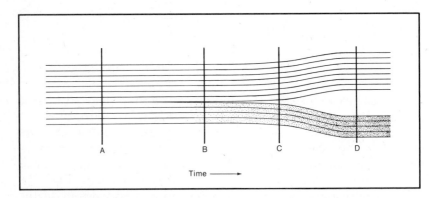

Fig. 6.2.   Schematic representation of the splitting of one species into two. Because the process is gradual, there is no one point in time when one can say that two species had just come into existence. Each line represents a different population. In *A,* there is only one species. In *D,* there are two. In *B* and *C,* the divergence between two groups of populations has started but is far from complete.

Homo erectus became *Homo sapiens?* Clearly not, since evolution is a more-or-less gradual process of change.

However, just because the changes from generation to generation were on the whole small, it would not be appropriate to classify our remote ancestors and ourselves in the same species. It is useful to identify the differences between our remote ancestors and ourselves by means of different species names as it is useful to distinguish daytime from nighttime, or childhood from adulthood. Evolutionists use, therefore, a somewhat arbitrary criterion to distinguish species in organisms that lived at different times. Whenever the morphological differences are about as large as among living species, organisms are classified in different species. Problems may arise when there is a more or less continuous sequence of organisms that gradually change from the characteristics of one species into those of another; but the fossil record is usually quite. spotty and hence these problems rarely arise. Species with different names but in the same line of descent are sometimes called *chronospecies* (from *chronos,* meaning "time" in Greek). In this sense *Homo erectus* and *Homo sapiens* are different chronospecies.

Reproductive Isolating Mechanisms

The definition of species uses reproductive isolation as the criterion to identify species in sexual organisms. Individuals naturally able to interbreed are members of the same species; otherwise they belong to different species. It should not be surprising that reproductive isolation be used as the fundamental criterion to define species—reproduction is a basic and indispensable function of life. The biological properties of organisms that prevent interbreeding are called *reproductive isolating mechanisms* (RIMs). Before discussing RIMs we should point out that geographical isolation is not a RIM, because it is not a biological property of organisms. Geographical separation sometimes makes it impossible for organisms, or for groups of organisms, to interbreed. House mice in different islands, salamanders in different creeks, or trees in different mountain ranges, cannot interbreed owing to their physical separation, but not necessarily because they are biologically incompatible.

A classification of RIMs is shown in Table 6.1. Reproductive isolating mechanisms may be classified into prezygotic and postzygotic. *Prezygotic* RIMs impede hybridization between members of different populations and thus prevent the formation of hybrid zygotes. *Postzygotic* RIMs reduce the viability or fertility of hybrids. Prezygotic and postzygotic RIMs serve all the same purpose —they forestall gene exchange between populations. But there is an important difference between them: the waste of reproductive effort is greater for postzygotic than for prezygotic RIMs. If a hybrid zygote is produced but is inviable (*hybrid inviability,* 2a in Table 6.1), two gametes have been wasted that could have been used in nonhybrid reproduction. If the hybrid is viable but sterile (*hybrid sterility,* 2b in Table 6.1), the waste includes not only the gametes but also the resources used by the hybrid in its development. The waste is even greater in the case of *hybrid breakdown* (2c in Table 6.1), because it involves the resources used not only by the hybrids but also by their progenies. One prezygotic RIM, *gametic isolation* (le in Table 6.1) may also involve reproductive waste when gametes fail to form viable zygotes. The other prezygotic RIMs avoid gametic wastage, but some energy may be wasted in unsuccessful courtship (*ethological isolation,* lc in Table 6.1) or in unsuccessful attempts to copulate (*mechanical isolation,* ld in Table 6.1). Natural selection promotes the development of prezygotic RIMs between populations already isolated by postzygotic RIMs whenever the populations coexist in the same territory and there is opportunity for the formation of hybrid zygotes. This occurs precisely because reproductive waste is reduced or altogether eliminated when prezygotic RIMs come about.

The reproductive isolating mechanisms listed do not all occur between any two species, but usually two or more isolating mech-

Table 6.1
CLASSIFICATION OF REPRODUCTIVE ISOLATING MECHANISMS (RIMS)

1. *Prezygotic,* which prevent the formation of hybrid zygotes.

    a. *Ecological isolation:* populations occupy the same territory, but live in different habitats and thus do not meet.
    b. *Temporal isolation:* mating or flowering occur at different times, whether in different seasons or at different times of day.
    c. *Ethological isolation* (from the Greek *ethos,* meaning "behavior"): sexual attraction between females and males is weak or absent; also called *sexual isolation.*
    d. *Mechanical isolation:* copulation or pollen transfer is forestalled because of the different size or shape of genitalia or the different structure of flowers.
    e. *Gametic isolation:* female and male gametes fail to attract each other; or the spermatozoa or pollen are inviable in the sexual ducts of animals or in the stigmas of flowers.

2. *Postzygotic,* which reduce the viability or fertility of hybrids.

    a. *Hybrid inviability:* hybrid zygotes fail to develop or at least to reach sexual maturity.
    b. *Hybrid sterility:* hybrids fail to produce functional gametes.
    c. *Hybrid breakdown:* the progenies of hybrids (F_2 or backcross generations) have reduced viability or fertility.

anisms are involved in the reproductive isolation between species rather than a single one. Some RIMs are more common in plants (e.g., temporal isolation), others in animals (e.g., ethological isolation); but even among closely related species, different sets of RIMs are often involved when different pairs are compared. This is another example of the opportunism of natural selection (Chapter 5, p. 189): the evolutionary function of RIMs is to prevent interbreeding, but how this is accomplished depends on environmental circumstances as well as on the genetic variability available.

Examples of Prezygotic Isolating Mechanisms

Ecological isolation   At least six species of mosquitoes exist in the *Anopheles maculipennis* group, some of which are involved in the transmission of malaria. Habitat separation contributes to their reproductive isolation. For example, some species, (*A. atroparvus, A. labranchiae*) live in brackish water, others (*A. maculipennis*) in running fresh water, and still others (*A. melanoon, A. messeae*) in stagnant fresh water. Habitat isolation is common in plants because some may grow in only one or another kind of soil, and either in humid or in dry habitats.

Temporal isolation   A remarkable example is provided by three species of the orchid *Dendrobium*. Their flowers open at dawn and wither by nightfall, so that fertilization can occur only within a period of less than one day. Flowering is brought about by a meterological stimulus such as a sudden storm on a hot day. The same stimulus acts on all three species, but the lapse between the stimulus and flowering is eight days in one species, nine in another, and ten or eleven in the third; interspecific fertilization is thus effectively impeded. Temporal isolation exists between pairs of closely related species of cicadas; in each pair one species emerges every thirteen years, the other every seventeen years, so that the two species of a pair have the opportunity to form hybrids only once every $13 \times 17 = 221$ years (Fig. 6.3).

Ethological isolation   This is often the strongest RIM keeping related sympatric (living in the same territory) species of animals from interbreeding; ethological isolation does not, of course, occur in plants. In most organisms with separate sexes, mating requires first that males and females search for each other, and come together. Then, complex courtship and mating "rituals" usually take

Fig. 6.3. The orchid *Dendrobium nobile*. The genus *Dendrobium* includes about 900 species native to the Philippines, Japan, China, Malaya, Burma, and India.

place before copulation or release of the sex cells. These elaborate rituals vary from species to species and play a significant role in species recognition. If some component element of the search-courting-mating sequence of events is disharmonious in the two sexes, the process is interrupted. Courtship and mating rituals have been carefully studied, particularly in some mammals, birds, fish, and various invertebrates.

The efficacy of ethological isolation in keeping species apart is remarkable. *Drosophila serrata, D. birchii,* and *D. dominicana* are three species morphologically very similar living in Australia, New Guinea, and nearby islands. Although the three species occur sympatrically in many places, no hybrids are known to occur in nature. Experiments to test the strength of ethological isolation were conducted in the laboratory by placing ten females of one strain

Table 6.2
SEXUAL ISOLATION BETWEEN THREE CLOSELY RELATED SPECIES OF *DROSOPHILA* (After Ayala, 1965.)

Females	Males	Number of females tested	Number inseminated	Percent inseminated
D. serrata	*D. serrata*	3,841	3,466	90.2
D. serrata	*D. birchii*	1,246	9	0.7
D. serrata	*D. dominicana*	395	5	1.3
D. birchii	*D. birchii*	2,458	1,891	76.9
D. birchii	*D. serrata*	699	7	1.0
D. birchii	*D. dominicana*	250	1	0.4
D. dominicana	*D. dominicana*	43	40	93.0
D. dominicana	*D. serrata*	163	0	0.0
D. dominicana	*D. birchii*	537	20	3.7

and ten males of another in a culture for several days. When the two strains were of the same species a large majority of the females (77 to 93 percent) were inseminated, but few inseminations took place (from 0 to 4 percent) when males and females were of different species (Table 6.2).

These experiments also illustrate a point made earlier, namely that several RIMs are usually combined in keeping species apart. Hybrid inviability and hybrid sterility on top of ethological isolation maintain the reproductive isolation between these three species of *Drosophila*—from the rare interspecific inseminations very few hybrid zygotes develop into adult flies, and these are always sterile. The stimuli involved in species recognition during courtship, in *Drosophila* as well as in other animals, are chemical (olfactory), visual, auditory, and tactile. Specific substances, known as *pheromones,* have been chemically identified in some moths and butterflies and have been demonstrated to play an important role in species recognition in those organisms as well as in other animals, including mammals.

Mechanical isolation Differences in shape and size of the genitalia may make copulation impossible in animals; variations in flower configuration may impede pollination in plants. Mechanical isolation makes difficult interspecific pollination, for example, between two species of sage, *Salvia mellifera* and *S. apiana* from California. The flowers of *S. mellifera* have typical two-lipped flowers

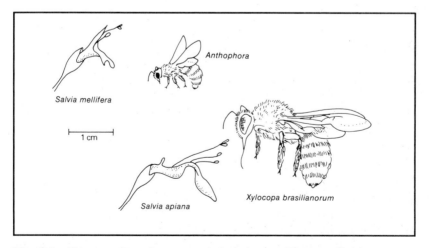

Fig. 6.4. Two species of sage and their bee pollinators. *Salvia mellifera* has small flowers and is pollinated by small or medium size bees such as *Anthophora*. *Salvia apiana* is larger and is pollinated by large bees such as *Xylocopa brasilianorum*.

with stamens and style in the upper lip, while those of *S. apiana* have long extended stamens and style and a specialized floral configuration (Fig. 6.4). The normal pollinators of *S. mellifera* are small or medium-size bees that carry pollen from flower to flower on their backs. On the other hand, *S. apiana* is normally pollinated by large carpenter bees and bumblebees that carry the pollen on their wings and other body parts. The pollinators of *S. mellifera* fail to pollinate *S. apiana*, and vice versa.

Mechanical isolation is common in animals. Interspecific copulation obtained in laboratory experiments between animals of different species may even result in injury or death of the participants, as in the case of matings between *Drosophila pseudoobscura* females and *D. melanogaster* males.

Gametic isolation   This is particularly important in some marine animals that release their eggs and spermatozoa into the surrounding water, where fertilization takes place; the attraction between the gametes of different species may be reduced or absent. For example, when eggs and sperm of two species of sea urchins, *Strongylocentrotus purpuratus* and *S. franciscanus* are simultaneously released, most of the fertilizations are homogametic, i.e., between eggs and sperm of the same species. In animals with internal fertilization, spermatozoa may be unable to function in the female sexual ducts of different species. In the plant kingdom, pollen grains of one species may fail to germinate on the stigma of another species.

Examples of Postzygotic Isolating Mechanisms

Hybrid Inviability   Prezygotic mechanisms are widespread, but they occasionally break down, making possible the formation of hybrid zygotes. Often, however, these fail to develop into adult individuals. For example, fertilization is possible between goat and sheep, but the hybrid embryos die in early developmental stages. Hybrid zygotes may abort soon after fertilization or at any stage of the life cycle: J. A. Moore found various degrees of inviability in hybrids between closely related species belonging to the leopard frog (*Rana pipiens*) group.

Hybrid Sterility   Sometimes hybrid zygotes develop into adults, even into vigorous adults, such as mules (progeny of the cross between horse and donkey), but these adults are sterile. Partial or complete hybrid sterility is common in interspecific hybrids, and it sometimes occurs in hybrids between populations that have not yet become completely different species (for example, between the subspecies *Drosophila equinoxialis equinoxialis* and *D. equinoxialis caribbensis* cited later in the chapter). Hybrid sterility often operates on top of other isolating mechanisms, as between *Drosophila serrata, D. birchii,* and *D. dominicana.* As shown in Table 6.2, interspecific matings are rarely successful even when the conditions are made most favorable in the laboratory; when hybrid zygotes are formed, most fail to develop, and the few that develop are sterile.

Hybrid Breakdown Hybrids are sometimes partially or completely fertile, but gene exchange may nevertheless be staved off because the second generation or backcross progenies are poorly viable or sterile. For example, hybrids between the cotton species, *Gossypium barbadense, G. hirsutum,* and *G. tomentosum* are generally vigorous and fertile, but F_2 progenies die in seed or early in development or develop into weak plants.

The Process of Speciation

Species are reproductively isolated groups of populations. The question of how species come about is, therefore, equivalent to the question of how reproductive isolation arises between groups of populations. Two theories have been advanced to explain the origin and development of reproductive isolation between populations. One theory considers isolation as an accidental by-product of genetic divergence: populations that become genetically more and more different (as a consequence, for example, of adaptation to different environments) may eventually be unable to interbreed because their gene pools are disharmonious. The other theory regards isolation as a product of natural selection: whenever hybrids are less fit than nonhybrid individuals, natural selection will directly promote the development of RIMs because gene variants interfering with hybridization have greater fitness than those favoring hybridization, since the latter will often be present in poorly fit hybrids.

The two theories of the origin of reproductive isolation are not mutually exclusive. Reproductive isolation may indeed come about as an incidental by-product of genetic divergence between separated populations. Consider, for example, the evolution of many endemic species of plants and animals in the Hawaiian archipelago. The ancestors of such species arrived in the Hawaiian islands long ago, perhaps several millions of years before the present. There they evolved and became adapted to the local conditions. Although natural selection did not directly promote reproductive isolation between the species evolving in Hawaii and the continental populations from which the colonizers came, reproductive isolation has nevertheless become complete in many cases.

In fact, however, the process of speciation commonly involves both kinds of phenomena postulated by the two theories of speciation. Reproductive isolation usually starts as an incidental byproduct of genetic divergence, but is completed when it becomes directly promoted by natural selection. Speciation may occur in a variety of ways, but generally two stages may be recognized in the process (Fig. 6.5).

Stage I The onset of the speciation process requires, first, that gene flow be somehow interrupted (completely, or nearly so) between two populations of the same species. Absence of gene flow makes it possible for the two populations to become genetically differentiated as a consequence of their adaptation to different local conditions or to different ways of life (and also as a consequence of genetic drift, which may play a lesser or greater role depending on the circumstances). The interruption of gene flow is necessary because otherwise the two populations will in fact share in a common gene pool and fail to become genetically different. As populations become more and more genetically different, reproductive isolating mechanisms may appear. This occurs because different gene pools are not mutually coadapted; hybrid individuals will have disharmonious genetic constitutions and reduced fitness in the form of reduced viability or fertility.

Two characteristics of the first stage of speciation should be noted: (1) reproductive isolation appears primarily in the form of postzygotic RIMs; and (2) these RIMs are a by-product of genetic differentiation—reproductive isolation is not directly promoted by natural selection at this stage.

Genetic differentiation, and therefore the appearance of post-zygotic RIMs, is often a gradual process. As with other gradual processes, it is somewhat arbitrary to decide whether or not populations have already initiated the process of speciation. Populations are considered to be in the first stage of speciation if RIMs have appeared between them; if the populations are only slightly genetically differentiated and there is no clear indication of reproductive isolation, they are not considered to be in the first stage of speciation. Local populations of a given species are often genetically somewhat different, but they are not thought to be in the first stage of speciation if their differentiation is small and does not result in the appearance of RIMs.

Stage II This stage encompasses the completion of reproductive isolation. Assume the disappearance of the external conditions

Fig. 6.5. Generalized model of the process of speciation. *A*. Local populations of a single species are represented by circles; the arrows indicate that gene flow occurs between populations. *B*. The populations have become separated into two groups between which there is no gene flow. The two groups gradually become genetically different, which is indicated by the shading of the populations on the left. As a consequence of their genetic differentiation reproductive isolating mechanisms (usually *postzygotic*) arise between the two groups. This is the first stage of speciation. *C*. Individuals from different population groups are able to intermate. However, owing to the preexisting reproductive isolating mechanisms, little if any gene flow takes place, which is represented by the small lines crossing the arrows that connect the two population groups. Natural selection favors the development of additional reproductive mechanisms, particularly prezygotic ones which avoid matings between individuals from different population groups. This is the second stage of speciation. *D*. Speciation has been completed because the two groups of populations are fully reproductively isolated. There are now two species which may coexist in the same territory without gene exchange.

that interfered with gene exchange between two populations in the first stage of speciation. This might occur, for example, if two previously geographically separated populations expanded and came to exist together in the same territory. Two outcomes are possible: (1) a single gene pool comes about, because the loss of fitness in the hybrids is not very great (or because one population is eliminated by the other through ecological competition); and (2) two species ultimately arise, because natural selection favors the development of reproductive isolation.

The first stage of speciation is reversible: if it has not gone far enough it is possible for two previously differentiated populations to fuse into a single gene pool. However, if matings between individuals from different populations leave progenies with reduced fertility or viability, natural selection would favor genetic variants promoting matings between individuals of the same population. Consider the following simplified situation. Assume that at a locus there are two alleles, A_1 and A_2, and that A_1 favors matings between individuals of the same population while A_2 favors interpopulational matings. Then A_1 will be present more often in progenies from intrapopulational crosses, that is in individuals with good viability and fertility, while A_2 will be more often present in interpopulational hybrids. Because these have low fitness, A_2 will decrease in frequency from generation to generation. Generally, natural selection will result in the multiplication of alleles that favor intrapopulation matings and in the elimination of alleles that favor interpopulation matings. Natural selection, therefore, will favor the development of prezygotic RIMs, which avoid the formation of hybrid zygotes.

The two characteristics of the second stage of speciation, then, are: (1) reproductive isolation develops mostly in the form of prezygotic RIMs, and (2) the development of prezygotic RIMs is directly promoted by natural selection. These two characteristics of Stage II stand in sharp contrast to the characteristics of Stage I.

Nevertheless, speciation may take place without the occurrence of Stage II. As explained, populations in the absence of gene exchange may develop complete reproductive isolation if the process of genetic differentiation continues long enough. However, whenever the opportunity for gene exchange arises after previous genetic differentiation, natural selection will accelerate the development of reproductive isolating mechanisms; Stage II, then, speeds up the speciation process. In fact, speciation often involves both stages.

In the following sections we consider particular modes of speciation, classified into two classes, "geographic" speciation and "quantum" speciation.

Geographic Speciation

One mode of speciation, perhaps the most common, is known as *geographic speciation*. The general model of the speciation process advanced in the previous paragraphs applies well to geographic speciation. Stage I starts owing to geographic separation between populations. Terrestrial organisms may be separated by water (such as rivers, lakes, and oceans), mountains, deserts, or any kind of territory uninhabitable for the population. Freshwater organisms may be kept separate if they live in different river systems or unconnected lakes. Marine populations may be separated by land, by waters of greater or lesser depth than the organisms can tolerate, or of different salinity.

As a result of natural selection, geographically separate populations become adapted to local conditions, and thus they become genetically differentiated. Random genetic drift may also contribute to genetic differentiation, particularly when populations are small, or when populations are derived from only a few individuals as when one or a few individuals colonize a new territory (an island, a lake, or a new mountain range, for example). If geographic separation continues for some time, incipient reproductive isolation may appear, particularly in the form of postzygotic RIMs; the populations will, then, be in the first stage of speciation.

The second stage of speciation starts when previously separated populations come into geographic contact, at least over part of their distributions. This may happen, for example, by topographic changes on the earth's surface, or by ecological changes in the intermediate territory that make it habitable for the organisms, or by migration of members of one population into the territory of the other. Matings between individuals from different populations may then take place. Depending on the strength of the existing (postzygotic) RIMs, and on the extent of hybridization, the two populations may fuse into a single gene pool or may develop additional (prezygotic) RIMs and become separate species.

The two stages of the process of geographic speciation may be illustrated with a group of closely related species of *Drosophila* that live in the American tropics (Fig. 6.6). The group consists of 15 species, six of which are siblings. One of the sibling species is *D. willistoni*, which consists of two subspecies: *D. willistoni quechua* lives in continental South America west of the Andes; *D. willistoni willistoni* lives east of the Andes in continental South America and also in Central America, Mexico, and the Caribbean. These two subspecies do not meet in nature. They are separated by the Andes because the flies cannot live at high altitude. There is incipient reproductive isolation between them, particularly in the form of hybrid sterility. When males and females from different subspecies are intercrossed in the laboratory, the result depends on the direction of the matings:

♀ *D. w. willistoni* x ♂ *D. w. quechua* → fertile female and male progeny
♀ *D. w. quechua* x ♂ *D. w. willistoni* → fertile females but sterile males

If these two subspecies would come into geographic contact and intercross, natural selection would favor the development of prezygotic RIMs because the male progenies of all crosses between *quechua* females and *willistoni* males are sterile. In other words,

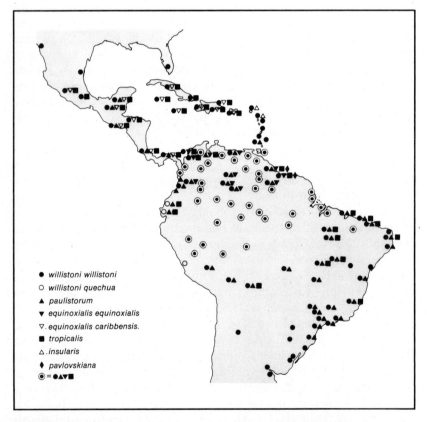

● willistoni willistoni
○ willistoni quechua
▲ paulistorum
▼ equinoxialis equinoxialis
▽ equinoxialis caribbensis.
■ tropicalis
△ insularis
♦ pavlovskiana
◉ = ●▲▼■

Fig. 6.6. Geographic distribution of six closely related species of *Drosophila*. *D. willistoni* and *D. equinoxialis* each consist of two subspecies. The subspecies represent populations in the first stage of geographic speciation.

the subspecies are two groups of populations in the first stage of speciation.

The first stage of speciation is also found in another species of the group. *Drosophila equinoxialis* consists of two geographically separated subspecies: *D. e. equinoxialis* inhabits continental South America; *D. e. caribbensis* lives in Central America and the Caribbean islands. Laboratory crosses between these two subspecies always produce fertile female progenies but sterile males, independently of the direction of the cross. Thus, there is somewhat greater reproductive isolation between the two subspecies of *D. equinoxialis* than between the two subspecies of *D. willistoni*. Natural selection in favor of prezygotic RIMs would be stronger in the case of *D. equinoxialis* because all hybrid males are sterile.

It is worth noting that there is no evidence of prezygotic RIMs between the subspecies of *D. willistoni* or of *D. equinoxialis*; in particular, they have no ethological isolation, the most effective RIM in *Drosophila* as well as in many other animals. When males and females from two different localities are placed together in the laboratory, the probability of matings between flies from different strains is the same whether the two strains belong to the same subspecies or to different subspecies (Table 6.3). Reproductive isolation between the subspecies is, therefore, far from complete, and thus they are not considered different species.

Stage II of the speciation process can also be found within the *D. willistoni* group. *Drosophila paulistorum* is a species consisting of six semispecies, or incipient species, two or three of which are sympatric in many localities (Fig. 6.7). The semispecies exhibit hybrid sterility similar to that found in *D. equinoxialis*: crosses between males and females of two different semispecies yield fertile females but sterile males. But two or three semispecies have come into geographic contact in many places, and there the second stage of speciation has advanced to the point that ethological isolation is complete or nearly so. When females and males from two different semispecies are placed together in the laboratory, the results depend on the geographic origin of the flies. When both semispecies are from the same locality, only homogamic matings take place; but when they are from different localities, heterogamic as well as homogamic matings occur, indicating that ethological isolation is not yet complete in the latter case. The semispecies of *D. paulistorum* thus provide a remarkable example of the action of natural selection during the second stage of speciation; reproductive isolation has been completed where the semispecies are sympatric but not elsewhere, because the genes involved have not yet fully spread throughout each semispecies. Whenever this occurs, the

Table 6.3
LACK OF ETHOLOGICAL ISOLATION WITHIN AND BETWEEN SUBSPECIES OF *DROSOPHILA*

In each experiment 12 females and 12 males of each of two strains are placed in an observation chamber and the matings are observed. Each line summarizes the results from several experiments. The results show that the probability of matings between females and males of two different strains is about the same whether the two strains are of the same or of different subspecies. (From Ayala and Tracey, 1974, and Ayala et al., 1974.)

Populations tested		Types of matings				Total number of matings	Percent homogamic matings**
A	B	A × A	B × B	A × B	B × A		
D. w. willistoni	*D. w. willistoni**	30	31	22	27	110	55.5
D. w. willistoni	*D. w. quechua*	156	167	107	113	543	59.5
D. e. equinoxialis	*D. e. equinoxialis**	74	45	28	57	204	58.3
D. e. caribbensis	*D. e. caribbensis**	28	11	6	17	62	62.9
D. e. equinoxialis	*D. e. caribbensis*	91	194	163	60	508	56.1

*Tests involving two different local populations of the same subspecies.
**A × A plus B × B.

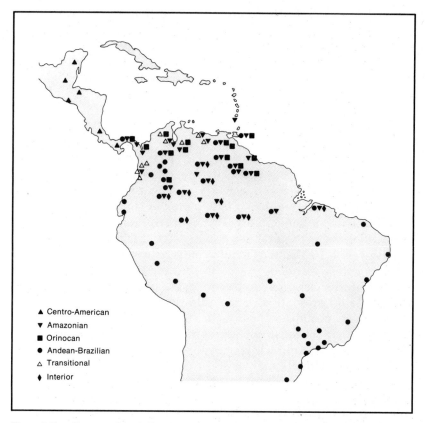

Fig. 6.7. Geographic distribution of the six semispecies of *Drosophila paulistorum.* The semispecies are populations in the second stage of geographic speciation. Speciation has been virtually completed in places where two or three semispecies coexist without interbreeding.

Legend:
▲ Centro-American
▼ Amazonian
■ Orinocan
● Andean-Brazilian
△ Transitional
◆ Interior

semispecies of *D. paulistorum* will have become fully different species.

The final result of the process of geographic speciation can be observed in the species of the *D. willistoni* group. *Drosophila willistoni, D. equinoxialis, D. tropicalis,* and other species of the group coexist sympatrically over wide territories with completely separate gene pools. Hybrids are never found in nature, are extremely difficult to obtain in the laboratory, and are always sterile. Complete reproductive isolation among the six sibling species of the group has been achieved with very little morphological differentiation. The species can be identified in the males by small differences in the configuration of their genitalia, but there is no reliable way to identify the species of the females by external examination. Species identification of females can be done, however, by examining the giant chromosomes of the salivary glands, and in other ways. As we shall see later in this chapter, the sibling species are genetically quite different in spite of their morphological similarity.

Quantum Speciation

In the case of geographic speciation, Stage I involves gradual genetic divergence of geographically separated populations. The development of postzygotic RIMs as by-products of genetic divergence usually requires a long period of time—thousands, perhaps millions, of generations. However, there are other modes of speciation where the first stage of speciation, and the appearance of postzygotic RIMs, may require only relatively short periods of time. *Quantum speciation* refers to these modes of speciation that involve an acceleration of the process, particularly of the first stage. "Rapid speciation" and "saltational speciation" are terms also used to encompass these modes of speciation.

One form of quantum speciation is *polyploidy,* the multiplication of entire chromosome complements (Chapter 3, p. 107). Polyploid individuals may arise in one or two generations. Polyploid populations are reproductively isolated from their ancestral species, and thus they are new species. As stated in Chapter 3, there are two kinds of polyploids: autopolyploids, resulting from multiplication of sets of chromosomes of one single species; and allopolyploids, resulting from the combination of chromosome sets from different species. Polyploidy is a common mode of speciation in plants, but it is rare in animals. Allopolyploid plant species are much more common than autopolyploids.

Assume that two diploid plant species have the same number of chromosomes. We can represent the chromosome complement of one species as *AA,* and the other as *BB.* The chromosomal constitution of an interspecific hybrid may be represented as *AB;* the hybrid will usually be sterile because of abnormal pairing and segregation at meiosis (Chapter 2, p. 29–34). However, chromosome doubling may occur spontaneously as a consequence of either abnormal mitosis or abnormal meiosis. In mitosis, if the chromosomes double but the cell fails to divide, a cell results with four sets of chromosomes (*AABB*). Branches and flowers derived from that cell are tetraploid and, since there are two chromosomes of each kind, normal meiosis is possible, which produces gametes with the constitution *AB.* Union of two such gametes at fertilization produces a tetraploid individual (*AABB*), which may be fertile because meiosis occurs normally owing to the presence of two chromosomes of each kind. Self-fertilization in plants makes possible the formation of a tetraploid individual following abnormal mitosis in a single cell. Polyploidy may also originate from abnormal meiosis in a hybrid, when all chromosomes go to one gamete, which will then have the constitution *AB*. Union of two such gametes produces a tetraploid zygote, *AABB,* from which a fertile individual may develop.

The origin of autopolyploidy is similar to that of allopolyploidy, except that the abnormal mitosis or meiosis occurs in a nonhybrid individual.

Polyploid individuals are reproductively isolated from their diploid ancestors. Progenies of the cross *AABB* x *AA* will be *AAB,* and meiosis will be abnormal because there is only one of each of the *B* chromosomes and they find no homologous chromosomes with which to pair. On the other hand, since the polyploid *AABB* is fertile, it may multiply by self-fertilization and produce a polyploid population, which is in fact a new species.

In polyploidy, the suppression of gene flow that is required for the onset of the first stage of speciation is due not to geographic separation, but to cytological irregularities. Reproductive isolation in the form of hybrid sterility does not require many generations but follows immediately owing to chromosomal imbalance. If diploids and their fertile polyploid derivatives occur near each other and hybridization occurs, natural selection will favor the development of prezygotic isolating mechanisms (Stage II) that avoid interfertilization and wastage of gametes.

Examples of polyploid species were given in Chapter 3. Tobacco (*Nicotiana tabacum*) and New World cottons (*Gossypium hirsutum* and *G. barbadense*) are additional examples of polyploid cultivated species. Nearly half (about 47 percent) of the species of flowering plants and a majority of ferns (about 95 percent) are polyploids.

Modes of "quantum" speciation not involving polyploidy are known to occur in plants. H. Lewis has suggested that several species of the annual plant genus *Clarkia* have arisen by rapid speciation events. One instance of quantum speciation involves the two diploid species, *Clarkia biloba* and *C. lingulata*. Both species are native to California, although *C. lingulata* has a narrow distribution, being known from only two sites in the central Sierra Nevada at the southern periphery of the distribution of *C. biloba*. The two species are outcrossers (reproduce by crossing between two different plants), although they are capable of self-fertilization and are similar in external morphology, except for differences in flower shape. However, their chromosomal configurations differ by a translocation, several paracentric inversions, and an extra chromosome in *C. lingulata* that is homologous to parts of two chromosomes of *C. biloba* (Fig. 6.8). The narrowly distributed species, *C. lingulata,* has arisen from *C. biloba* by a rapid series of events involving extensive chromosomal reorganization. Chromosomal rearrangements such as translocations, fusions, and fissions reduce the fertility of individuals heterozygous for the arrangements. The first stage of speciation may be, thus, accomplished through chromosomal rearrangements without extensive allelic differentiation. Self-fertilization facilitates the propagation of the rearrangements. Once there is a population of individuals exhibiting some reproductive isolation from the rest of the population owing to the chromosomal rearrangements, natural selection favors the development of additional RIMs.

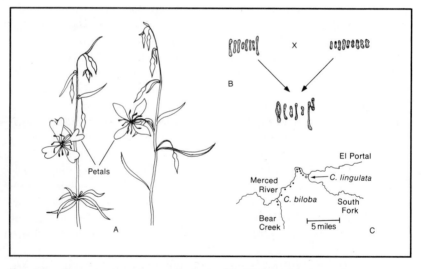

Fig. 6.8.   Two species of annual plants: *Clarkia lingulata* has arisen from *C. biloba* by quantum speciation. *A.* Flowering branches of the two species showing a difference in petal shape, which is bilobed in *C. biloba, left,* but not in *C. lingulata. B.* Paired chromosomes, at meiotic metaphase, *left,* of *C. biloba* (eight pairs), *right, C. lingulata* (nine pairs), and of the F_1 hybrid between the two. The chromosomes differ by at least two reciprocal translocations and a fission (or fusion). *C.* A small portion of the Merced River Canyon, in the Sierra Nevada of California, west of Yosemite Valley, showing the southernmost populations of *C. biloba* (black dots at left) and the two known populations of *C. lingulata* (asterisk at right).

Rapid speciation initiated through chromosomal rearrangements has also occurred in animals, for example in some flightless Australian grasshoppers such as *Moraba scurra* and *M. viatica.* Incipient species differing by chromosomal translocations are found in adjacent territories. According to M. J. D. White, who has named this mode of speciation *stasipatric speciation,* a translocation establishes itself at first in a small colony by genetic drift. If members of this colony possess high fitness, they may subsequently spread and displace the ancestral form from a certain area. The ancestral and the derived population may then coexist contigually, their individuality maintained by the low fitness of the hybrids formed in the contact zones, since the hybrids are translocation heterozygotes. The first stage of speciation is thus rapidly accomplished, and natural selection favors the development of additional RIMs (Stage II). This mode of speciation (or a variant of it sometimes called *parapatric* speciation, because it occurs between contiguous populations) seems to be common in several animal groups, particularly in rodents living underground and having little mobility, such as mole rats of the group *Spalax ehrenbergi* in Israel and pocket gophers of the group *Thomomys talpoides* in the southern Rocky Mountains of the United States (Fig. 6.9).

The speciation process may also be initiated by genetic changes in only one or a few loci, when these changes result in a change of host in the case of parasites, or in general when they result in a change of ecological niche. Many parasites use their host as a place for courtship and mating; organisms with different host preferences may as a consequence become partly reproductively isolated. If the hybrids are poorly fit because they are poor parasites in either one of the two kinds of host, natural selection will favor the development of additional RIMs. Guy Bush has argued that this is a common mode of speciation among parasitic insects, a large group comprising perhaps several hundred thousand species.

A model of speciation intermediate between quantum speciation and geographic speciation has been proposed by Hampton L. Carson to account for the extensive speciation of *Drosophila* in Hawaii. According to this model, speciation starts with the arrival of one or a few inseminated females to a new locality, where they rapidly multiply owing to lack of competitors. Eventually the population expansion is curtailed by limited resources, and a population crash takes place. During the population flush, natural selection is greatly relaxed and the gene pool of the population becomes largely changed and reorganized by chance. During the population crash, however, natural selection is intense. Ultimately, a gene pool very different from the original one may result. This

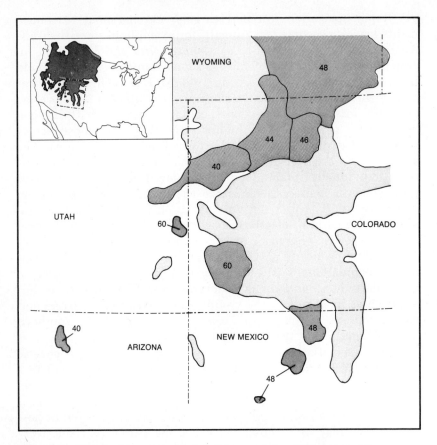

Fig. 6.9. Distribution of pocket gophers of the *Thomomys talpoides* complex in the southern Rocky Mountains. The number of chromosomes in different populations ranges from 40 to 60, although individuals from different populations are morphologically indistinguishable. Differences in chromosome numbers keep individuals from different populations from interbreeding in the regions where they are adjacent to one another.

flush-and-crash model of speciation is similar to the geographic model in that the derivative species starts as a population geographically separated from the parental species. The model is similar to the quantum speciation model in that speciation occurs in a geologically brief time. The flush-and-crash model of speciation may be common in the colonization of islands and other habitats previously unoccupied by potential competitors of the colonizing organisms.

Genetic Differentiation During Geographic Speciation

Evolution occurs through changes in the genetic constitution of populations. We may, therefore, ask how much genetic change takes place during the speciation process. Because the process of speciation involves two stages, the question should be answered separately with respect to each stage. Moreover, we must discuss different modes of speciation separately, since they may require different amounts of genetic change.

Answers to the question of how much genetic differentiation occurs in the process of speciation have not been available until recent years. This may seem surprising because evolutionists have been interested in the question since the birth of genetics at the beginning of the twentieth century. This apparent paradox becomes understandable when we consider the kind of information required to answer the question. To find out the proportion of genes at which two populations or species differ, one must be able to study a "random" sample of genes, i.e., a sample of genes that are not likely to be on the average either more similar or more different between the two populations than the rest of the genome. Yet, the classical methods of Mendelian genetics ascertain the presence of genes by studying segregation in the progenies of individuals differing in phenotype. Whenever a gene is not variable, the classical methods of Mendelian genetics do not make it possible to know that there is a gene at all; thus, only genes that are different can be studied, and it cannot be known how many or what proportion of the genes are identical. This difficulty was also encountered in attempts to determine the amount of genetic variation in natural populations, as discussed in Chapter 3 (p. 79). Moreover, the applicability of segregation studies in severely limited in the study of species differences by the inviability or sterility of most interspecific hybrids.

The advances of molecular genetics and the development of the techniques of gel electrophoresis have made possible in recent years the estimation of genetic differentiation between populations, as they made possible the estimation of genetic variation in populations. Because of the correspondence between genes and proteins (polypeptides), we may study genes by looking at the proteins they encode. Genes that are identical in two populations code for identical proteins as well; if the proteins are different, the genes are also different. Estimates of genetic differentiation between populations can be obtained by studying a sample of proteins, selected for study without knowing whether or not they are different in the populations. The genes coding for the proteins may be assumed to be a random sample of all the structural genes in the organism with respect to the differentiation between the populations. The results from the study of a moderate number of gene loci can, therefore, be extrapolated to the whole genome.

Gel electrophoresis provides estimates of allelic frequencies in populations. Data obtained for many gene loci in two populations can be summarized in a variety of ways. One method uses two parameters: (1) *genetic identity*, represented by I, which estimates the proportion of genes that are identical in two populations; (2) *genetic distance*, represented by D, which estimates the number of allelic substitutions per locus that have occurred in the separate evolution of two populations. There is an allelic substitution when one allele is replaced by a different allele, or when a set of alleles is replaced by a different set. The method recognizes that not all observed allelic substitutions are complete: an allele may have been partly replaced by a different one, but the original allele may still exist in greater or lesser frequency; and some alleles from a set may have been replaced by different alleles.

Genetic identity, I, may range in value from zero (no alleles in common) to one (the same alleles and in the same frequencies are found in both populations). Genetic distance, D, may range in value from zero (no allelic changes at all) to infinity; D can be greater than one because each locus may experience complete allelic substitutions more than once as evolution goes on for long periods of time.

CLOSER LOOK 6.1   Calculating Genetic Identity and Genetic Distance

Electrophoretic studies provide data in the form of genotypic frequencies that can be readily converted into allelic frequencies. Assume that A and B are two different populations, and K a given gene locus. Assume that i different alleles (1, 2, 3, etc.) are observed in the two populations, and let us represent the frequencies of the alleles in population A as a_1, a_2, a_3, etc., and in population B as b_1, b_2, b_3, etc. The genetic similarity between the two populations at this locus may be measured by I_K, defined as

$$I_K = \frac{\Sigma a_i b_i}{\sqrt{(\Sigma a_i^2 \cdot \Sigma b_i^2)}}$$

where the symbol Σ means "summation of"; $a_i b_i$ represents the products $a_1 b_1$, $a_2 b_2$, $a_3 b_3$, etc.; a_i^2 means a_1^2, a_2^2, a_3^2, etc.; and b_i^2 means b_1^2, b_2^2, b_3^2, etc. The formula for I_K calculates the normalized probability that two alleles, one taken from each population, are identical.

Let us consider some simple examples. First assume that only one allele is observed with frequency one in both populations. Then $a_1 = 1$, $b_1 = 1$, and therefore

$$I_K = \frac{1 \times 1}{\sqrt{1^2 \times 1^2}} = \frac{1}{1} = 1$$

Not surprisingly, the value of I_K is 1, indicating that the two populations are identical at this locus.

Assume now that we observe two different alleles, the first having a frequency of 1 in A, and the second a frequency of 1 in B. Then $a_1 = 1$, $b_1 = 0$, $a_2 = 0$, $b_2 = 1$, and therefore

$$I_K = \frac{(1 \times 0) + (0 \times 1)}{\sqrt{(1^2 + 0^2) \times (0^2 + 1^2)}} = \frac{0 + 0}{\sqrt{1 \times 1}} = 0$$

The value of I_K is zero, indicating that the two populations are genetically completely different at this locus.

Now consider the case where two alleles exist in both populations with frequencies $a_1 = 0.2$, $a_2 = 0.8$ ($a_1 + a_2 = 1$), and $b_1 = 0.7$, $b_2 = 0.3$ ($b_1 + b_2 = 1$). We, then, obtain

$$I_K = \frac{(0.2 \times 0.7) + (0.8 \times 0.3)}{\sqrt{(0.2^2 + 0.8^2) \times (0.3^2 + 0.7^2)}} = 0.605$$

The value of I_K is intermediate between one and zero, as we would expect since the two populations share common alleles although not in identical frequencies.

Assume, finally, that both populations are polymorphic at the locus, but share no alleles in common; for example $a_1 = 0.4$, $a_2 = 0.6$, $a_3 = 0$, $a_4 = 0$ ($a_1 + a_2 + a_3 + a_4 = 1$), and $b_1 = 0$, $b_2 = 0$, $b_3 = 0.1$, $b_4 = 0.9$ ($b_1 + b_2 + b_3 + b_4 = 1$). Then

$$I_K = \frac{(0.4 \times 0) + (0.6 \times 0) + (0 \times 0.1) + (0 \times 0.9)}{\sqrt{(0.4^2 + 0.6^2 + 0^2 + 0^2) \times (0^2 + 0^2 + 0.1^2 + 0.9^2)}} = 0$$

We find that I_K is equal to zero, as expected since the two populations are genetically completely different at this locus.

In order to estimate the genetic differentiation between two populations, several loci need to be studied. Let I_{ab}, I_a, and I_b be the arithmetic means (i.e., the averages), over all loci, of $\Sigma a_i b_i$, Σa_i^2, and Σb_i^2, respectively. Then, the *genetic identity, I,* between the two populations may be measured, as proposed by M. Nei, by

$$I = \frac{I_{ab}}{\sqrt{I_a \cdot I_b}}$$

And the *genetic distance, D,* between the populations may be measured by

$$D = -\log_e I$$

Assume that the four examples given above of differentiation at single loci actually correspond to four different loci studied in two populations. Then, we have

$$I_{ab} = \frac{1 + 0 + 0.38 + 0}{4} = 0.345$$

$$I_a = \frac{1 + 1 + 0.68 + 0.50}{4} = 0.795$$

$$I_b = \frac{1 + 1 + 0.58 + 0.82}{4} = 0.850$$

Therefore,

$$I = \frac{0.345}{\sqrt{0.795 \times 0.850}} = \frac{0.345}{0.822} = 0.42$$

and

$$D = -\log_e 0.42 = 0.87$$

That is, it is estimated that 0.87 allelic substitutions per locus (or 87 allelic substitutions for every 100 gene loci) have occurred in the separate evolution of the two populations. More than four gene loci need to be studied in order to obtain an acceptable estimate of genetic differentiation between any two populations, but the four loci are sufficient to show how genetic identity and genetic distance are calculated.

The statistics I and D can be used to measure genetic differentiation during the speciation process. First, we shall consider geographic speciation. The *Drosophila willistoni* group of species was used as a model of geographic speciation because both stages of the process can therein be identified. This group of species has been extensively studied using electrophoretic techniques. The results are summarized in Table 6.4. Five levels of evolutionary divergence are represented in the table. The first level involves comparisons between populations living in different localities but without any reproductive isolation between them; the genetic identity is 0.97, indicating a very high degree of genetic similarity.

The second level involves comparisons between different subspecies (such as *D. w. willistoni* compared with *D. w. quechua,* and *D. e. equinoxialis* compared with *D. e. caribbensis*). These populations are in the first stage of speciation and exhibit postzygotic RIMs in the form of hybrid sterility. They also show a fair amount of genetic differentiation, $I=0.795$, and $D=0.230$; complete allelic substitutions have occurred, on the average, about once for every four gene loci (23 for every 100 gene loci).

The third level of evolutionary divergence in Table 6.4 involves comparisons between the incipient species of the *D. paulistorum* complex. These are populations in the second stage of speciation exhibiting some prezygotic, as well as postzygotic, RIMs. It is interesting that these populations apparently are not genetically more differentiated than those in the first stage of speciation. The values of I and D are not significantly different for comparisons between subspecies or between incipient species. This means that the second stage of speciation does not require much genetic change, which is perhaps not surprising. During the first stage of speciation, reproductive isolation comes about as a byproduct of genetic change: a fair amount of genetic change needs to take place over the whole genome before postzygotic RIMs develop. However, during the second stage of speciation, natural selection directly favors the development of prezygotic RIMs; only a few genes—affecting courtship and mating behavior, for example —need to be changed to accomplish it.

The fourth level in Table 6.4 involves comparisons between sibling species (such as *D. willistoni* compared with *D. equinoxialis*). In spite of their morphological similarity, these species are genetically quite different; about 58 allelic substitutions have occurred, on the average, for every 100 loci. Species are independently evolving groups of populations. Once the process of speciation is completed, species will continue to diverge genetically. The results of this gradual process of divergence are also apparent in the comparisons between morphologically different species of the *D. willistoni* group (fifth level in Table 6.4). On the average, somewhat more than one allelic substitution per gene locus has occurred in the evolution of these nonsibling species.

Table 6.4
GENETIC CHANGE IN GEOGRAPHIC SPECIATION

Genetic differentiation between populations of the *Drosophila willistoni* group at various levels of evolutionary divergence. Levels of comparison 2 (subspecies) and 3 (incipient species) represent, respectively, Stage I and Stage II of the process of geographic speciation. *I* estimates the degree of genetic similarity, *D*, the degree of genetic differentiation. The values given are the means and standard errors for several comparisons. (From Ayala, 1975.)

Level of comparison	I	D
1. Local populations	0.970 ± 0.006	0.031 ± 0.007
2. Subspecies	0.795 ± 0.013	0.230 ± 0.016
3. Incipient species	0.798 ± 0.026	0.226 ± 0.033
4. Sibling species	0.563 ± 0.023	0.581 ± 0.039
5. Morphologically different species	0.352 ± 0.023	1.056 ± 0.068

Using the techniques of gel electrophoresis, comparisons between populations at various levels of evolutionary divergence have been carried out during the past few years in many kinds of organisms. Evolution is a complex process determined by the environmental conditions as well as by the nature of the organisms, and thus the amount of genetic change corresponding to a given level of evolutionary divergence is likely to vary from organism to organism, from place to place, and from time to time. The results of electrophoretic studies confirm this variation, but also show some general patterns. With few exceptions, the genetic distance between populations in either the first or the second stage of speciation is about 0.20 (most comparisons fall between 0.15 and 0.30) for organisms as diverse as insects, fishes, amphibians, reptiles, and mammals. Moreover, the results obtained with many such organisms are consistent with the conclusions derived from the study of the *Drosophila willistoni* group: the first stage of the geographic speciation process requires a fair amount of genetic change (of the order of 20 allelic substitutions for every 100 gene loci), while little genetic change is required during the second stage (the amount of genetic change involved in the second stage is often so small that, as in the *D. willistoni* group, it is not detectable with the methods of gel electrophoresis).

Genetic Differentiation During Quantum Speciation

How much genetic change takes place in the quantum mode of speciation? It is clear that when a new species arises by polyploidy no genetic changes other than the chromosome duplications are required. The new species has the alleles present in the parental species and no others. However, because polyploid species usually start from only one individual, they possess at the beginning less genetic variation than the parental species (see "founder effect"; p. 117).

Other modes of quantum speciation start with chromosomal rearrangements that cause either partial or total hybrid sterility. As in the case of polyploidy such rearrangements do not necessarily involve changes in allelic constitution, although there is often a reduction of genetic variation because the derivative population starts from only one or a few individuals. The first stage of speciation is, therefore, accomplished with little or no genetic change at the level of the individual genes.

What about genetic change in the second stage of quantum speciation? The second stage of speciation is similar in geographic and in quantum speciation. In both cases, the populations already exhibit postzygotic RIMs and are developing prezygotic isolation by natural selection. If in the case of geographic speciation the second stage requires genetic changes in only a small fraction of the genes, the same should be true in quantum speciation. Therefore, examining populations in the second stage of quantum speciation provides a good test of the conclusion derived with respect to the second stage of geographic speciation. Since no genetic change is required in the first stage of quantum speciation, we should find that populations in the second stage of quantum speciation are genetically very similar to each other. Experimental results confirm such prediction (Table 6.5).

The first comparison in Table 6.5 is between two annual plant species, *Clarkia biloba* and *C. lingulata,* discussed earlier as examples of quantum speciation. These species remain genetically quite similar: $I = 0.880$ and $D = 0.128$, indicating that only about 13 allelic substitutions for every 100 gene loci have occurred in their separate evolutions.

The second comparison is also between two annual plants, *Stephanomeria exigua* and a population derived from it only very recently. Leslie Gottlieb has shown that the original and the derivative populations differ by one chromosomal translocation and by their mode of reproduction—the original species reproduces by outcrossing while the derivative population reproduces by self-fertilization. As expected, the two populations are genetically very similar (about six allelic substitutions for every 100 loci) in spite of exhibiting postzygotic reproductive isolation, since hybrids have considerably reduced fertility and few hybrids are produced owing to the different system of reproduction of the two populations.

The third and fourth comparisons in Table 6.5 involve rodents. *Spalax ehrenbergi* is a species of mole rat consisting of four groups of populations differing in the number of chromosomes (52, 54, 58, and 60). The populations are largely allopatric (living in different territories) although they enter in contact one with another in narrow zones at the edge of their distributions and some hybridization takes place there. The difference in chromosome number due to chromosomal fusions or fissions provide effective postzygotic RIMs. Moreover, some ethological isolation has developed—laboratory tests show greater preference for matings between individuals of the same chromosomal type, although they appear morphologically indistinguishable. As shown in Table 6.5, these four populations in the second stage of quantum speciation are, on the average, genetically very similar: only about two allelic substitutions for every 100 gene loci have taken place in their separate evolution.

Thomomys talpoides is a species of pocket gopher, consisting of more than eight populations differing in their chromosomal arrangements, living in the north central and northwest United States and neighboring areas of southern Canada (Fig. 6.9). As in the case of *Spalax,* the populations of *Thomomys* are mostly allopatric but are in geographic contact at the margins of their distributions. The chromosomal reaarangements keep the populations from interbreeding in the zones of contact. Nevertheless, the average genetic distance between these populations is quite small ($D = 0.078$, or about eight allelic substitutions for every 100 loci).

In conclusion, then, quantum speciation can occur with little change at the level of the genes; that is, neither Stage I nor Stage II requires substantial allelic evolution. This result, in turn, confirms the conclusion reached earlier, namely that the second stage of

Table 6.5
GENETIC DIFFERENTIATION IN QUANTUM SPECIATION

As in Table 6.4, I estimates the degree of genetic similarity and D the degree of genetic differentiation. Little genetic differentiation is observed between species or incipient species arisen by quantum speciation. (After Ayala, 1975.)

Populations compared	I	D
Plants:		
Clarkia biloba versus *C. lingulata**	0.880	0.128
*Stephanomeria exigua***	0.945	0.057
Rodents:		
*Spalax ehrenbergi***	0.978	0.022
*Thamomys talpoides***	0.925	0.078

*Comparison between two recently arisen species.
**Comparisons between incipient species, i.e., populations completing the second stage of speciation.

speciation—when natural selection promotes the development of prezygotic RIMs—does not require major genetic changes, even in the case of geographic speciation.

Species in Asexual Organisms

The organisms so far discussed in this chapter are those reproducing sexually. There are microorganisms, such as the bacteria and blue-green algae, which do not reproduce sexually but rather by fission; populations of asexually reproducing organisms are clones made up of genetically identical individuals. The concept of species advanced earlier in this chapter does not apply to these organisms. Bacteria and blue-green algae are classified into different species following criteria such as external morphology, chemical and physiological properties, and genetic constitution.

It should be noted that gene exchange may occur among bacteria and blue-green algae by various means known as transformation, transduction, and bacterial conjugation. However, gene exchange in these microorganisms does not involve mutual exchange between two entire genomes. Rather, the exchange is usually unidirectional; one or a few genes from one individual are incorporated into the genome of another individual, either replacing similar genes in the receiving individual or simply being added to the existing ones. It deserves notice that in these microorganisms, genes from one individual can sometimes be incorporated into an individual classified in a different species and in fact quite different from the former in its biological properties.

QUESTIONS FOR DISCUSSION

1. Is the process of speciation irreversible? What about the formation of races or subspecies?

2. Give examples from plants and animals of the various reproductive isolating mechanisms. Consider pairs of closely related species and enumerate the reproductive isolating mechanisms that, in your opinion, are keeping the species separate.

3. What are the main differences between the geographic model of speciation and the quantum model?

4. Discuss the reasons why polyploidy is rare in animals. Give examples of polyploid animal species.

5. What circumstances favor the stabilization of the progeny of interspecific hybrids so that they will breed true and give rise to new adaptive populations?

Recommended for Additional Reading

Dobzhansky, T. 1970. *Genetics of the evolutionary process.* New York: Columbia University Press.

An excellent discussion of speciation and of most of the topics so far discussed in this book. Advanced.

Dobzhansky et al. 1977. *Evolution.* San Francisco: W. H. Freeman.

A fairly complete presentation of current evolutionary theory, including the problem of speciation. Moderately advanced.

Grant, V. 1971. *Plant speciation.* New York: Columbia University Press.

An extensive treatment of the origin of plant species. Moderately advanced.

Mayr, E. 1963. *Animal species and evolution.* Cambridge, Mass.: Harvard University Press.

Thorough discussion of speciation, covering most of the topics presented in this book up to this point. Advanced.

White, M. J. D. 1978. *Modes of speciation.* San Francisco: W. H. Freeman.

Moderately advanced discussion of speciation, with emphasis on animal speciation.

7

Reconstructing the Tree of Life

Under appropriate circumstances, the processes of speciation can be very prolific, producing vast arrays of distinctive species. There are over one and a half million living species. Many millions more have become extinct. Some species resemble each other quite closely, and some groups of animals share features in common so as to give them a sort of family resemblance. Mosquitoes, for example, all look pretty much alike; and although different groups of insects are rather different from each other when viewed in detail, insects as a whole tend to resemble one another when compared with other distinctive groups of animals, such as clams or fish. There are many more species of insects than of any other comparable group of animals, even though insects are nearly all restricted to the terrestrial habitat; only a relatively few insects live in truly aquatic environments. In the oceans, the group with the most species is the snails (though the nematodes, a poorly known worm group, may have about as many). Organisms seem to live in every conceivable habitat on the earth's surface, from the greatest ocean depths in deep-sea trenches to Himalayan mountain peaks, and from subzero arctic plains to hot springs. A major reason for the great diversity of living things is the great diversity of environmental conditions found on earth, as Darwin observed long ago.

The Hierarchy of Classification

The sheer number of species of organisms poses problems for life scientists. No one person is capable of learning the characteristics of all living species, much less of comprehending the patterns of resemblance among them all, so long as they are treated as individual entities. In order to make clearer the patterns of resemblance among species, naturalists have had to devise a system of classification.

Classifications were begun in classical times or even before, and by the mid-eighteenth century rather elaborate ones existed. At that time a Swedish naturalist named Carolus Linnaeus (Fig. 7.1) devised a system of classification so thoroughgoing and superior for its time that the tenth edition of his work, *Systema Naturae* (1758), was subsequently taken as the starting point for modern animal classification.

Fig. 7.1. Carolus Linnaeus (1707–1778), Swedish naturalist, the father of modern taxonomy.

The rules of classification are slightly different for animals and plants, but the general systems are similar. Both classifications are hierarchical. Animal species that resemble each other closely are classed into a genus; genera that resemble each other closely are classed in the same family; families that resemble each other into an order; orders into a class; and classes into a phylum. Phyla are then grouped into a kingdom, such as Metazoa (animals) (Fig. 7.2). Plants are classed in a similar hierarchical scheme. Hierarchical classifications are extremely efficient ways to organize large bodies of data or large numbers of units; governments, armies, and even colleges are organized on such a plan. Hierarchies may represent the only logical architecture for organizing great complexity.

Placing species within the hierarchy is only part of the problem of classification. There must also be some system of identifying species and formal groups of species so that scientific facts about them can be communicated and accumulated in an orderly manner. Popular names of species are not adequate; they vary from language to language, or even from region to region in the same language area. At any rate, most species do not have popular

CATEGORIES	TAXA	
Kingdom	Metaphyta	Metazoa
Superphylum	—	Deuterostomia
Phylum	—	Chordata
Class	Angiospermae	Mammalia
Order	Fabales	Primates
Family	Fabacea	Hominidae
Genus	*Pisum*	*Homo*
Species	*Pisum sativum*	*Homo sapiens*
Individual organism	A garden pea	A human

Fig. 7.2. The hierarchical classification used in taxonomy. The categories are ranks within the hierarchy; the taxa are groups of organisms classed in the categories. Thus humans are a taxon ranked at the level of species; primates are a taxon ranked as an order.

names. Therefore a special system of nomenclature has been adopted by international agreement, which identifies each species or group of species unequivocally. The smallest grouping of species is the genus. The complete species name consists of the name of the genus followed by the species name itself. For humans, for example, the species name is *Homo sapiens,* and no other animal species may bear this name. By international convention the generic name is capitalized and the generic as well as the species name are italicized.

The practice of classifying living things is called *taxonomy* (or sometimes *systematics*). Each species, genus, family, or other grouping that is named is called a *taxon*, while each level within the classification—the species level, generic level, and so on—is called a *taxonomic category. Homo sapiens* is a taxon in the category of species, while *Homo* is a taxon in the category of genus.

Form and Phylogeny

Living things all share the same genetic code and have in common many biochemical features. It is inconceivable that precisely the same code and biochemistry would evolve more than once. Therefore living organisms must have all descended from some common ancestor from which they inherited these characteristics. The incredible diversity of life forms must have sprung from this common ancestral form, radiating and branching repeatedly. One of the tasks of the evolutionist is to discover the lines of descent of organisms in order to reconstruct the ancestral-descendant pathways among living and extinct animal groups. Such work aims to determine the geneology of life, which can be expressed in a family tree that depicts the pattern of relatedness among taxa (Fig. 7.3).

It is, however, very difficult to establish the precise lines of descent, termed *phylogenies,* for most organisms. A direct method of tracing phylogenies has been to trace series of fossils that resemble each other but show a sequence of changes leading through time from an ancestral to a descendant form. Relationships among the fossils are thus judged by their relative ages and their morphological resemblances and differences. This works well when abundant fossils are available in a continuous record, but unfortunately the fossil record is quite incomplete. Most animals have no easily fossilizable hard parts, and only a small fraction of animals with shells or bones are actually preserved as fossils. For most lineages we have to employ more indirect methods of phylogenetic reconstruction.

Fig. 7.3. A family tree for Darwin's finches, several species of birds from the Galapagos Islands that have radiated from a common ancestor. (After Lack, 1953.)

Adaptive Patterns

When fossils are lacking, the degree of resemblance among living organisms may be taken as an indication of their relationship. This works in a general way, but the fossil record shows that resemblance alone is not an accurate measure of closeness of relationship. Lineages that are closely related may change so as to resemble each other only distantly—a phenomenon known as *divergence* (Fig. 7.4). Divergence indicates that the allied lineages are becoming adapted to separate modes of life, usually in dissimilar environments. It is common to find lineages that split frequently, producing a large number of divergent lineages, which may also split, so that the evolutionary pattern is one of a burst of diverging lineages, a *radiation* (Fig. 7.4). This usually indicates that a lineage has been able to exploit a novel mode of life; such radiations are commonly called *adaptive radiations*. The fossil record shows that different radiating lineages may evolve at very different rates. Some diverging lineages soon look quite dissimilar while others in the same group still resemble each other closely, even though they may all be related equally. On the other hand, some lineages that are only distantly related and that begin quite distinct morphologically may evolve so as to resemble each other closely. This is *convergence* (Fig. 7.4) and indicates that the separate lineages are becoming adapted to similar modes of life, usually in similar environments.

Other patterns of resemblance between different lineages include *parallelism,* when two or more independent lineages change in similar ways so that their degree of resemblance remains about the same (Fig 7.4); and *iteration,* when a given trait appears a number of separate times. Parallelism suggests a single environmental change that evokes similar responses in distinct lineages, while iteration suggests that a given environmental opportunity has appeared repeatedly, perhaps due to some cyclical environmental change, evoking similar evolutionary responses each time. Clearly, the patterns of morphological resemblance may be misleading as criteria of relatedness among lineages, even though in general closely related organisms resemble each other closely.

When comparing different taxa of organisms, it has proven useful to distinguish between characters that resemble each other because they are descended from a common ancestor, called *homologous* characters, and those that resemble each other because they have similar functions, called *analogous* characters. The forelimbs of chickens and humans are homologous and have similarities that are due to their common ancestral condition in primitive reptiles

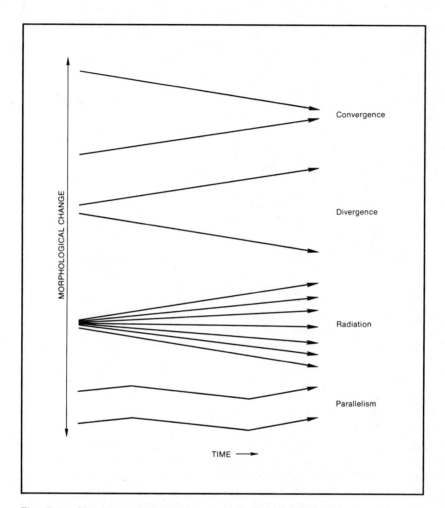

Fig. 7.4. Some common patterns of evolution displayed by independent lineages. Convergence occurs when two (or more) lineages evolve towards a similar state. Divergence occurs when two lineages evolve so as to become less similar. Radiations are divergences involving more than two lineages. Parallel evolution occurs when two (or more) lineages change similarly so that despite evolutionary activity they become neither more nor less similar.

(Fig. 7.5). The eyes of a mammal and the eyes of an octopus, however, have been evolved independently but they are analogous, having similar function. When using morphological similarity as a guide to phylogeny we must be careful to avoid taking analogous structures for homologous ones.

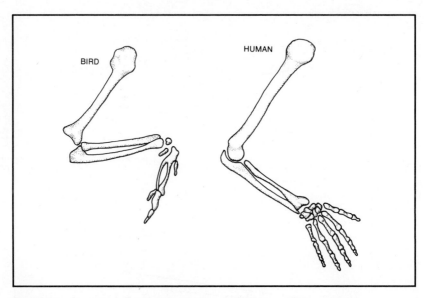

Fig. 7.5. Homology in the forelimbs of humans, *right,* and birds, *left;* the bones resemble each other because they have been inherited from a common reptilian ancestor, although they have been modified to serve different functions.

Mosaic Evolution

There are other difficulties in using morphology as a guide to relatedness. One is that different parts of animals often evolve at different rates. A good example is provided by the lungfishes, a branch of a rather primitive fish group (lobe-finned fishes) from which land animals evolved (Fig. 7.6). About 390 million years ago these fishes were rather fast swimmers that fed by biting or slashing, as indicated by their sharp teeth. However, their diet shifted and their dentition evolved to a grinding type during the next 40 million years; this shift also involved changes in their general head structure, associated with the mechanics of feeding in the new manner. Their bodies did not change much, however, until the evolution of the head was almost completed. Then the bodies changed, presumably because rapid swimming for prey was no longer required, and structural adaptations for a more sluggish swimming mode appeared. Thus the head and body parts of these fish evolved separately. This is *mosaic evolution.* Organisms may display independent evolution of different morphological characters.

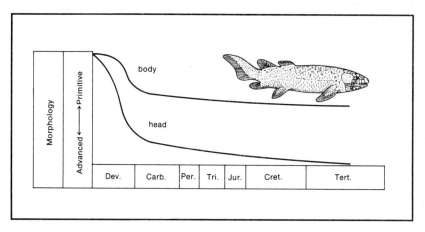

Fig. 7.6. Mosaic evolution in lungfish. Steepness of curve indicates morphological change. Note that the head evolved rapidly before the body characters changed very much. (After Westoll.)

A famous forged fossil—Piltdown Man, believed for a while to be a true stage in the evolution of man from apelike ancestors—illustrates the problems that are posed by mosaic evolution. The forger joined the jaw of an orangutan with a human cranium to give a small-jawed, large-brained "ape-man." This forgery was then planted in ancient sediments and "discovered." We know now that while the jaw and brain size of the lineage leading to man did indeed display mosaic evolution, man's apelike ancestors actually had large jaws and small brains. Unless we have some theory to explain which parts of organisms evolved when, it is difficult to reconstruct hypothetical intermediates between known animals. Mosaic evolution may have occurred so that no animal ever actually existed that was precisely halfway in every character between the end points. This is probably the usual case. It is often difficult to judge where any given fossil type falls among the many branches of a phylogenetic tree.

Recapitulation and Larval Resemblances

Phylogenetic reconstructions were once widely based upon the hypothesis of recapitulation. In its simplest form, this hypothesis states that as organisms develop from fertilized eggs through embryos and into adults, they retrace in their life histories the outlines of their evolutionary histories. The sequence of form they display during development is supposed to reflect the sequence of adult forms in their evolutionary past—"ontogeny recapitulates phylogeny," as the notion was commonly phrased. This idea was promoted in the last century by a famous follower of Darwin named Ernst Haeckel. It has now been disproven, at least in its classic form.

As we have seen in earlier chapters, what is inherited is a genome that contains a set of structural genes and a regulatory system that is coded so as to develop into an adult of a given sort, as specified by the genes. The entire life cycle, including the developmental stages, is inherited. Embryonic or young organisms usually face quite different environments than do adults. Therefore selection pressures are different during different stages of the life cycle, and adaptations may be found in, say, embryonic stages that are not useful in the adult.

Marine invertebrates provide many examples. Many invertebrates that live on the seafloor as adults have larvae that live afloat in seawater. This is particularly true in tropical latitudes where food is continuously available in the water. In latitudes nearer the poles, however, where food availability is highly seasonal, fewer invertebrates have such larvae. These invertebrates tend to develop on the seafloor within large eggs that contain adequate food as yolk. Invertebrate species that are closely related but that live in different regions—tropical or polar—frequently have such distinctly different developmental stages, even though the adults resemble each other closely. Clearly this is because the early life stages have each become adapted to differences in the conditions that each must face, while the adults have similar modes of life.

Generally speaking, the developmental stages of species belonging to the same major animal group resemble each other somewhat; starfish larvae tend to resemble each other more than they resemble snail larvae, and so on. Yet the effects of convergence and divergence in larval forms is often greater than upon adults, and one must use caution in inferring phylogenies from early stages of life histories.

Genetic Phylogenies

When lineages diverge from a common ancestor into two or more descendant species, the gene pools of the descendants diverge from each other as well. If we compare a sample of the genes of two allied species, we gain some idea of how closely the species are related; the differences in the genes form an estimate of the relatedness of the species. There are a number of ways that genetic comparisons between different species may be made. The structure of the chromosomes may be compared; the similarities between DNA strands may be estimated by hybridization or other techniques; proteins coded by the same gene locus but in different species may be assayed and compared, either by such techniques as electrophoresis, immunology, or by actually determining the sequences of amino acids in the proteins. These modern methods will become increasingly more important in evolutionary studies in the future and therefore their strengths and limitations should be understood.

Chromosome Phylogenies The chromosomes in the nuclei of eukaryotic organisms may change in number, size, and shape during the course of time owing to the various sorts of chromosomal mutations discussed in Chapter 3, including inversions, translocations, duplications, and deletions, as well as fusion and fission of entire chromosomes. Since these changes are inherited, they can be used to infer phylogenies when their sequence of occurrence can be determined. One method involves the use of inversions that have accumulated in chromosomes. The inversions can be detected by microscopic examination of bands that are visible on giant chromosomes that occur in certain cells of flies.

Drosophila flies have been extensively studied phylogenetically using this technique. Suppose we label eight characteristic bands in one fly species ABCDEFGH in order. In a second species, the bands occur as AEDCBFGH. It is clear that there has been an inversion of BCDE, but we cannot be sure whether the first or the second species is ancestral. If in a third species the sequence is AEDFBCGH, we can conclude that it differs from the second by an inversion of CBF, but again we cannot be sure which species is the ancestor. The first and third species, however, cannot be derived one from the other without an intermediate step. The sequence of species is either 1→2→3 or 3→2→1 or 3←2→1, but it cannot be 1→3 or 3→1 without going through 2. If we can determine from other evidence whether 1, 2, or 3 is oldest, we have obtained a phylogeny for these three species. By piecing phylogenies together in this way, the family tree of over 100 species of Hawaiian *Drosophila* has been reconstructed. In the Hawaiian archipelago, new species are often evolved after the invasion of an island by members of an ancestral species; the isolation of the invaders from their parent species and the novel environmental conditions on the island form an ideal condition for the appearance of new characters and the divergence of the invader from the parent. The probable island of origin and the direction of colonization of many fly species has been determined after working out which species are ancestral and which are descendant by chromosome phylogeny. The probable colonizing routes of one group of Hawaiian vinegar flies worked out in this way are shown in Figure 7.7.

DNA Hybridizations As two species undergo divergent evolution their DNA sequences become less similar. A method for determining the degree of difference between the sequences of DNA in separate species has been developed in recent years. It is laborious and costly but can be used for any species whether or not the species has giant chromosomes. The method involves the dissociation of the complementary DNA strands in the double helix in each of the species being compared, and then joining together a single strand from each species to obtain hybrid-paired DNA strands.

Using some experimental tricks, the single strands of two species are cut up into small segments and then given an opportunity to re-pair. The DNA of one species is labeled with a radioisotope,

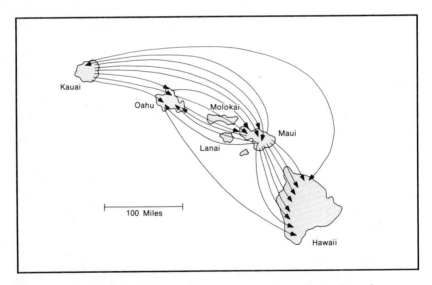

Fig. 7.7. Speciation of the "picture-wing" group of species of *Drosophila* in Hawaii. The arrows indicate the minimum number of interisland migrations required to account for the diversification of this group according to the gene arrangements on their chromosomes. (After H. C. Carson, D. E. Hardy, H. T. Spieth, and W S. Stone, 1970.)

such as tritium. By measuring the radioactivity of the rejoined pairs in a variety of experiments it is possible to estimate the numbers of DNA segments that are similar in the two species—similar enough that hybrid pairing can occur. The higher the similarity of DNA, the greater the degree of relationship is presumed to be. For example, in humans only 10 percent of DNA is similar to that of chickens, but over 90 percent is similar to that of gibbons, while the DNA of humans and chimpanzees appears 100 percent similar.

In order to obtain even more accurate estimates of DNA similarities, hybridization studies have been refined further. Single strands of DNA will pair even when they are not precisely alike— as between humans and chimpanzees. If the hybrid DNA segments are dissociated again, they will separate at rates that are inversely proportional to their similarity, in other words to the degree of complementarity of their nucleotide sequences. Using this technique, it has been estimated that humans and chimpanzees differ by 2.4 percent of their nucleotides. Figure 7.8 shows a family tree based on the percentage of nucleotide difference among a number of species of primates, including man, as estimated by this method.

Techniques have been recently developed for obtaining the actual nucleotide sequence in DNA. These techniques can provide the most exact measure of genetic distance between species since the exact number of nucleotide differences becomes known. However, nucleotide sequencing techniques have not been used to any extent as yet to determine the overall degree of genetic differentiation between groups of organisms.

Electrophoretic Methods   The principle underlying the techniques of electrophoresis was explained in Chapter 3. Recall that electrophoretic studies tend to underestimate the true amount of genetic variability present in a species. Nevertheless, species may be compared on the basis of electrophoresis if it is assumed that the degree of electrophoretic similarity that they display is proportional to their genetic similarity. Species can be grouped in a pattern of electrophoretic similarity that should be close to their phylogenetic pattern.

Electrophoretic similarities between two populations are usually measured by coefficients I and D that are sensitive to differences in allele frequencies (see Chapter 6). Genetic differences detected by electrophoresis are sometimes rather independent of morphological differences. Figure 7.9 depicts the pattern of electrophoretic resemblance among morphologically similar species of vinegar flies from the New World tropics. Most of these species are

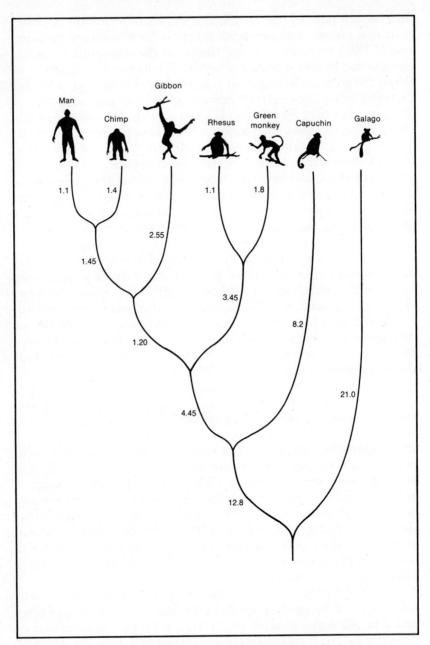

Fig. 7.8. Similarity in the DNA of a number of species of primates based on DNA hybridization techniques. The numbers on the branches estimate the percentage of nucleotide-pair substitutions that have occurred during evolution. The pattern of similarity suggests a family tree. (After Kohne, Chiscon, and Hoyer, 1972.)

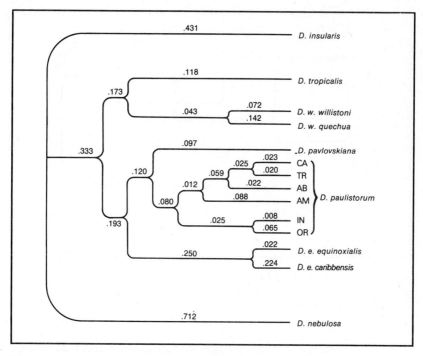

Fig. 7.9. Pattern of electrophoretically determined differences in proteins among a group of closely related species and subspecies of *Drosophila* from the American tropics. The numbers are estimates of the electrophoretically detectable nucleotide substitutions that have occurred during evolution and suggest the phylogenetic relationships of the taxa. The six populations labeled *D. paulistorum* are incipient species. (After Ayala, 1975.)

so similar morphologically that they can hardly be told apart, but breeding experiments have demonstrated that they are infertile when crossed; they are true sibling species (see Chapter 6). One species (*Drosophila nebulosa*) is somewhat more distinctive morphologically. There is quite a range of electrophoretically detectable genetic differentiation between this species and the others. The pattern of differentiation in Figure 7.9 may be taken as an indication of the phylogenetic pattern.

Immunological Techniques The immune reaction can also be used as an indication of relatedness among species. A purified protein from one species is injected into a second, rather different species. For example, a human protein might be injected into a rabbit. The rabbit develops antibodies against this foreign protein.

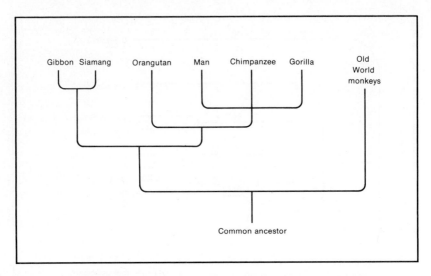

Fig. 7.10. Pattern of immunological differences in the albumin proteins of a number of primate species, which suggests their phylogenetic relationships. (After Sarich and Wilson, 1967.)

Those antibodies react against the human protein and will also re-act against similar proteins coming from animals closely related to humans. In general, the more closely the animals are related to humans, the stronger the immune reaction involving the rabbit antibodies. Thus a number of animal species may be ranked in relatedness according to evidence from immune reactions. Figure 7.10 depicts the inferred phylogenetic relationships among a number of primate groups based on immune reactions to albumin proteins.

Amino Acid Sequences   A rather direct way of judging the genetic similarity among species is to determine the actual sequence of amino acids present in proteins of one species and then to compare them with sequences of amino acids of the same proteins in other species. Recall that each protein is composed of one or more polypeptide chains, and that each chain is composed of a sequence of amino acids. The sequences are termed the *primary structure* of the polypeptides and are controlled by the nucleotide sequences of the genes that code for the polypeptides. During evolution, mutations arise that change the nucleotide sequences so that descendant populations have amino acid sequences in many of their proteins different from those of their ancestors. The degree of dissimilarity between the amino acid sequences of polypeptides in two species, then, is an indication of their genetic divergence from a common ancestor.

To determine the sequence of amino acids, the polypeptide is first broken into small segments by an enzyme that breaks the bonds between certain amino acids at specific sites. The various small segments are then separated, chiefly by chromatographic methods, and their sequences are determined by using an apparatus called a *sequinator*. Next, a second sample of the original polypeptide is broken into different small segments by using a different enzyme, one that breaks bonds at different sites. These segments are also separated and sequenced. Finally the complete sequence of the polypeptide is pieced together by matching up overlapping portions of the different segments. The whole process is quite laborious.

Insulin was the first protein to be sequenced, reported by F. Sanger and E. O. P. Thompson in 1953. About 500 different sequences are now known, and more are reported each year. The protein cytochrome *c,* which is involved in cell respiration, has been sequenced in a number of diverse organisms. These studies show that cytochrome *c* has evolved only very slowly; humans, moths, and yeast all have identical sequences of amino acids in large proportions of cytochrome *c.* This makes cytochrome *c* a good molecule with which to study the phylogeny of organisms that are only distantly related—organisms belonging to separate classes or phyla, for example. Figure 7.11 depicts the phylogeny of 20 organisms as inferred from cytochrome *c* sequences, including two animal phyla (Chordata and Arthropoda), several classes of chordates, and some lower organisms not in the animal kingdom (yeasts for example). To establish the scale of relatedness in the diagram, the sequence of amino acids for each organism was translated into its genetic code. Then the minimum number of mutations necessary to derive each code sequence from each of the others was calculated. These are called *mutation distances*.

Mutation distances provide a conservative estimate of the amount of evolutionary change at the cytochrome *c* locus among all the organisms. It is not a precise measure of the change because there may have been back mutations that restored some earlier mutational changes back to the original sequence. Furthermore there is no assurance that cytochrome *c* evolution exactly mirrors the average evolution of the organisms as a whole—why should one among many thousands of proteins be precisely average? Important evolutionary trends may have occurred in some lineages without any cytochrome *c* changes but with much change in the sequence of other loci—a mosaic evolution on the molecular level.

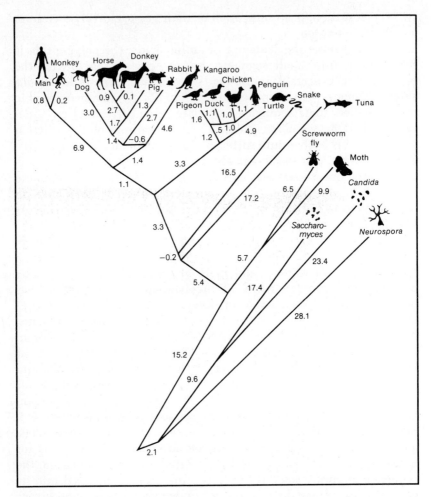

Fig. 7.11. Pattern of differences in the protein sequence of cytochrome c in organisms ranging from yeast to man. The numbers are estimates of the nucleotide substitutions in this protein that have occurred during evolution. This pattern agrees well with the pattern of phylogenetic relationships worked out by classical techniques of comparative morphology and from the fossil record. (After Fitch and Margoliash, 1967.)

The True Phylogenies

None of the methods of determining phylogenies is perfectly foolproof. Indeed, most of these methods are subject to enough error so that we accept their results only tentatively, as a sort of hypothesis or model of what the phylogeny might be like, but certainly subject to revision as new evidence comes to light. The best we can do at present is to employ evidence from all the different methods that we can use. Incorrect models will tend to become inconsistent with new lines of evidence sooner or later, so that the more evidence we have, the more closely constrained the phylogenetic models tend to become. Even when all evidence is used, there are often several alternate phylogenies that are equally plausible. This is especially true for taxa in higher categories, such as phyla or classes.

Probably the best hope for establishing true phylogenies lies in the technique of amino acid sequencing of homologous proteins. Although a phylogeny based on any one protein may be in error, there are a great many proteins, and when the sequences of large numbers of proteins have been learned in representatives of all living phyla and classes we should be nearly certain of their relationships. For extinct groups, however, we shall have to continue to rely upon morphological resemblances and their patterns of appearance and change in the fossil record.

The Art of Classification

When he proposed his great system of classification, Linnaeus did not know that patterns of resemblance among organisms reflected their evolutionary histories. Classification was based on morphological resemblances alone. After Darwin's theories became accepted, however, evolutionists became concerned about the classification of organisms because they believed (as Darwin did) that classification should reflect the evolutionary history of life. To construct a classification under such a stipulation is an exercise in phylogenetics, an attempt to trace patterns of descent. Evolutionists usually prefer that the organisms placed in each taxon be more closely related to each other than to organisms in other taxa of the same rank or category. Thus all species of a genus should be descended from a common ancestor that lies on the branch from which they sprang. Taxa of this sort are *monophyletic* —all from one lineage. Taxa that include species that are of diverse ancestry are *polyphyletic*. These terms are illustrated in Figure 7.12.

Another property of evolutionarily oriented classifications is that the ranking of taxa—whether a given monophyletic group of animals should constitute a genus, family, or order, for example— reflects their ecological significance insofar as possible. Thus a group of organisms that has become quite distinctive and ecologically dominant in some wide environmental domain should be accorded a significantly high ranking, with due regard to its phylogenetic history.

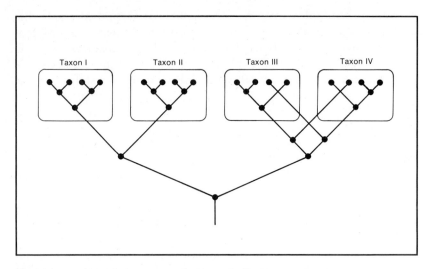

Fig. 7.12. Monophyly and polyphyly. Taxon I and Taxon II are monophyletic, being descended from common ancestors (*a* and *b*, respectively). Taxon III and Taxon IV are not descended from common ancestors, however, and are therefore polyphyletic.

Classification has proven to be exceptionally difficult. Since the true phylogenies of most groups are not known for certain, taxonomists tend to rely upon their own opinions as to the most plausible phylogeny. Since opinions differ, there are frequently a number of different classifications in use, each implying a different phylogenetic history. As old phylogenies are modified by new evidence, old classifications must be modified as well. Therefore classifications are not at all immutable and stable but are altered frequently.

Even if we had a true phylogenetic tree, so that the patterns of relatedness were known for sure, it would still be possible to form many different classifications, based on different rules and approaches. Some taxonomists will divide an order into numerous families or a family into numerous genera; they are "splitters." Other taxonomists will divide the same order into only a few families or a family into few genera; they are "lumpers." For example, living species of the cat family, the Felidae, have been split into 28 genera by some taxonomists and lumped into a single genus by others.

As a final complication, there is an involved International Code of Zoological Nomenclature that specifies rules and conditions for determining the valid names of families, genera, species, and similar ranks. The code specifies that the first names that are valid are those used by Linnaeus in his standard edition, or the first names proposed thereafter if Linnaeus did not name them, assuming that the names meet certain eligibility requirements. As old publications come to light, names based on later works must be changed if the older names are eligible.

Numerical Taxonomy

The instability of classification is vexing to biologists, who sometimes scarcely know how to classify an organism without consulting a specialist (and if they seek a second opinion they may end up with two separate answers). There have been various attempts to bring more order to the system of classification. One approach represents a return to the system of classifying organisms strictly on the basis of their morphological resemblance without regard to their descent—a *phenetic* (based on appearance) rather than a genetic system. For each organism, a great many morphological characters are measured, counted, or otherwise encoded numerically. The numerical scores of one organism may be compared to those of another, most efficiently by using a computer. A numerical index of similarity may then be calculated between them. Whole arrays of species may be compared in this fashion. Those that are morphologically most similar can then be clustered together by the computer, and the cluster can be taken to represent a taxon. Species clusters may be grouped into genera; a number of genera closely resembling each other may be grouped into a family, and so on.

Such numerical taxonomies are clearly subject to all the difficulties raised by divergent, convergent, and iterative evolution —morphology does not accurately reflect descent. The proponents of numerical taxonomy have argued that since true phylogenies are unknown and the phylogenetic hypotheses are changeable anyway this should not be of great concern, and that morphology and phylogeny are probably very close at any rate. The stability introduced by numerical methods would make up for any small departures from a phylogenetic arrangement.

Whether this is true is not yet clear, but there is one important study that bears upon the problem. Fossil bivalve mollusks (clams) of the family Lucinidae are fairly common, and the lines of descent of many genera may be traced with some confidence during the last 60 million years. From this record and from conventional morphological comparisons, a family tree has been worked out for 42 genera and subgenera of these clams. Figure 7.13 depicts the phenetic relationship as represented in a morphological tree (*phenogram*). Numbers along the scale at the left indicate the morphological similarity; 1.00 indicates perfect similarity. Taxa 11 and 12, near the center, are quite similar, for they cluster near 1.00, while Taxa 10 and 36, on the far right, are much less similar. At the 0.00 level of similarity there are four large clusters (two are shaded in Figure 7.13). If morphology closely reflects phylogeny in these taxa, then the taxa in these clusters should be monophyletic descendant groups. Unfortunately they are not. For example, the right-hand cluster *D* contains taxa that belong to lineages otherwise represented in clusters *A* and *C,* plus several other genera that are not at all closely related. The convergence of species from different ancestries upon similar morphological characteristics has been so common that the phenetic classification is not very phylogenetic.

The methods of numerical taxonomy are nevertheless of great usefulness to evolutionists. Consider the foregoing example. If the bivalves had been classified only by traditional methods, we would not understand the extent of the convergence; if they had been classified only by methods of numerical taxonomy, we would not understand their phylogeny. Taken together we have a good idea of the pathways of descent and also learn that some characteristics arose independently but repeatedly among them. This of course suggests that powerful selective forces have been at work and stimulates us to study the adaptive significance of the important convergent characters.

Cladistics

Another method intended to stabilize classification and improve taxonomic practice is called *cladistics.* The term refers to the branching or splitting of evolving lineages (*cladogenesis*) into two descendant groups or *clades.* When paleontologists trace a lineage through time they sometimes find that the morphology changes although no splitting occurs. Eventually the characters of the early populations are so altered that the later populations look quite different and would ordinarily be placed in a different species from the early ones ("chronospecies," see Chapter 6). In Figure 7.14, the species *Athleta petrosa* has been divided into four successive subspecies, but the characters have in fact changed rather gradually so that the subdivision of the lineage is largely arbitrary—each generation was parent to the next, obviously, so that fertility was continuous through time. How is one to determine subspecies or species within such a lineage?

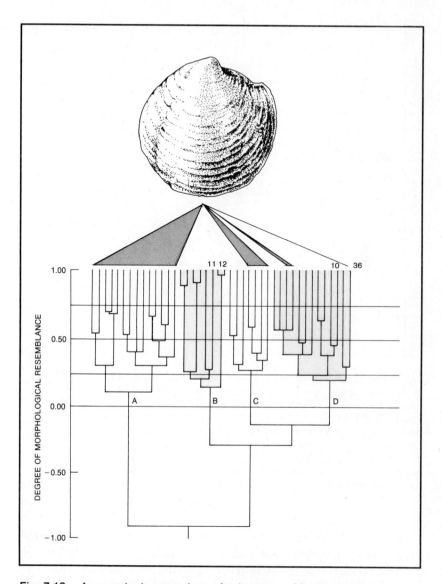

Fig. 7.13. A numerical comparison of subgenera of fossil clams of the family Lucinidae. They have been clustered by computer into four major groups (*A* to *D*) according to their phenetic (morphological) resemblances. However, when the evolutionary history of this family is worked out from the fossil record, it turns out that the clusters differ significantly from the phylogenetic relationships. For example, allied members of the *Lucina* lineage (pictured) are scattered over three of the four major groups, as shown by the lines which connect them to a point below the figured specimen. (After S. Bretsky, 1970.)

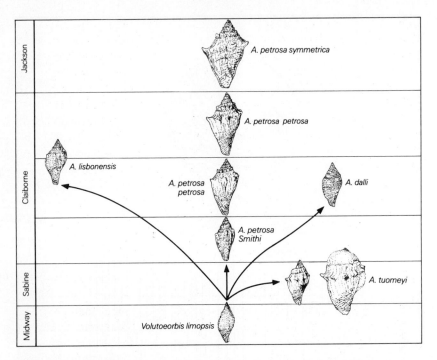

Fig. 7.14. Evolution of size, shape, spinosity, and ornamentation in Eocene marine gastropod lineages. The length of time represented by the fossiliferous units (the oldest is on the bottom) is about 25 million years. The *A. petrosa* lineage is broken up into a number of subspecies, but these are intergrading through time like chronospecies. The rate of evolution, both in rate of speciation and in rate of morphological change, was clearly greatest during the late Midway-early Sabine interval, when three distinct species originated. (After Fisher, Rodda, and Dietrich, 1964.)

Cladists have taken the view that the only event during the evolution of lineages that provides a truly objective datum for classification is the branching or splitting of a lineage into two daughter lineages. The contemporary sister lineages resulting from such a split are evolutionarily independent and should be placed in separate taxa. Cladists require that sister lineages be of the same rank, and that the mother lineage be placed in a different taxon from either daughter. They also require that each taxon be monophyletic in the strictest sense, with the last common ancestor of all members of the taxon included in it.

The consistent application of the requirements or rules of cladistics removes much of the subjectivity from classification, once a phylogenetic tree is established. Removing subjectivity is clearly a great advantage, but there are also disadvantages that arise from the strict formalism of cladistics. In Figure 7.15, a phylogenetic tree is broken up into species numbered according to traditional (A) and cladistic (B) methods. In method A, an ancestral lineage splits to give rise to a second lineage that has split twice to produce two more lineages; all the lineages are recognized as species. These same lineages would be classified differently cladistically (B). The ancestral lineage would be broken into two species and the second lineage into three species, even though the different species in each of these lineages are essentially identical. The occurrence of cladogenesis has forced the subdivision of an otherwise homogeneous lineage.

The One True Classification?

The need for a stable classification is great, but new systems that have attempted to classify organisms "objectively" involve as many difficulties as traditional methods. Numerical taxonomy too often lumps lineages that are rather unrelated. Cladistics too often dismembers lineages that are similar or even identical. Both systems create biologically artificial groupings of taxa and neither

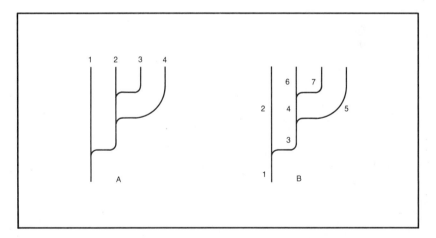

Fig. 7.15. Some effects of cladistic taxonomy. A, a conventional assignment of four morphologically distinctive monophyletic lineages to separate taxa. B, a cladistic classification of the same pattern. The branching events require that ancestors and each descendant be placed in separate taxa, thus dismembering lineages (such as 1–2 and 3–4–6) that may be morphologically and genetically identical.

solves the problems of ranking taxa into categories. A major problem is that these systems lack adequate methods of evaluating the biological significance of the taxa with which they deal. Reconstructing the tree of life involves more than connecting taxa by lines that depict their relatedness; it involves determining the size, length, and distinctiveness of the branches as well. Classification should mirror all such properties, or as many as possible.

When the phylogenies of living organisms have finally been worked out by molecular and other techniques, the organisms can be classified without fear of future changes due to revisions of their patterns of relatedness. Disputes over classification may nevertheless continue, for the relative importance of the various branches and twigs of the phylogenetic tree will still be debatable. There can be a true phylogeny, at least in principle, but there cannot be one true classification, because the rules of classification are made by people to serve their convenience and not by nature or by history. We may eventually have several stable classifications, each based on a different set of rules to serve a special purpose and each displaying a particular aspect of the tree of life.

At present we know what evolution is capable of, because the results are all around us in the biota of the modern world and the fossil record. We are not yet sure of the pathway by which evolution achieved many of these results, even in very important cases (see Chapter 11). As phylogenies become well-established we shall understand the course of these pathways and will be able to investigate such important questions as why one pathway was followed rather than another. It is impossible to predict what new discoveries and insights will emerge from such investigations, but the prospects are exciting.

QUESTIONS FOR DISCUSSION

1. Distinguish taxonomic categories from the taxa themselves. Give examples of each.

2. Give three reasons why morphological resemblances need not reflect phylogenetic relationships.

3. Can we ever hope to have a true phylogeny? Why?

4. Can we ever hope to have a true classification? Why?

5. How can we improve existing phylogenies and classifications? Give examples.

Recommended for Additional Reading

Hennig, W. 1966. *Phylogenetic systematics*. Translated by D. D. Davis and R. Zangerl. Urbana, Ill.: University of Illinois Press.

An account of the rationale underlying cladistic taxonomy by the principal founder of the method. Moderately advanced.

Linnaei, C. (Linnaeus Carolus). 1758. *Systema nature per regna troa naturae etc*. 10th ed. Reprinted, 1956. London: British Museum (Natural History).

The foundation ,for all animal nomenclature, this classic is still fascinating to look through. Notice the way that animal phyla are classed, especially the worms (Vermes), and also the classification of primates, which includes man (the first species entry, p. 20).

Sneath, P. H. A., and Sokal, R. R. 1973. *Principles of numerical taxonomy*, 2nd ed. San Francisco: W. H. Freeman.

A clear statement of the aims and methods of this invaluable technique. Moderately advanced.

8

Transspecific Evolution

Living organisms are very diverse in form. Complexity is an important aspect of their differences. Some organisms, such as bacteria, are composed of only a single cell that contains a relatively small amount of DNA. Others, such as insects or mammals, are vastly more complex, with numerous kinds of complicated cells (mammals may have 200 cell types) that total in the billions. We know from the fossil record that simple cells evolved early in the history of life, complex cells later, and multicellular organisms still later. Mammals have existed for only 200 million years or so; human beings have evolved only within the last few million years. There is thus some sense of increasing complexity to the grand evolutionary pattern.

The Concept of Evolutionary Progress

This general trend of increasing complexity has led to the concept of evolutionary progress. For change to qualify as progress, two conditions must be fulfilled: (1) the change must be *directional;* and (2) the change must be *for the better.* In order to consider evolutionary changes as directional we must be able to order them in a linear sequence, such that the organisms in the early part of the sequence more nearly resemble those in the middle than those in the later part, with respect to some property or feature that we are interested in. In order to judge whether or not change is a betterment, we must have some standard of reference. The standard need not have any moral or ethical overtones. The standard might be some aspect of size, or number, or it might be an aspect of complexity.

If we compare the organisms of two billion, one billion, and one-half billion years ago with each other and with living organisms, we find that more complex forms are present at each successively younger time. However, when we look in detail at life history we find that evolutionary change is sometimes in the direction of simplification. Animals that evolve to be of minute body size, as when they become colonial or parasitic, frequently become simplified in structure relative to their ancestors. Increasing complexity is therefore not a pervasive trend, and progress by this criterion is not *uniform*—it does not apply to every sequence of evolutionary change. In fact, there seems to be no criterion by which evolutionary change is uniformly progressive.

Grades and Clades

Some patterns of evolutionary change clearly do involve progress with respect to particular features, while others need not show such progress. *Anagenesis* (Fig. 8.1) is a change in some feature within a group of related organisms over time. The change need not be in complexity, but it must extend over a series so that later forms become progressively more different from early forms. Sustained anagenetic changes should lead to truly novel types of organisms. *Cladogenesis* (Fig. 8.1) is the splitting of an ancestral lineage into two (or more) daughter lineages. Since daughter species diverge from a common founding species, change occurs in at least one of them. A widespread episode of cladogenesis may thus produce large numbers of new lineages and greatly multiply the numbers of different species that are present. Since the new lineages need not differ very much from their ancestor and need not display any sustained trends, there may be very little net change in the quality of the organisms involved. On the other hand, some species appear to differ markedly from their immediate ancestors. *Stasigenesis* (Fig. 8.1) is applied to lineages that display no significant change over an appreciable amount of time but simply remain evolutionarily stable, without showing progressive change.

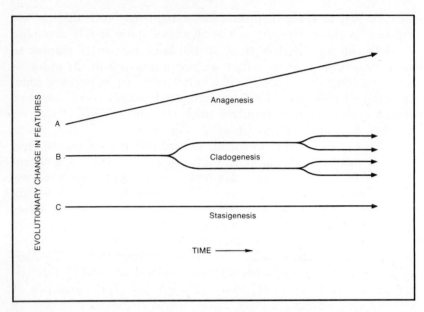

Fig. 8.1. Evolutionary patterns within lineages. *A,* anagenesis, a progressive evolutionary change in some features. *B,* cladogenesis, splitting of a lineage into daughter lineages, which need not involve significant progressive changes. *C,* stasigenesis, the mere persistence of a lineage without significant change.

A sustained anagenetic episode leads to descendants with novel characters and abilities beyond those of their ancestors. A suite of such characters and abilities constitutes a *grade.* To evolve from one grade to another requires net evolutionary progress. Reptiles are cold-blooded; mammals evolved from reptiles and became warm-blooded during the process; they have an entirely new functional ability and have achieved a new grade. Birds have also advanced to the warm-blooded grade, quite independently of mammals. That is, they had a different cold-blooded ancestor than mammals. Therefore, a grade may be polyphyletic—it may be occupied by animals with diverse ancestral backgrounds.

By contrast, a *clade* is a single branch of the tree of life and must be monophyletic (Fig. 8.2). It may be a bare branch of only one lineage, or it may have many branchlets resulting from cladogenetic events. All members of the clade, however, must have the founder of the clade as a common ancestor. To the extent that our classification truly reflects evolutionary history, all genera, families, orders, classes, and phyla represent clades.

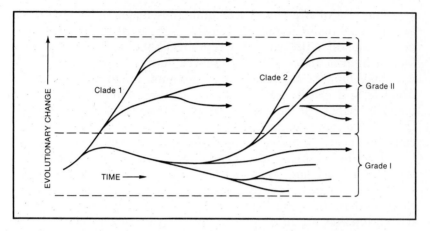

Fig. 8.2. Clades and grades. Clades are monophyletic branches (which may undergo much or little diversification). Grades are levels of functional or morphological complexity which may be monophyletic or polyphyletic. In the diagram, Grade 1 is monophyletic, Grade 2 polyphyletic (having been achieved independently by Clades 1 and 2).

The Origin of Novel Adaptive Types

Genetic changes within populations producing varieties that are different from the parental population have been well studied in the field and laboratory. As seen in Chapter 6, populations that display all stages of differentiation from complete similarity and fertility to incipient genetic isolation to complete infertility can be found in nature.

At high taxonomic levels, however, there are gaps between taxa that are not filled by living intermediates and which in fact may never have been bridged by intermediate forms. For example, take a species of insect—say a butterfly—and human beings. There has never been a butterfly man or any other kind of form intermediate between butterflies and humans. To be sure, both butterflies and humans have descended from a remote common ancestor, most likely a small wormlike marine animal resembling a flatworm. The ancestors of butterflies and humans each underwent a long divergent history involving much anagenesis. These lineages are now so different that even the first division of their fertilized eggs occurs on a different plan. Each represents a separate, major branch of the phylogenetic tree, and like tree branches they are not connected by any crossties.

The tree of life consists of numerous major branches—there are about 25 major living subdivisions (phyla) of the animal kingdom alone, all with gaps between them that are not bridged by known intermediates. The differences between species on different major branches are commonly profound. Furthermore it is unusual to find fossils that are the immediate common ancestors of major branches. In fact, there are no extinct fossil groups known that are the common ancestors of two or more living phyla, and the common ancestral stocks of only a few classes (out of many score) have been found. Most taxa at these high levels appear abruptly in the fossil record, and we do not know their immediate ancestors.

Partly because of this pattern, some evolutionists have thought that the processes of genetic change that we can study within lineages are insufficient to account for *macroevolution,* that is, the appearance of truly novel types of organisms. They felt that there must be some additional undiscovered processes at work to develop new phyla or classes. Today it is generally conceded that the processes producing new species are also sufficient to produce genera, families, or even phyla. However, the various processes differ in importance and outcome, depending upon the circumstances in which they operate.

CLOSER LOOK 8.1 An Explanatory Model of the Origin of Novel Taxa

The biological environment is not a simple continuum. It is patchy, with conditions changing more abruptly between patches than within them. The ocean and the land, or the plains and the mountains, are large patches that pose very different adaptive problems for their inhabitants. Within each large patch are many smaller patches with different local conditions—the shore zone or deep water in the sea, rocky shores or sandy beaches along the shore, above or under rocks on the rocky shore, and so on. This great *environmental heterogeneity* or patchiness provides not only a host of adaptive problems, but also a wide variety of opportunities for many different types of animals to coexist, each adapted to its own distinctive habitat. Even within a given environment, there are commonly many possible life modes; animals living near the surface of the open ocean may be swimmers or floaters, for example. We depict the potential modes of life on the accompanying diagram as *adaptive zones* and *subzones*.

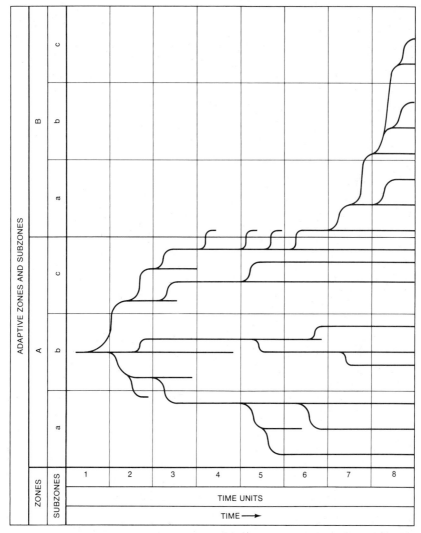

At times *1* and *2*, a clade inhabits zone *A*, but zone *B* is uninhabited by this clade or by any potential competitors. The clade diversifies so as to be represented in all subzones *a*, *b*, and *c* of zone *A* by time *3*; the environmental discontinuity between zones *A* and *B* is so great, however, that the latter zone remains unoccupied. It finally happens that some characters are evolved by the lineage in subzone *c* that happen to be useful for life in zone *B* as well. These are *preadaptations;*

they have arisen entirely for life in zone *A* and only by chance are adaptive to the requirements of zone *B*. Preadaptive characters would normally appear in populations inhabiting subzones most like the zone for which they are preadaptations, of course. Once the ability to invade the open zone becomes established, a pioneering population may appear there, owing to the pressure of population growth and overcrowding in the old zone. However, adaptation of this invader to the open zone will ordinarily be well below average, since the pioneers have been evolved for life in their old zone. Therefore extinction is a common fate of pioneering populations (as for example those that invade zone *B* at times 4 and 5 in the diagram). Selection will nevertheless be at work among the pioneers and in some cases appropriate gene combinations and arrangements can be developed so that adaptation improves. These new adaptations—*postadaptations*—are directly evolved for life in the open zone, and in combination with the preadaptive features form the basis for a new mode of life for the pioneers. Thus a new adaptive type has appeared and may eventually radiate to become diverse and successful in the new zone.

Clearly, the period of actual invasion and the beginnings of postadaptation are critical times in the development of a new adaptive type. The difficulties inherent in a novel adaptive zone will cause a severe winnowing among the pioneers, which might stand little chance for success except that, at the same time, selection is relaxed in other areas. Competition and predation pressures might be significantly reduced within the invaded zone, for example. Nevertheless the population size of the ill-adapted pioneers might be quite small at times, so that genetic drift could come into play to produce a variety of novel gene combinations in frequencies that they could not attain in large populations. Postadaptive gene combinations may appear among such novelties.

The Punctuational Model of Speciation

Commonly, fossil species appear abruptly at a given geological time with no indication of their immediate ancestries (although there is often a record of older allied species, we often cannot trace an allied form into a new species). Perhaps this is not too surprising, since the geographic model of speciation, believed to be particularly common (see Chapter 6), implies that descendant species evolve someplace else than where their main ancestral populations are found. Therefore an ancestral-descendant fossil sequence would not ordinarily be found in the same area. The founding population, perhaps atypical of the ancestral population in any event, might be so small as to escape preservation or at least to escape notice. Founding fossil populations have been identified in a few well-studied cases.

The surprising thing about fossil species is not that they often appear without clear ancestry, but that many of them do not change their morphologies in a significant way throughout their entire durations, which average a few million years for most animal groups. Most commonly, they remain morphologically stable. Since they appear abruptly and last millions of years without change, it follows that most of the morphological change in these species was associated with their origin. In them, evolution proceeded by bursts separated by periods of stasis when little change occurred, rather than by a gradual shift in characters through time. It is as if these species achieved morphological equilibria soon after they arose, permitting them to endure for significant lengths of time even in the face of some environmental changes. They may give rise to new species in turn through cladogenetic events involving the establishment of new morphological equilibria. Niles Eldredge and Stephen Jay Gould, who explored these data explicitly for their evolutionary consequences, call this the mode of *punctuated equilibria.* It may well be the most common pattern of speciation. They point out that trends in morphological change within higher taxa may be recorded in successions of such species. Thus a directional, anagenetic trend may result, not from gradual change within an evolving lineage, but from the accumulation of a number of abrupt changes as new species branch off and as ancestral lineages eventually become extinct. This has been termed *species selection* by Steven Stanley.

Why do so many fossil species display punctuational patterns? Ernst Mayr suggested long ago that species' gene pools become highly integrated because of genetic coadaptation; they are mixtures of genes, the products of which get along very well with each other as well as with the outside world. When conditions change for a population (because the population migrates or the environment changes), selection may dictate that new or rare alleles with rather different products become frequent. If the products of these new alleles do not mix well, the harmonious integration of the old gene pool may be destroyed. Genes at many loci may have to be replaced by alleles whose products function well with the products of the critical new genes. Thus the gene pool undergoes a sort of genetic revolution, being reconstituted into a new gene association to rebuild harmonious coadaptation. Such a revolution could underlie the sudden bursts of change that punctuate periods of stasis in species' histories.

Origin of Genus

Most evolutionary change does not involve the appearance of a totally novel body plan, but rather the development of more-or-less minor variations on a previous anatomical theme. A well-studied case of evolution that has produced new genera from a single founding population occurs among Hawaiian honeycreepers (family Drepanididae, shown in Figure 8.3). As the Hawaiian archipelago is less than 10 million years old, the evolution of these birds must have occurred within this time, and probably within a much shorter period. At some point, a single lineage colonized the islands and clearly prospered. Occasionally but infrequently, inter-island migration has occurred. Populations migrating to new islands have been effectively isolated from their parental populations and subjected to somewhat different conditions. Therefore the honeycreepers evolved different habits on different islands and some became incapable of interbreeding, thus achieving the status of separate species.

Subsequently, some species did manage to redisperse between islands and to come into contact with their close relatives. Although there had been some divergence between species there was naturally much overlap in form and habits as well. Thus individuals of the separate species might compete with each other for some resources when the species occurred together. In such cases, those individuals of two species that are most similar to each other compete most strongly. This could limit their success and thus lower their fitness. Therefore there is a tendency for individuals that are least similar to another competing species to be fitter than those most similar to it, and the competing species may evolve so as to partition rather than to share their common resources. Thus they evolve so as to resemble each other less. The morphological shifts that accompany such evolution are termed character displacement (see Chapter 5). Hawaiian honeycreepers are inferred to have undergone character displacement frequently. In this way distinctive new species arose.

Among the honeycreepers, the inferred pattern was migration, divergence accompanied by the establishment of reproductive barriers, migration reestablishing sympatry, and then greater divergence through character displacement. The shape of the beaks is

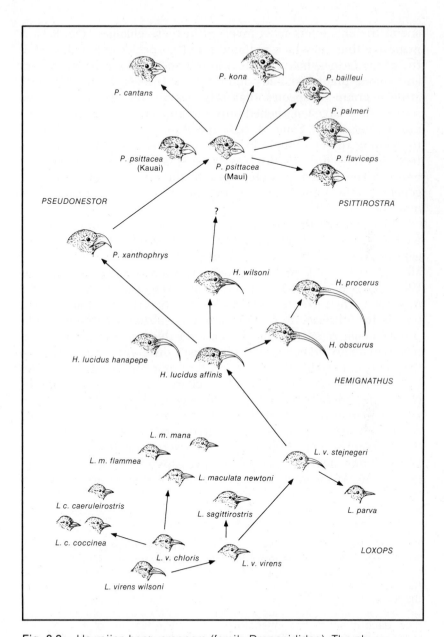

Fig. 8.3. Hawaiian honeycreepers (family Drepanididae). The phylogenetic relations inferred between species of one subfamily are indicated by the arrows. New genera are derived from a species of a preceding genus that leans toward a different mode of life. The new genus represents the successful adoption of this mode, a different adaptive zone. In the case of these birds the chief differences are based on feeding modes and are reflected in beak morphology. (After Bock, 1970.)

one of the characters most involved in these changes (Fig. 8.3). A primitive lineage, which radiated into the species group classed as the genus *Loxops,* had a small thin bill and fed by probing in bark and among plant leaf buds and pods for insects. From this stock arose a group of species that rely heavily on nectar as a food source; they have lengthened bills and are classed in a separate genus (*Hemignathus*). Some of these species evolved stouter bills used in opening insect tunnels in wood, and this led to the ability to feed on the plants themselves (as displayed by species of *Psittirostra*). Here, then, is a clear case of the evolution of distinctive genera by the same processes responsible for the appearance of new species.

If we survey the genera that have been studied in any detail, we find several basic histories that are responsible for creating the gaps between genera. These are summarized in Figure 8.4. First of all a gap is created when a lineage enters a new adaptive zone (see Closer Look 8.1). The insectivorous honeycreeper stocks entered a new zone when they evolved to become herbivores. Another pattern is for a lineage to radiate into new subzones without undergoing a major zonal shift. In this case the descendants of species on different main branches will often look different simply because they have inherited the characters that first separated the species that founded the branches. The characteristic groups can be classed as genera. A third case is for a generalized ancestor to give rise to increasingly specialized descendants, so that the original adaptive zone is progressively partitioned. The descendants of early branches may inherit their characteristic features and be grouped into distinct genera. Finally, gaps may occur because of extinctions. If members of a large group of species are rather ev-

Fig. 8.4. The evolution of genera. *A,* a preadaptation species lineage crosses an adaptive threshold into a new adaptive zone and subsequently diversifies; the cluster of species that it produces inherits the features of the colonizing lineage and is separated from the ancestral stock as a new genus. *B,* divergence of sister lineages with accompanying cladogenesis leads to two clusters of species, each of which has inherited the distinctive features of their founding sister lineage. They are therefore classed as genera. *C,* a variable, generalized lineage partitions its niche into a number of more specialized niches as it becomes diversified; the descendants can be clustered into genera on the basis of features inherited from their founding sister lineages. *D,* a pattern of extinction creates morphological gaps that allow the clustering of species into genera.

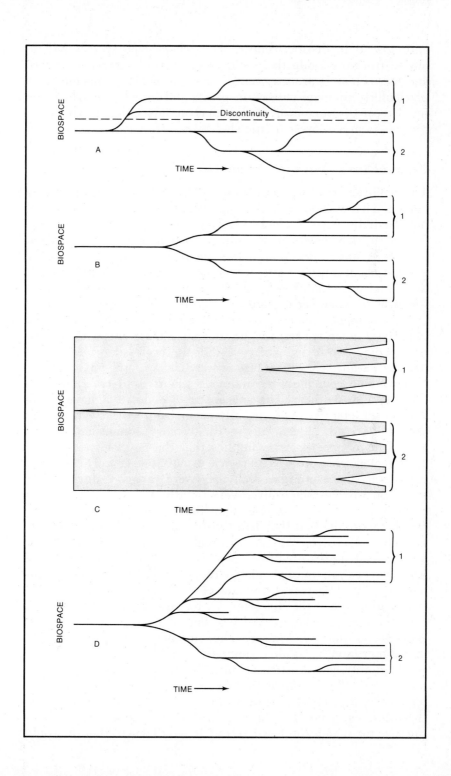

enly spaced in distinctiveness, then when some species are removed from the group there will be clusters of surviving species, and these clusters are sometimes classed as genera. This may be a particularly common pattern in plants and insects, for which the fossil record is very poor. If the survivors are grouped with care, the genera may represent true clades.

Origin of Taxa Above the Genus Level

The evolutionary origins of taxa in the higher categories are poorly known. The possibility of the gaps being caused by extinction of intermediates becomes increasingly unlikely. A possible explanation of the gaps between, say, classes is that they have originated as gaps between genera, and subsequent anagenetic changes have brought about an increasing morphological distance between their descendants, so that they are grouped in very different higher taxa —classes or whatever is appropriate. This implies that taxa in increasingly higher categories simply require more time to develop than those in lower categories in order to achieve their greater distinctiveness. However, when we consult the fossil record this is not the pattern we find. Most orders, classes, and phyla appear abruptly and commonly have already acquired all the characters that distinguish them. The anagenetic episodes that have led to very novel adaptive types usually occur before these types enter the fossil record. Indeed, this is true of most families and genera, and of many species. This implies that the novelties evolved in some region wherein fossils are not usually preserved (or at least not usually discovered), or that they were very rare during their evolution, or that they evolved very rapidly.

It was once common to explain the lack of ancestors of many taxa by assuming that they had lived in areas that happened to be unknown. The deep-sea and the higher mountain environments are not well represented by fossils, for example, and if novel taxa had first evolved in such places their primitive members might never be found. However it is now clear, from studies of the adaptive significance of the characteristic features of taxa, that many of them evolved in some of the best-known ancient environments, such as on shallow seafloor. The "unknown environment" explanation cannot be applied generally.

Certainly, many early members of novel lineages may have been rare; the adaptive zone model certainly suggests that pioneering populations could be small. However, if it is assumed that these early novel lineages evolved at the same rates as have lineages for which the fossil record is well known, then it would take many hundreds of millions of years to develop the great degrees of morphological difference that they exhibit. It is not reasonable to expect these lineages to have small population sizes (and certainly not to be poorly adapted) for such a long time. Furthermore, even a rare lineage, if skeletonized, should appear in the fossil record sooner or later if it persisted for such a long period. We are forced to the conclusion that most of the really novel taxa that appear suddenly in the fossil record did in fact originate suddenly.

As we have seen, even species commonly appear abruptly, which can be at least partly explained as an abrupt shift in coadaptation affecting both the genetic and morphological levels when they face new environmental conditions. Such shifts may account for the development of species and genera such as the Hawaiian honeycreepers. However, this is probably not the whole story of abrupt speciation, and it will usually not account for the origin of truly novel morphologies as in higher taxa. Major changes in the architecture of organisms must be due chiefly to changes in the patterns of gene expression. This involves an alteration in the pattern of gene regulation rather than the addition or substitution of new structural genes. Relatively abrupt shifts in morphology at the species level can also originate through changes in gene regulation.

CLOSER LOOK 8.2   The Punctuational Model of
Speciation: Some Evolutionary Consequences

If most morphological changes are associated with speciation events during the branching off of new lineages, as implied by the punctuational model, then relatively few changes occur within gradually evolving lineages that do not give off branches. There are many interesting consequences of this situation. Steven Stanley has explored them in some detail.

For example, since it is the more diverse clades that have undergone more speciation events, it is to be expected that more evolutionary novelties will be introduced in them than in clades with few species. Indeed, since most novelties are associated with the branching pattern of speciation, we should expect clades with histories of frequent speciation to be the most divergent morphologically and to furnish truly novel adaptive types. On the other hand, clades that have not speciated much will appear to be evolutionarily stagnant. The gradual changes within evolving lineages (*phyletic evolution*) unaccompanied by any significant genetic reorganization, will often be ineffective in producing novel morphologies. Phyletic evolution may not often be involved with shifts from one adaptive zone to another, but chiefly with adjustments centering around the ancestral adaptive modes. Lineages known as "living fossils"—brachiopods such as *Lingula,* which appear essentially unchanged since the Ordovician, and lizards such as *Sphenodon,* which have survived since well back in the Mesozoic—can be explained as lineages that have undergone little speciation and therefore little morphological change. They happened to have escaped extinction by being unusually adaptable, or by surviving in refugia wherein environmental changes have been minimal.

A similar line of reasoning has been applied by Stanley to help explain the advantages of sex. Since sexually reproducing females have their fitnesses diluted when their eggs receive a 50 percent genetic contribution from males, it is a puzzle as to why selection has spread sexuality so widely. Species with parthenogenetic females (which do not require fertilization to reproduce) are well known in many phyla, but sexual reproduction is nevertheless overwhelmingly predominant. One possibility is that since only sexually reproducing lineages can speciate in the usual sense of the word, only they can produce a rich supply of novel adaptive types through branching. They are thus better able to weather the effects of extinction and to occupy rapidly any empty adaptive zones that become available. They have therefore become much more common than asexual groups.

It is interesting to keep the punctuational model in mind when reading through Chapters 10 and 11. It may provide insights into the historical patterns of evolution.

Gene Regulation and Clades and Grades

The effects of gene regulation can be easily appreciated by considering the differentiation of the cells in an animal body. Even the simplest living animals have several different types of cells, while the more complex may have 200 or more. Each type of cell has a different morphology or cell phenotype (Fig. 8.5) as well as a different function. Muscle cells and nerve cells look quite different

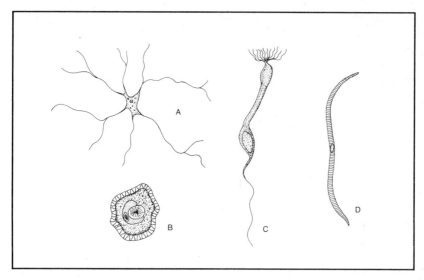

Fig. 8.5. Some metazoan cell types. *A*, a connective cell from a snail. *B*, a pharyngeal cell from a tunicate (a sea squirt). *C*, a sensory cell from a snail. *D*, a muscle cell from a jellyfish.

and certainly perform different tasks. The cells are associated in tissues, such as nervous or muscle tissue. Tissues in turn are associated in organs, such as a muscle, a brain, or a liver. Organs are associated in organ systems, such as the nervous system or digestive system, which have major physiological responsibilities. Thus the entire organism is composed of a harmoniously integrated and functioning set of cells, tissues, organs, and organ systems.

It is the development of an organism from an egg that is encoded in the genes. The numbers and kinds of cells that appear, their association in tissues, the construction of organs—all are encoded and inherited in the genome. The actual work of development is largely done by enzymes acting with materials (food) imported from the environment. The cells have all originated by simple cell division (mitosis) from a fertilized egg; the chromosomes are duplicated precisely in all cells (except animal gametes and plant spores and pollen which have undergone meiosis). Although all cells possess the same genes, different sets of genes are active in different cells, producing different products that form different phenotypes. The activation and deactivation of sets of genes is carried out through the regulatory part of the genome. The number of cells of each type produced and the position of the cells within the anatomy of an organism must also be controlled so as to conform to the requirements of organ function and of the body plan of the organism. Again, it is generally in the regulatory portion of the genome that this information resides.

A regulatory gene may control the expression of numerous structural genes. It is reasonable to suppose (and there is supporting evidence) that regulatory genes control integrated sets, or *batteries,* of structural genes, the products of which form biosynthetic pathways all serving a particular requirement. The batteries of genes need not be mutually exclusive. Indeed, some gene products must be employed in many biosynthetic pathways, so those genes are activated in many different cells and under many different circumstances in the same cell, in association with various sets of genes. The activating substance, perhaps a hormone or a protein, may be produced by previous biosynthetic activity. The products of one gene battery may be programmed to effect the switching on of the next gene battery so that the entire time sequence of gene activity in a cell be orderly; the entire process being started by fertilization. The regulatory circuits for the different cell phenotypes must be encoded in each cell, but normally only the circuits appropriate to that cell function. This normal situation may be upset by certain conditions. In mutations induced in *Drosophila* flies by X-radiation, for example, legs or wings may be developed in body segments where they are normally absent. The many genes required to produce these complex structures have been switched on where they are supposed to be nonactive.

When organisms are evolving so as to increase in complexity—as when they are developing new cell types, functions, and organs—it is clearly necessary that the genetic regulatory apparatus evolve. For example, the evolution of a multicellular organism from a unicellular one required new regulatory functions. The earliest multicellular forms required, first of all, a set of genes that encoded the growth and metabolism of a cell (just as in a unicellular organism). In addition, they needed to specify the required number and positions of cells, and to regulate cell interactions. Furthermore, multicellular organisms with several cell types need to encode several patterns of cell growth and metabolism rather than one. Thus an increase in the complexity of organisms involves a growth in the complexity of the regulatory part of the genome. No doubt growth in the number of structural genes often occurs as well. Finally, the substitution of new alleles at structural loci must accompany most or all evolutionary change in order to maintain appropriate levels of physiological response as environmental interactions are altered.

It is not yet known just how important a role is played by the addition of new structural genes during the evolution of complexity. There is evidence to suggest that a majority of the structural gene functions in the cell of a higher organism such as a human are also required in the cells of unicellular forms; after all, every cell has certain basic metabolic functions of energy uptake, respiration, growth and maintenance of physiological activity, and so on. The chief differences between organisms reside not so much in the kinds of enzymes encoded by structural genes but in the pathways by which the enzymes are produced and associated so as to produce different cell phenotypes. One could make a rough analogy with the construction of a building with bricks, wood, and certain other materials. The same materials may be used to construct a cottage, a sprawling ranch home, or an apartment building; the plans are, however, quite different. In the long run of evolution, there is no doubt that many new structural gene loci *have* evolved. The contributions of new loci, relative to new regulatory arrangements, are not yet known.

Gene Regulation and New Body Plans

If the evolution of new grades does involve the regulatory genome, what of the evolution of new body plans or of body-plan modifications within an established grade? Horses and rhinoceroses are at the same grade, having diverged from a common ancestor some 55 million years ago. Their distinctive anatomies suggest regulatory gene evolution. One way to demonstrate this is to contrast the rates of evolution of structural genes in mammals, which display great morphological diversity, with those in a group such as frogs, which do not display much morphological differentiation.

Each of these groups contains thousands of species (Table 8.1), yet frogs are so compact a group anatomically that they are classed in only one order, while mammals are divided into at least 16 orders. Differences in the structural genes coding for the protein albumin have been studied in hundreds of species of frogs and mammals by immunological techniques (see Chapter 7). The albumin gene seems to have evolved no faster in mammals than in frogs.

Such data as are available for other genes in these groups (from studies using electrophoretic methods and even amino acid sequencing) display the same pattern. Some mammal species that are placed in separate orders are as similar in their hemoglobin sequences as some frog species that are placed in the same genus. The great morphological diversity of the mammals has been achieved, not by allele substitutions at structural gene loci alone, but by evolution of the regulatory portion of the genome as well.

This is not too surprising, actually, since as we have seen, anatomical patterns are expected to reflect patterns of gene expression as controlled by regulatory genes.

Human beings are no exception to the rule of large regulatory gene involvement in mammalian evolution. Chimpanzees and humans are quite similar to each other when their structural genes are compared. About half of their genes are identical when compared by electrophoresis, and proteins of both species that have been sequenced have proved to be closely similar. This is roughly the amount of genetic difference found between sibling species of vinegar flies, which are morphologically nearly identical, at least to the human eye. Yet chimpanzees and humans have quite distinctive appearances. Their impressive morphological differences have resulted chiefly from the evolution of the regulatory gene system after they diverged from their last common ancestor.

Although we still have little experimental evidence on regulatory gene systems, what we have is sufficient to support inferences as to the evolutionary effects of these systems. First and perhaps most importantly, a single mutation at an important regulatory locus could switch the gene activities of tens or hundreds of structural genes, causing them to operate in a new context relative to the rest of the genome. They can switch on earlier or later during development, altering the pattern of cell differentiation or producing more or fewer cells of certain types. They can also cause cell division in different directions or along new growth gradients so as to produce different tissue geometries. The effects of one or a few mutations to regulatory genes could thus be a fairly spectacular change in morphology.

Table 8.1
RATES OF EVOLUTION IN SOME FEATURES OF FROGS AND OF PLACENTAL MAMMALS
(After Wilson, 1976.)

Property	Frogs	Placental mammals
Number of living species	3,050	4,600
Number of orders	1	16–20
Age of the group (millions of years)	150	75
Rate of morphological evolution	Slow	Fast
Rate of albumin evolution	Standard	Standard
Rate of change in chromosome number	Slow	Fast
Rate of change in number of chromosomal arms	Slow	Fast

Evolution of Novelty

The evolutionary change in human ancestry provides an example of the sorts of effects arising from changes in the timing of gene expression during ontogeny. During the rise of the human body style from apelike ancestors, fitness was promoted by a constellation of anatomical modifications that permitted erect locomotion (see Chapter 12). This trend was presumably highly adaptive to a new mode of life, which for these protohumans was probably hunting and gathering of food in bands along forest margins or in relatively open country. It happens that juvenile apes have rather erect postures and other features—flat faces, high foreheads, and so on— that developed among the adult ancestors of humans. Presumably the young of the apelike human ancestors also had these juvenile features, which happened to be preadaptive to the new mode of life. Selection did not have to laboriously assemble a correlated set of novel genes to solve many of the problems associated with the prehuman shifting of life modes. It merely had to favor the extension of those preadapted features, already present in the juvenile portion of the life cycle, into adult life stages (Fig. 8.6). The regulatory circuitry for these features was already in place.

Fig. 8.6. Chimpanzees, baby and adult, illustrating that infant apes resemble adult humans more closely than do adult apes. Evidently humans have acquired many of their adult characteristics by neoteny, deriving them from the juvenile characteristics of their ancestors. This notion is supported by the fossil record. (After Naef, 1926.)

That characters can shift from one part of the life cycle to another was inferred on morphological grounds more than a hundred years ago. There are many terms describing the different types of such shifts, but few of them are in common use. These character shifts are complicated because the key reference event in the life cycle is the onset of reproduction, and this event itself is commonly shifted into earlier or later developmental stages. In general, developmental events can either be accelerated (can appear earlier in ontogeny) or retarded (can appear later). If the onset of sexual maturity is retarded, then all other characters appear to accelerate (since they then appear earlier in relation to sexual maturity), and vice versa. One set of terms to describe different combinations of developmental events is given in Figure 8.7. *Neoteny* is the most commonly described case of developmental shift. The shifting of the expression of some juvenile ape characters into human adulthood, reviewed above, is a case of *neoteny*.

Perhaps a reason that neoteny occurs so commonly is that it represents a way to escape the specialized features that tend to appear late in ontogeny. Organisms start out as single cells and their most complicated morphology develops last in most cases. It is difficult for selection to discover a pathway leading from an extremely specialized adult that is "locked in" to a complex of habits

Developmental timing	Occurs earlier	Occurs later
Somatic (nonreproductive) features	Acceleration	Neoteny
Reproduction	Progenesis	Hypermorphosis
	☐ Recapitulation	☐ Paedomorphosis

Fig. 8.7. Terms applied to evolutionary changes in developmental timing. When characters are somatic (nonreproductive) features and they shift to occur earlier in development they naturally appear earlier relative to the onset of reproduction. The same effect is also found if reproduction shifts to occur later in development. These two sorts of change are termed *recapitulation*. When somatic characters shift later or reproduction begins earlier, then the somatic characters appear later relative to reproduction. These types of change are termed *paedomorphosis*. (Usage after Gould, 1977.)

because so many genes are coadapted to support them. Through neoteny, simpler characters found in juvenile stages become adult characters; the former complex adult characters can then be discarded from the life cycle altogether. A lineage can thus alter its evolutionary direction.

QUESTIONS FOR DISCUSSION

1. Why do organisms within evolving lineages tend to become more complicated over long periods of time?

2. Why don't all descendant lineages become more complicated than their ancestors?

3. How does evolution of regulatory genes raise the possibility of speciation by saltation?

4. Criticize the thesis that ontogeny recapitulates phylogeny.

5. Describe the circumstances under which a genome might respond to selective processes by changes in the regulatory genes rather than in the structural genes.

Recommended for Additional Reading

Gould, S. J. 1977. *Ontogeny and phylogeny.* Cambridge, Mass.: Harvard University Press.

A new review of the history and problems of individual development in evolution. Advanced.

Mayr, E. 1970. *Populations, species and evolution.* Cambridge, Mass.: Harvard University Press.

Rensch, B. 1960. *Evolution above the species level.* New York: Columbia University Press.

Simpson, G. G. 1953. *The major features of evolution.* New York: Columbia University Press.

Works by major contributors to the synthetic theory that deal at least in part with transspecific evolution. Moderately advanced.

9

Evolution and Ecology

Evolution has produced the varied organisms, living and extinct, that have appeared in over three and a half billion years of life history. This is chiefly owing to the ability of natural selection to adapt organisms to environments, both physical and biological, that they encounter. The relations between organisms and their environments, so basic to evolution, are also studied by the science of ecology. Evolution is in large part an ecological process—"ecogenetic" to be more precise.

The Architecture of the Biosphere

The ecological systems in which living organisms are found are so complex and varied as to be beyond the comprehension of any one person. Since organisms have been formed by evolutionary processes, the ecological systems are themselves products of evolution in great measure. It is clearly useful to organize or classify these systems in some way so that they can be understood in this light. When taxonomists are faced with the problem of classifying the vast natural diversity of organisms they resort to a hierarchical scheme (see Chapter 7) that has permitted great advances in our understanding of evolutionary patterns and pathways. Similarly, ecologists have classed the ecological systems into a hierarchical scheme, with increasingly inclusive ecological units on each higher level. The main outlines of such a scheme are depicted in Figure 9.1.

Individual Organisms

Individual organisms comprise the lowest level of the ecological hierarchy. The environments that different individual organisms face are extremely varied. For example, for some bacteria, the environment is the interior of a cow's rumen, while for the cow the environment is pastures or rangelands. Ecological relations are carried out by physiological and behavioral responses to the environment, ultimately based on the information encoded in the genotype. Genotypes respond to environmental factors during ontogeny, so that different phenotypes may be produced from the same genotype under different circumstances; there is a range of potential reactions inherent in a given genotype.

We will call the reactions that are actually realized by an individual phenotype its *functional range,* which is a portion of the reaction range inherent in its genotype. The functional range of an individual is bounded by the limits of its tolerances and requirements (Fig. 9.2). For any individual, there is a certain temperature

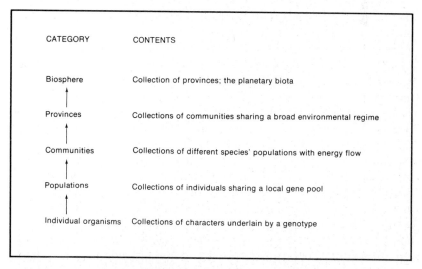

CATEGORY   CONTENTS

Biosphere   Collection of provinces; the planetary biota

Provinces   Collections of communities sharing a broad environmental regime

Communities   Collections of different species' populations with energy flow

Populations   Collections of individuals sharing a local gene pool

Individual organisms   Collections of characters underlain by a genotype

Fig. 9.1   The hierarchy of ecological units. The categories form a nested hierarchy containing increasingly inclusive units. The units are actual individuals, populations, communities, and so on.

range within which it may function. The individual requires food of certain types in certain amounts, and it also needs a place of refuge from predators.

Organisms may exist wherever the environmental factors lie within the limiting bounds of the functional range. When one or more factors are out-of-bounds they are termed *limiting factors*. Fitness is a measure of how well an organism functions relative to other individuals belonging to the same species in the environment that it actually encounters.

Populations

Populations are composed of individuals of the same species that share a local gene pool. Members of a given population have a greater chance of breeding with individuals in the same population than with those from other populations. Thus there is much intermixing of genes within a population and less gene flow between populations. Populations have ecological characteristics that individuals cannot possess. For example, individuals are born once, exist in a certain place at any time, and die once. Populations, however, have birthrates, patterns of dispersion, and mortality rates—all based upon the reproductive regimes, functional ranges, and life expectancies of the individuals that make them up. Other such group properties include density of individuals, rates of emigration and immigration, age structure, reproductive potential, and of course the frequencies of genes and gene arrangements in the gene pool. Units in each succeeding higher level of the ecological hierarchy display properties that are not found in units on the preceding level, because they are formed by measures or processes involving aggregations of those lower units. These are termed *emergent properties*.

No two individuals will have precisely the same functional ranges. The functional ranges found within a population embrace those of all its individuals and may be considerably broader than the tolerances of any individual. This will depend on the range of phenotypes that develop from the range of genotypes assembled from the gene pool. The environment required by a population has properties that the environment of an individual does not. For example, a population requires a certain space and some diversity of habitats to accommodate a number and variety of individuals in order to permit the expression of the emergent population properties. The population functions include those of the individuals plus the emergent properties of the group. This concept of population functions is usually termed the *niche* (Fig. 9.2).

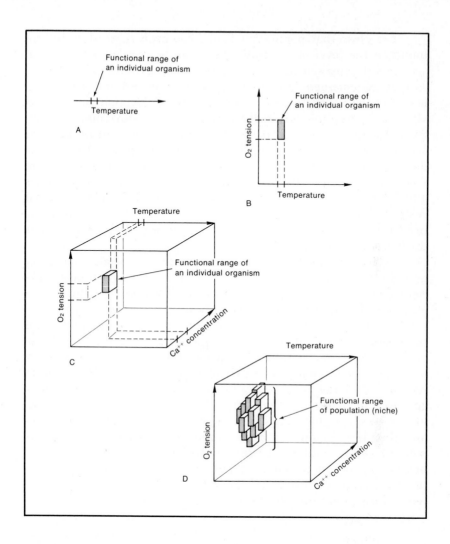

Fig. 9.2 Functions of some ecological units. The tolerances of requirements of an organism for any given factor can be represented as a segment of a line that scales that factor (*A*); for any two factors as a plane geometrical figure (*B*); and for any three factors as a solid figure (*C*). Real environments contain many more factors, so the three dimensional figure (*C*) is a much simplified way of representing the functional range of an individual, which can extend into as many dimensions as there are factors. In *D,* the functional range of a population is represented as the ranges of all its individual members. Actually there are even more environmental factors and therefore dimensions in *D* than in *C,* owing to emergent factors, but they cannot be diagrammed. The population functions are called the *niche.*

It is clear that population functions, even those that are not found on the level of the individual, can be regulated by natural selection. Group selection is not a dominant evolutionary process, so that the emergent properties of populations are not usually selected at the population level but arise from selection among individuals. An example will make this important point much clearer. An important population parameter (and an emergent property at the population level) is the rate of reproduction. It takes a great deal of energy to reproduce, energy that must be acquired from food, so that populations that have high reproductive rates require a high level of food resources. Furthermore, such populations must restrict their energy output for other functions (such as escaping predators) when they use large proportions of their energy budgets for reproduction.

Suppose that a population suffers high mortality of early adults. This is the case with many fish populations that are exploited for food by humans. In this case, fish that reproduce early and that have large broods are highly fit. The individuals that reproduce later might actually be more robust, have longer natural life expectancies, and be better competitors or hunters, but these attributes will not contribute much to their fitness because they are fished when young. The exploited population will contain an increasingly large percentage of individuals that reproduce early, and thus it will evolve a higher reproductive potential. If exploitation stops, however, the offspring of the early reproducers may then be outcompeted by more robust, longer-lived individuals which reproduce later but which will then become the fitter. The percentage of these latter types will then increase and reproductive potential will fall. Population properties have been altered through natural selection for individual fitness.

Communities

Communities are collections of different species' populations that occupy a distinctive range of habitat conditions. The inhabitants of a sandy desert basin in the American Southwest form a community. Small mountains bordering the basin will possess a different habitat association and therefore will have a different biota, representing another community. Along a seacoast, exposed rocky shores support a distinctive association of marine life, sandy beaches another, and the muddy flats of quiet bay shores still another. These are communities. Within such associations, populations interact to create the emergent properties at the community level of the ecological hierarchy. Common sorts of interactions are described in Table 9.1.

Table 9.1
TYPES OF SPECIES INTERACTION WITHIN NATURAL COMMUNITIES

	One population neutral		
Population 1	0	0	0
Population 2	0	–	+
	Neutralism	Amensalism	Commensalism

	One population negative *(the other not neutral)*	
Population 1	–	–
Population 2	–	+
	Competition	Parasitism, predation

	One population positive *(the other not negative or neutral)*
Population 1	+
Population 2	+
	Mutualism, protocooperation

The transfer of energy in the form of food is a particularly characteristic community property. Plants or single-celled organisms that derive energy from the sun to create living matter through photosynthesis lie at the base of the energy chain. It costs energy to synthesize the molecules in plant bodies, and many of them will release energy when broken down. Herbivores derive their energy from eating photosynthetic organisms. Some carnivores derive theirs from eating herbivores, while large or fierce predators may eat smaller or less active ones. Energy thus passes from the sun to plants to herbivores and to the carnivores that eat them and so up the chain of prey and predator species (Fig. 9.3). Energy resources are termed *trophic resources,* and each level in the prey-predator chain a *trophic level* (see Chapter 5).

The photosynthesizing organisms at the base of the trophic chain generally do not require biosynthetic compounds in order to grow. They are termed *autotrophs* ("self-feeders"). Animals or single-celled forms that do require organisms or biosynthetic products as food are *heterotrophs* ("different-feeders"). The rate at which autotrophs produce new protoplasm is termed *primary productivity,* and this production forms the basis of support for all the world's heterotrophs. However, much of the energy that is assimilated by plants is used by the plants themselves, in respiration and reproduction. Therefore only a fraction of the energy used by plants (often 15 to 20 percent) is actually available to herbivores.

Similarly, herbivores respire, forage for plants, reproduce, perhaps even frolic, and this all requires energy. Only 15 or 20 percent of the energy that they assimilate is available to their

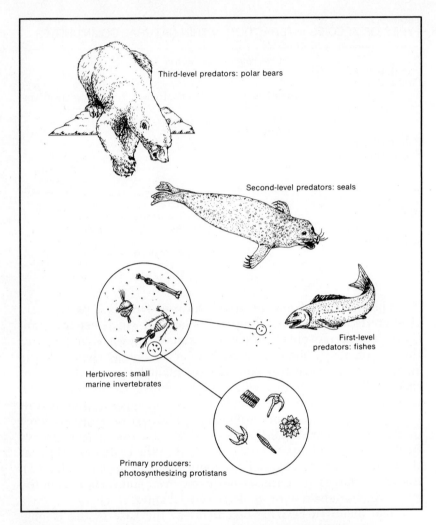

Fig. 9.3 Trophic levels in an arctic marine community. The auto-trophs are minute floating unicellular organisms, eaten by in-vertebrate herbivores, which are eaten in turn by fish, and so on up the chain of predators to the polar bear, which bridges into the ter-restrial environment.

predators. Thus a lesser amount of life can be supported at each higher trophic level. The amount of life present is commonly measured as *biomass* (living weight). It is clear that the biomass must decline by 80 percent or more at each trophic level, so that when the levels are plotted by biomass they form a trophic pyramid (Fig. 9.4), narrowing rapidly towards the top. Because of this rapid re-

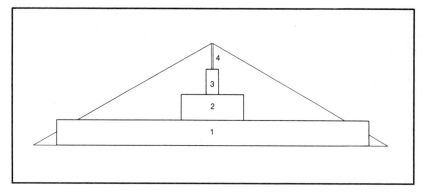

Fig. 9.4   The trophic pyramid. A reduction in biomass occurs at each successive trophic level. In the system pictured, 80 percent of energy intake is employed for metabolism, and 20 percent is stored and thus available to predators, on each trophic level. It is easy to see why the number of trophic levels is limited.

duction in the energy available at each higher tropic level, there are limits to the number of levels that are possible in nature— seven is the usual limit, although many communities possess only two or three. The number of species occupying any given trophic level also varies among communities, some being rich in species and others poor.

Instead of catching living prey, many species feed chiefly on organic remains that, while often somewhat decayed, still retain a supply of energy-furnishing compounds. A large number of species feed on plant remains. On land, a numerous and diverse insect fauna feeds on decaying plant litter on forest floors, and earthworms derive much of their food from plant humus. In the oceans, a rich bottom fauna of invertebrates feeds on the remains of large plants, while many species that burrow into the seafloor mine the fine organic detritus there. The detritus is derived chiefly from the small floating marine photosynthesizers (such as diatoms) with contributions from the decay of just about everything else living in the sea, plant and animal alike. Scavengers also exist in nearly all habitats, feeding on animal carcasses.

All these feeding types that eat dead organisms or their degraded products have a special place in communities, different from the plant-herbivore-predator chain. That chain can be termed the *consumer* chain. Feeding on the chain that leads from dead plants or animals to scavengers, and from detritus to detritivores, does not put predation pressure on the populations that serve as food; the individual prey have already succumbed to some other source of mortality. In effect this latter chain serves to retain within the community a lot of energy (still locked up in detritus) that would otherwise escape. This latter chain is therefore called the *recuperator* chain; it saves energy losses and supports a rich assemblage of animal populations, which often provide food items for predators on the consumer chain (Fig. 9.5). The diversity of animals within communities is thereby enhanced greatly. In some communities the recuperator chain is more important in energy transfer between trophic levels than is the consumer chain.

The environmental aspects of communities are much broader than those of any given population niche, because a range of conditions must be present to support all the constituent populations. Like individuals and populations, communities have functional ranges. They are composed of the functional ranges of all the populations that constitute the community, plus all the functions associated with the emergent community properties. The functional range of the community is the *ecosystem* (Fig. 9.1). This term in-

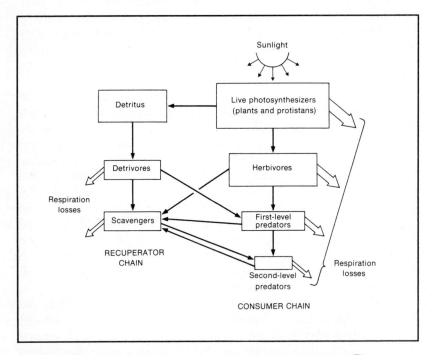

Fig. 9.5 Consumer and recuperator chains in an ecosystem. The relative sizes of feeding types are not to scale; the proportion of biomass and the number of species found on the two chains varies enormously between communities.

cludes the organisms and the physical-chemical aspects of their environment as well.

Provinces

Biotic provinces are regions of the world, usually of subcontinental extent, within which communities maintain characteristic species compositions. For example, the cool temperate shores of western North America include a number of distinctive communities, such as those on rocky shores, sandy beaches, and muddy tidal flats. Around Tasmania a similar climate is found and a suite of communities is developed that appears very similar in general aspect to the suite of communities in western North America.

However, whenever two similar communities are compared in detail, the species found are totally different. What is similar is the setting of climate and habitat. Species with analogous habits in the different communities look very much alike—they are similar adaptive types—and the numbers of species and their trophic proportions are closely similar in the different communities. However, the barriers to migration are so great between these two regions, separated as they are by the Pacific Ocean and lying on opposite sides of the tropics, that species cannot migrate from one to the other. The regions lie in different provinces. If migration became possible, it is likely that many species could become established in both provinces since conditions are so similar. It appears that the similarity of conditions in the two different areas has evoked similar adaptive responses leading to great similarities on the individual, population, and community levels. Organisms introduced by humans across natural barriers often flourish in new provinces; rabbits and wild oats, both of European origin, flourish in Australia and western North America, respectively.

Climatic differences can create barriers to migration and climatic changes commonly form provincial boundaries. The rich tropical rain forests of Central America give way to the subtropical communities of northern Mexico and the southern United States, and these to the cool-temperate rain forests of the Northwest, and eventually to the tundra of the Far North. Plants and animal species adapted to a given climatic regime are replaced by other species as one climate gives way to another (Fig. 9.6).

In the sea an analogous situation is found. On the shallow marine shelves that border the world's continents, and in shallow waters off island coasts, the marine biota can be classed into over 30 distributional regions (Fig. 9.7). Some of these regions are sepa-

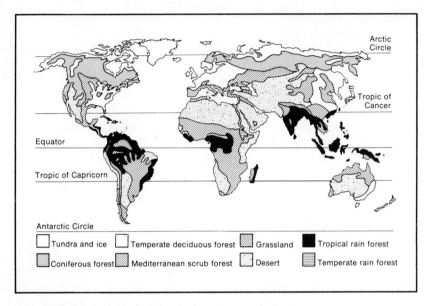

Fig. 9.6 Major floral regions of the land. (After G. G. Simpson, C. S. Pittendrigh, and C. H. Tiffany, 1957.)

rated by continents (as are the east and west coasts of the Americas) or by deep ocean basins (as are the eastern American coasts and the western coasts of Europe and Africa). Many shallow marine species cannot migrate across deep, wide ocean, or across land of course, so they are restricted in their distribution to their native coastlines. Along a given coastline, different distributional regions may be developed if there are corresponding climatic changes. The tropical communities of the Gulf of Panama are replaced northward in the temperate waters of Pacific Baja California by different species associations, even in analogous environments. There are six distinctive provinces between the equator and the Arctic Ocean in the northeastern Pacific Ocean (Fig. 9.7).

Thus today, latitudinal chains of climatically distinct marine provinces border the north-south trending continents. Owing to topographic barriers there is a different chain on each side of each major ocean. The latitudinal climatic changes are restricted to relatively shallow water overlying the continental shelves and shallow island slopes. The great bulk of the seafloor, the truly deep-sea environment, is relatively homogeneous climatically. Deep-sea species tend to have relatively restricted depth ranges but often range very widely geographically, so that at a given depth, similar associations are found over thousands of kilometers. Deep-sea provin-

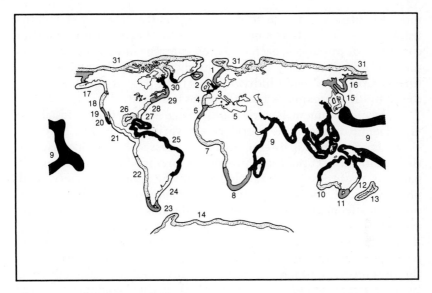

Fig. 9.7 Major biotic provinces of the marine realm, found in shallow water—chiefly on continental shelves, where about 90 percent of marine animals live. (After Valentine, 1973.)

ciality is therefore lower than that in shallow water. When submarine ridges form restricted basins, the deepest-water species cannot cross the shallower basin sills. There is therefore a certain amount of *endemism* (a certain number of native species) within deep basins.

The emergent properties of the provincial level have not been much studied as such, but they may be of considerable evolutionary importance. These properties include the number and relative sizes of communities and the pattern of population sizes, numbers, and sharing among different communities. Other important properties that vary from province to province include the total number of species present and the regime of primary productivity, both usually correlated with the climatic regime that characterizes each provincial environment. We may recognize the functions at the provincial level as the *provincial ecosystem*.

The Biosphere

When all provinces are grouped together they comprise the entire planetary biota, the functional aspect of which is the world ecosystem or *biosphere*. The biosphere functions on the solar energy received at the earth's surface, which supports the biomass of au-

totrophs, utilized in turn to support a smaller biomass of heterotrophs. The biosphere is a nested hierarchy of ecological units. It is
composed of provinces, each of which contains numbers of communities. Each community contains numbers of populations, each in
turn composed of many individuals. At the base of this hierarchy,
the individual organisms face their environment.

The environment is varied and changeable, and individual environmental interactions therefore are variable also. The outcome
of the environmental interactions is expressed by the fitnesses of
the organisms. Population properties are modified as changing
patterns of fitness lead to change in the quality, variety, and
proportions of individual organisms. Communities derive their
properties from their component populations, and as the environments change and populations evolve, communities may change in
the kinds, number, and proportions of the populations that comprise them. As the communities change, they change the aspects of
the provinces to which they belong. Provinces vary through time
in the kinds, numbers, and proportions of their communities. Finally, the entire biosphere changes as the kinds, numbers, and proportions of the provinces vary with shifts in the environment and
in changes in the functions of the world's organisms. The structure of the biosphere is determined by the nature of the environment and by the adaptations and evolutionary potential of the
organisms. It is an ecological structure but it is achieved through
evolutionary response to environmental opportunities.

Adaptive Strategies

Evolution of the functions of individuals and populations was the
central concern of the Darwinian model of natural selection and
remains a major concern of modern evolutionary theory. We have
dealt extensively with these topics in previous chapters. Here we
wish to review some aspects of functional ranges and niches that
are particularly important to a consideration of the development of
the structures and functions of emergent properties of ecological
systems.

All real environments are somewhat patchy in space and vary
somewhat through time. Yet there are some environments that are
relatively homogeneous in space and stable through time (as in the
deep-sea), while others are relatively heterogeneous in space (coral
reef complexes) or unstable in time (continental interiors in the
cool temperate zones). Organisms living in different places must
cope with different sorts of environmental variations. Adaptations
to patterns of environmental variability are called *adaptive strate-*

gies. Clearly, an individual living in a nearly constant physical environment requires only a narrow physiological tolerance, while an organism facing annual swings of 10°C must have a wide physiological tolerance for temperature. The ranges of tolerance developed by organisms in order to accommodate the environmental changes they usually face are components of their adaptive strategies.

Consider a predator that must hunt for widely dispersed prey. If it inhabits a very homogeneous environment it can travel widely in search of prey and still possess only a narrow range of physiological tolerances. On the other hand if it inhabits an environment extremely patchy with wide differences in conditions between patches, it must have broad tolerances in order to cross the patches that separate its prey. Here again the ranges of tolerance constitute adaptive strategy components.

One of the more difficult problems in ecology is to characterize the pattern of environmental variability. The concept of *environmental grain* has been developed in order to help visualize animal-environment interrelations associated with such patterns (Fig. 9.8). Recall the model of adaptive zones, in which the environment is considered as composed of large environmental patches (land, sea) subdivided into smaller and smaller patches; finally we can visualize relatively small areas of nearly uniform habitat conditions as basic patches, like tiles in a mosaic. Between these patches conditions change more abruptly than within them. Naturally, many species have their distributions limited by the boundaries between patches, since it is there that the environmental changes are the sharpest. Indeed, some species have such narrow tolerances that individuals must spend their entire lives within a single environmental patch type, unable to tolerate conditions in any neighboring patches. To them, the environment appears as very heterogeneous and "coarse." These species are said to have *spatially coarse-grained* adaptations.

By contrast, there are species that are broadly adapted, individuals of which may cross the boundaries between patches easily, hardly noting the environmental changes that are limiting for the coarse-grained forms. They perceive the same environment as nearly homogeneous, with only fine-scale changes. These more tolerant species have *spatially fine-grained* adaptations. Between these extremes are many species that can inhabit more than one patch but not a great many; these species perceive their spatial environments as of medium grain.

Environments vary in time as well as in space; seasons vary greatly in the colder climatic zones, and floods, droughts, forest

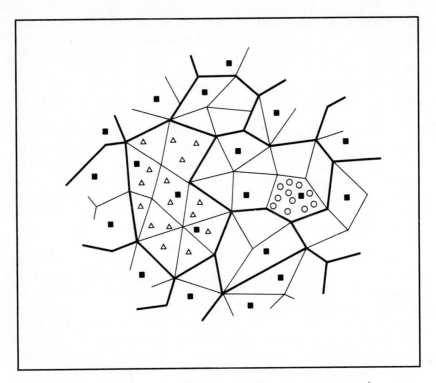

Fig. 9.8 Environmental grain. The mosaic pattern represents environmental patches that are relatively homogeneous internally but display significant differences in conditions across their borders. The heavy lines indicate greater environmental changes than the light lines. One population (open circles) is entirely restricted to one homogeneous patch; this population perceives the environment as coarse grained. A second population (triangles) ranges across several patches but is restricted by the changes that are indicated by heavy lines; it perceives the grain as medium. A third population (solid squares) ranges indiscriminantly throughout the entire region, perceiving the environment as fine grained.

fires, landslides, hurricanes, and other such events may disturb the usual conditions of life anywhere. Populations with narrow tolerances may find that a given environmental change exceeds their tolerance. They are perceiving their environment as temporally heterogeneous, or coarse, and can be called *temporally coarse-grained*. Other species' populations may tolerate the same changes easily, perceiving the environment as fairly stable; they are *temporally fine-grained*. Between the extremes are species that can tolerate the changes by adjusting their habits in some significant manner—hibernating (winter dormancy) or aestivating (summer dormancy), for example. They have temporal adaptations of medium grain.

All these grain-related adaptations are adaptive strategies. It is easy to see how selection would operate to develop different strategies. Two burrowing species might pursue different burrowing strategies, each successful in its own way. One species could solve the burrowing problem by becoming very efficient at burrowing into sediments of one type only, for which specialized adaptations would be evolved. The other species might solve the same problem by evolving the ability to burrow in nearly any kind of sediment, and thus would become a very generalized burrowing form. The generalized burrower might not be as efficient as the specialist in the latter's own milieu, but then it can live in many places where the specialist cannot. In a region full of many different kinds of specialized burrowers the generalist might suffer heavily from the competition. But if conditions change so that the sediment types required by specialists become rare, the specialists would suffer while the generalist could persist easily. Each strategy has its advantages and disadvantages.

The environment is comprised of numerous factors and it is common for species to be specialized for some and generalized for others. Indeed such mixed strategies can be coadaptive. Species that are highly specialized feeders (trophic specialists) often have generalized spatial strategies so as to be able to cross a variety of patches to search far and wide for prey. A generalized feeder may be able to find adequate food from its wide choice of items in one or a few patch types, and thus it may become a habitat specialist. In general, species have mixtures of strategies that are designed (by natural selection) for a particular mix of conditions.

Diversity Within Ecosystems

From the adaptive strategies of species, further properties emerge on the level of ecological systems. Some of the more important differences between the structures of community and provincial ecosystems seem to result from the different adaptive strategies of their component species.

One of the more obvious differences between ecosystems is that some contain many more species than others; communities in analogous habitats can differ in species richness by over an order of magnitude. We will refer to the number of species in an ecosystem as its *diversity*. The complex of factors that control diversity is still being unravelled. One factor that is well known to affect diversity is the degree of patchiness or spatial heterogeneity of the

environment. In environments that are spatially most heterogeneous, more species can coexist, other things being equal. In a
given region, a stretch of rocky coastline will support many more
species than a stretch of sand beach. Among the rocks there are
many more habitats; animals and plants can live under, behind, or
atop rocks, on rock faces exposed to or protected from waves, in
cracks and crannies, or in sand pockets interspersed among the
rocks. Along sand beaches the environment is relatively homogeneous and there is just one set of habitats—upon or within the
exposed sand bottom.

Spatial heterogeneity is one of the major influences on species
diversity patterns, but it is not the whole story because diversities
can vary between regions with equivalent spatial heterogeneities.
A tropical rocky shore is more diverse than a temperate one; a
tropical sand beach is more diverse than a temperate sand beach.
There is a striking gradient in diversity from the equator to the
poles (Fig. 9.9). Although the reefs and forests of the tropics derive
much of their diversity from biotic heterogeneity, this cannot be
the ultimate explanation for their species richness. The question
here is, what factors have permitted tropical organisms to generate

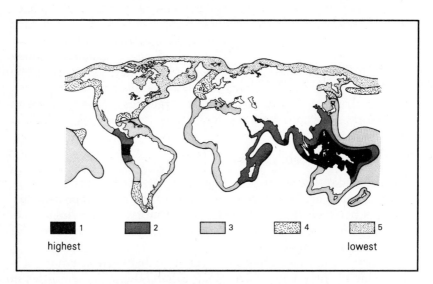

Fig. 9.9 Diversity pattern inferred for shallow-water marine communities. Generally speaking, highest diversities are found in the
inner tropics, especially in the more maritime regions. Lowest diversities are in high latitudes or where continentality is most severe.
(After Valentine, 1973.)

such biotic heterogeneity? Why are reefs and very diverse forests not present in temperate latitudes?

The answer to this last sort of question is not completely known. Diversity must be limited because some resources are used up by a certain number of species so that additional species cannot be supported. The two resources that are most commonly used up so as to limit the number of individuals are food and space. As we have seen, however, diversity patterns do not depend entirely on the amounts of these factors. Regions with similar productivities and similar spatial heterogeneities can have different levels of diversity. What does seem to be different about species found in analogous communities of different diversities is their modal adaptive strategies. Species in the more diverse communities tend to divide more finely the habitat and food resources than species in the less diverse communities. Thus factors that promote or inhibit the partitioning of resources may act to control diversity.

It has long been observed that there is a rough correlation between the variability of environments through time and their species diversities; the temporally less variable environments tend to have more species. To pick just one factor, consider variability in trophic resources—in nutrients and sunlight for photosynthesizers, and in prey species or detritus supply for heterotrophs. When the trophic resource base fluctuates, species must cope with food shortages at times. These are more easily dealt with by generalized feeders, for specialists that rely upon only a few food items will suffer if those few happen to become scarce. Additionally, habitat generalists are able to occur in a variety of habitats, permitting them to maintain relatively large populations and to be present in localities where conditions are least unfavorable— where food supplies are best, in this case. Thus fine-grained, generalized species are favored. Not many of these species, with their large voracious populations living in many habitats, can be packed into a given environment.

In stable environments, on the other hand, competition for resources can lead to coarse-grained strategies. The environment becomes highly partitioned. The smaller populations and restricted occurrences of these species do not put them in particular danger of extinction so long as their resources remain stable and thus dependable. The more stable tropics therefore have many more species than the more unstable arctic regions. This reasoning can be extended to other factors than just food supply. If, for example, greater temperature variations caused higher mortalities among arctic species, then their population would need to be larger to cope with the mortality. This could be accomplished by having fine-grained adaptations.

Stability of Ecosystems

Another way of approaching diversity regulation has been explored by ecological theorists. They have examined how populations interacting within a community structure might be most stable. The stability relevant to populations is best illustrated by imagining a population that is adapted to a given environment so that it can maintain a constant size through time. Now if we perturb the population from its equilibrium, perhaps by killing half of it off, resources will then be released and the population can increase toward its previous size. Often the population will actually rise above the level at which it can be sustained, since it can reproduce vast numbers of young with the available resources. When the food runs out, growth of all the young individuals cannot continue and many die. Thus the population size decreases again.

There are three classes of outcomes to this situation (Fig. 9.10). The population may gradually damp down the alternating size increases and decreases until it returns to its original constant size; this is *stability*. The population may instead continue to oscillate by the same amount indefinitely; this is *neutral stability*. Or the population may oscillate ever more wildly (in which case it will eventually become extinct); this is *instability*. It is clear that populations do tend to vary in size from season to season or year to year, some more than others. It is also clear that environmental conditions—food supply, temperature, and so on—vary also, often being far from optimal, so that populations do become perturbed from a steady state or from a normal regime. The question, then, is how is stability best preserved and instability prevented in nature?

One model, developed formally by Robert MacArthur, imputes stabilization of communities to a multiplicity of predator-prey interactions as based on the following scenario. If there is only one predator and one prey, the predator population may oscillate as much or more than its prey (Fig. 9.11); as its food supply grows it may grow, but as more individual predators appear more prey will be eaten until the prey population becomes rare; the predators will then starve and their population size will diminish rapidly. This permits the prey population to increase in size again, and the cycle may be repeated.

If one predator has two (or more) prey, however, it will tend to eat more of the most abundant one. If that population shrinks as

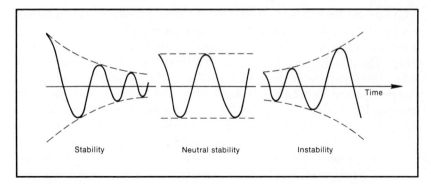

Fig. 9.10 Types of stability displayed by natural populations when disturbed from an equilibrium situation. Stable populations return to equilibrium (the figured population does so by a series of damped oscillations). Neutrally stable populations oscillate about their former equilibrium level indefinitely. Unstable populations oscillate more and more wildly, which will take them to size zero (extinction) eventually. (After May, 1973.)

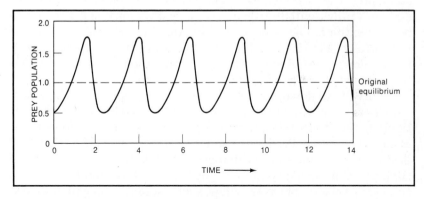

Fig. 9.11 Model of oscillation of prey population with a single predator when the prey is disturbed from an equilibrium. The result is commonly a neutral stability. (After May, 1973.)

a result of the predation (or from some other cause for that matter), the predators can turn most of their attention to the other population, now become the largest; in this way the first population will have a chance to recover from its decline without necessarily becoming extremely rare. If either prey population happens to grow unusually in size, perhaps because its own food supply happens to increase, then the predators will capture it in greater numbers and it will tend to even off its growth. Thus the populations are somewhat stabilized by their predator-prey relationships. If we add still another prey, or form a still more complicated predator-prey web (Fig. 9.12), stability should increase even further. The conclusion is that the greater the average number of trophic links (predator-prey interactions), the greater the stability. Since tropical communities have the most species and the most trophic links, they should be the most stable. Indeed tropical communities do appear to be more stable than, say, arctic communities, in that they change relatively little from year to year.

It has long been known that a single prey-predator system oscillated when modelled mathematically. It was not known whether a more diverse system would actually be more stable or even less stable. This question has been investigated by Robert May. Surprisingly, the more complex system is mathematically the less stable. Prey populations in single predator-prey systems, when perturbed from equilibrium, tend to oscillate in a neutral equilibrium. In a system with two prey and two predators feeding on either prey, when one prey is perturbed from equilibrium, the other prey leaves the equilibrium condition as well. Then in general both prey oscillate irregularly (unstably) until one happens to become extinct. The other then oscillates in a neutral equilibrium indefinitely (Fig. 9.13). Thus the two-prey, two-predator system breaks down into a one-prey system and then is no more stable than the usual one-prey, one-predator system.

This breakdown occurs because the population systems, which are naturally oscillating owing to seasonal or other changes in birth and death rates, have to be carefully balanced in order to maintain stability. Predators cannot eat too many prey or they will destroy prey population structures, but they must eat enough to support their own population, and so on. It is simply easier to achieve a balance between a few such oscillating population systems than between many. Adding a new oscillating subsystem (a new population) to a system composed of other oscillating subsystems cannot improve the stability of the whole system. At best the system will remain as stable (or unstable) as before; usually the system will be destabilized. In this view, tropical communities have more species than arctic ones because the tropical environment is more stable than the arctic with respect to the factors that cause populations to oscillate. Selection can be more effective in balancing populations that oscillate only a little than those that oscillate a great deal.

These two rather different ideas of diversity regulation may yet prove to be compatible; research is active on problems of diver-

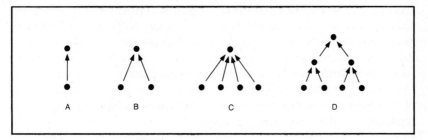

Fig. 9.12 Model of stabilization of communities by predator-prey interactions, following the ideas of R. MacArthur. *A,* a two-species system that is rather unstable (see Fig. 9.11). *B,* by adding a prey species stability is improved since the predator can switch its attentions to the more abundant prey when it expands or when the other declines. *C,* stability is further increased by adding even more prey. *D,* another pattern of predation, thought to be equally stable with *C.* In both *C* and *D,* there are four paths of energy flowing from the lower to the higher level, an indication of stability equivalence.

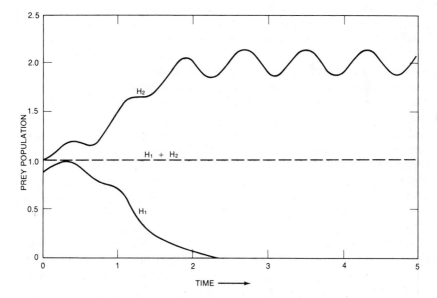

Fig. 9.13 Model of oscillations of two-prey populations with two predators when one prey population is disturbed from an equilibrium shared by all four populations. Both the prey populations oscillate unstably until one becomes extinct. The other then oscillates in a neutral stability as in the one-prey, one-predator system (After May, 1973.)

sity regulation and of community stability. They are of much more than purely intellectual interest, for human activity is tending to perturb both the world's populations of organisms and their physical environments. It therefore will be of great practical interest to learn the solution to these problems. With known solutions we can then direct our activities so as to minimize the possibility that species become extinct or community structures collapse owing to our ignorance of the governing principles.

Biogeography

The plants and animals in different regions of the world are very distinctive; zebras and lions are distinctly African, bison and puma distinctly North American, and so on. Darwin was struck with the faunal differences between parts of the world that he visited during the famous voyage of the *Beagle*. Not only the living organisms, but the fossils found in the Americas differed from those of Europe. These facts made sense if one assumed the doctrine of evolution. Species inhabiting distant regions are often isolated from each other by barriers to migration they cannot penetrate. Evolution therefore proceeds independently, and between regions isolated for very long, faunal differences can become very great.

In the past, the earth's environment has been considerably different from today, and the distribution patterns of many organisms have changed as environmental patterns have altered. During the ice ages of the last two million years or so, massive glaciers grew on land. The glaciers trapped water, as ice, that would otherwise have returned to the sea during spring melts. As a result, as glaciers waxed and waned, sea levels fell and rose through a hundred meters or more. When sea level was much lower than today, the Bering Straits were exposed as dry land and many native Asian animals that do not ordinarily cross marine barriers simply walked into North America. Humans may have migrated in this fashion. At other more remote times in the past, tropical waters were far more widespread than today and tropical marine communities lived in higher latitudes, many hundreds of miles from their present occurrence. Potential migration routes as well as provincial areas are thus created and eliminated through time.

Clearly, some animals (those with fine-grained strategies, for example) can range through many environmental changes that others find limiting. Therefore the ability to migrate is very different for different species. Furthermore, migration between favorable regions that are separated by a somewhat unfavorable barrier can be just a matter of time. If a species has one chance in 100 of crossing a barrier in a year, it would make it once every century on the average. Thus the spread of species is partially related to the strength of the barrier and to the time it is in operation. George Gaylord Simpson has distinguished three classes of migration routes. Most of a biota has a high probability of migrating along a *corridor*. For *filter routes,* some of the biota has a high migration probability but some forms are barred; thus migrants are filtered from among the available biota. *Sweepstakes routes* involve such strong barriers that there are only low probabilities of migration; nevertheless, given time, a few species will make it across, their successful migration being largely one of chance. They are the lucky sweepstakes winners.

Islands are somewhat isolated by definition, and many oceanic islands or island groups lie at considerable distances from continents or even from other island groups. Also, some new volcanic islands have appeared and old ones have become denuded of life by eruptions during comparatively modern times. Thus islands are natural laboratories in which migration and associated problems can be studied particularly well; there is even a special field termed *island biogeography*.

When the terrestrial biota of large numbers of islands of different ages, sizes, and degrees of remoteness are compared, some interesting generalities emerge. First of all, other things being equal, the larger the island the more species it supports. It is not certain whether this is a matter of size alone, whether the larger islands simply have more kinds of habitats (are more spatially heterogeneous), or whether the effect arises from a mixture of these two. Secondly, the more isolated islands have fewer species than islands near continents or other well-populated island groups. Presumably this is because they are so remote that migration is a sweepstakes for many species. Thirdly, very young islands have only a few species. When all these factors are taken into account, the number of species on an island can be fairly well predicted; there seems to be an equilibrium number of species that the island will contain. Since this is not an effect of immigration rate, it must be that the effects of immigration onto an island can be balanced by extinction, so that species diversity represents a dynamic equilibrium. Robert MacArthur and E. O. Wilson have examined data from many islands and have postulated the relationship depicted in Figure 9.14.

The division of the planet into provincial regions provides a way in which the number of species in the world at large can be raised far above any equilibrium levels established in communities. That is to say, each community may have a limit to the number of species it can accommodate at any one place. The total species found in a community may somewhat exceed this number

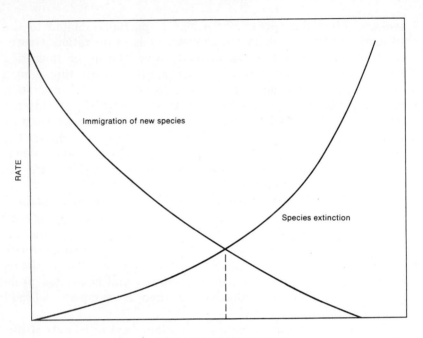

RATE

Immigration of new species

Species extinction

NUMBER OF SPECIES PRESENT, N

Fig. 9.14 A model of equilibrium species numbers (S) expected on an island. The number found will represent a balance between the rate of immigration (a function of distance from source, and other factors) and the rate of local extinction (a function of island size, diversity of habitat, and other factors). (After MacArthur and Wilson, 1967.)

if different species replace each other at different sites where the community appears. Nevertheless a given community has some species limits.

However, if the range of environmental conditions that support a given community in one province is inhabited by an entirely different set of species in another province, then the species diversity supported by those conditions is doubled. In the real world there is enough species overlap between provinces that adding a new province does not necessarily double the species numbers. On the other hand, some provinces contain ten times as many species as others. The large number of provinces that exist today raises the planetary diversity significantly, at least several times what it would be if only one terrestrial province and one marine province were present. This diversity makes the world a far more interesting and rewarding place to live.

QUESTIONS FOR DISCUSSION

1. What are the chief levels of categories in the ecological hierarchy?

2. If we knew all about the biological systems on one level—say the level of the individual—could we then predict the properties of systems on a higher level—say the community level? Why or why not?

3. Would you say that the producer or consumer chain contains more biomass of heterotrophs in the Arctic Ocean? The deep-sea? The rain forests of tropical Africa? A coral reef?

4. Would you predict that animals in the following regions have evolved primarily coarse-grained or fine-grained strategies, or that there are no significant differences: Arctic Ocean; deep-sea; rain forests of tropical Africa; Pacific coral reefs?

5. The Hawaiian archipelago has less marine diversity than many island systems of the tropical Pacific. Explain.

Recommended for Additional Reading

May, R. M., ed. 1976. *Theoretical ecology, principles and applications.* Philadelphia: W. B. Saunders.

Pianka, E. R. 1978. *Evolutionary ecology,* 2nd ed. New York: Harper & Row.

Smith, R. L. 1974. *Ecology and field biology,* 2nd ed. New York: Harper & Row.

Three works ranging from basic (Smith) to advanced (May) that consider ecology from evolutionary perspectives.

Scott, R. W., and West, R. R., eds. 1976. *Structure and classification of paleocommunities.* Stroudsburg, Pa.: Dowden, Hutchinson & Ross.

A collection of studies that deal with problems of identifying and interpreting communities in the fossil record. Intermediate level.

10

The Geological Record

Since the environment provides the opportunity for populations to evolve, it is important to gain some idea of the sorts of environmental changes that occur. It is useful to begin with those changes in the physical environment that are caused by physical factors, then to consider those caused by biological factors, and finally to consider biological changes themselves as environmental changes.

Evolution of the Environment

To discuss the origin and development of processes on earth we must have an appropriate time scale. Geologists have a traditional nomenclature for the historical periods of earth history, shown in Figure 10.1. Long after these periods were named, it became possible to determine the age of the rocks representing different geological periods by means of radiometric dating. Many varieties of the natural elements (*radioisotopes*) decay spontaneously at constant rates, creating lighter daughter isotopes. Many of the daughter isotopes are stable; that is, they are not subject to further decay. When radioisotopes deposited in minerals decay, the daughter isotopes commonly remain in the mineral structure. If the mineral is then analyzed and the amount of the parent and daughter isotopes determined, it is easy to calculate how long it has been since the mineral formed (providing that the rate of decay is known). For minerals that formed when volcanic lavas cooled, for example, we can determine how long ago the lava solidified. Ages of fossil-bearing rocks associated with dated volcanic rocks can then be estimated by extrapolation. The absolute ages of the boundaries between the important divisions of geological time are shown in Figure 10.1.

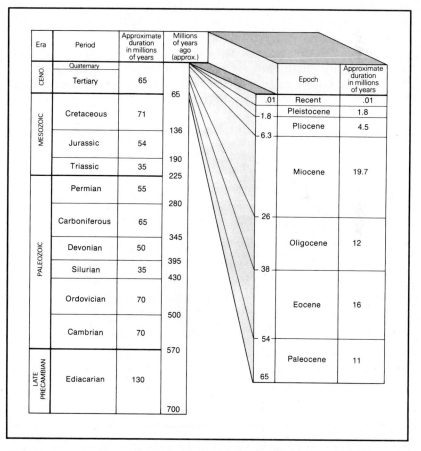

Fig. 10.1 The geological time scale. The traditional time subdivisions are eras, periods, and epochs (these last shown only for the Cenozoic era), based on fossiliferous rock sequences. A scale of absolute time, established by piecing together radiometric ages, is shown on the right.

The earth originated about 4.65 billion years ago, condensing from dust and gas that revolved around our newly formed sun. The earliest earth history is still largely unknown; the earliest known rocks are about 3.8 billion years old and have been much altered from their original condition. The early earth interior must have been much hotter than today's, for most of the interior heat is created by the decay of radioisotopes, and since they have been decaying steadily for 4.65 billion years there must have been much more radioactivity early in earth history. Under high temperature conditions, the material of the earth became differentiated into layers, with heavy material (as nickel and iron) "sinking" to form the earth's *core* and lighter material (aluminum and silicon compounds) "floating" to the surface to eventually form the earth's *crust*.

The earth's crust can be divided into two main types: oceanic crust, made up chiefly of basalt; and continental crust, lighter and thicker than basalt and therefore rising to higher elevations, com-

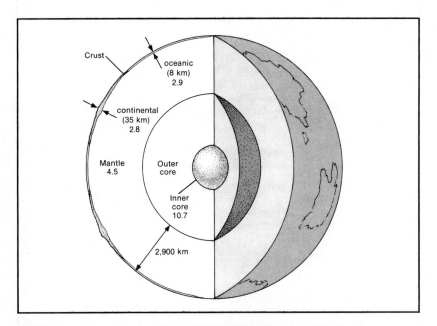

Fig. 10.2 Generalized cross section of the earth showing the major layers. The crust is thin (about 8 km beneath oceans, 35 km beneath continents) and light (density of 2.8 and 2.9 for continental and oceanic crust, respectively). The average density of the mantle is 4.5, and of the core, 10.7. The outer core behaves something like a fluid, the inner core like a solid.

posed of a complex of rocks that have the average composition of granite. Between the crust and the core is a region of rocks of intermediate densities, the *mantle* (Fig. 10.2). During early earth history, volatile materials escaped from the interior to form an early atmosphere (probably largely carbon dioxide) and hydrosphere (the oceans).

Seafloor Spreading

Even today, enough heat is generated within the earth to drive upward rising streams of molten rock, lighter partly because they are hotter, to the surface in volcanoes and in a system of elongate fractures or rifts that runs for thousands of kilometers beneath the oceans (Fig. 10.3). Since heat flow to the surface is high beneath the rift system, the crust bulges upward there to create a system of submarine ridges, on which the rifts are located. One such ridge, the Mid-Atlantic Ridge, bisects nearly exactly the Atlantic Ocean

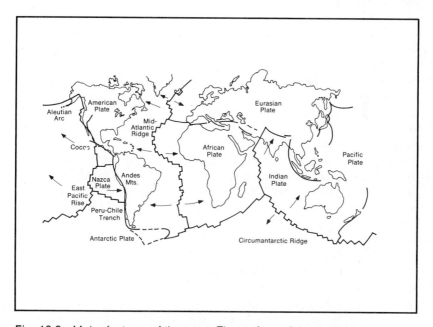

Fig. 10.3   Major features of the crust. The surface of the earth is divided into several major plates (named on the figure) and a number of minor ones by deep-sea ridges (heavy lines), trenches (arcuate lines), and transform faults (straight lines). Some plates bear continents (the American plate has two) and others do not (the Pacific plate has none). (After Vine, 1969.)

floor; two other ridges of major significance are the East Pacific Rise and the Circumantarctic Ridge.

As molten rock rises along the rifts it cools and solidifies; the rifts part slightly, and the sides of the rifts become coated with layers of new rock. When rifting (parting) recurs, the new fracture is within the newly intruded rock. The sides of the fractures thus spread away from each other as new rock solidifies between them. Although the invasion of any particular segment of a rift by molten rock is an episodic process, worldwide activity is fairly steady when viewed over millions of years. For example, the sides of the Mid-Atlantic Ridge have been spreading at the rate of a few centimeters per year for about 100 million years in low latitudes. The Atlantic Ocean was closed entirely during much of the Mesozoic era. North and South America, Europe, and Africa were then joined in a huge supercontinent called Pangaea (Fig. 10.6), which included all the present continents. Since rifting began along the site of the Mid-Atlantic Ridge, the trans-Atlantic continents have spread laterally away from the ridge as new seafloor appeared between them. Indeed, they are still spreading.

New seafloor cannot accumulate indefinitely, or else the earth would become hollow and expand like a balloon. Expansion seems in fact to have been modest if it has occurred. Instead, seafloor is returned to the earth's interior down zones of sinking, so that a long-term recycling process is involved. The layer that actually spreads at the ridges embraces the crust and upper portion of the mantle; it is a rather rigid layer termed the *lithosphere* (Fig. 10.4). Beneath it is a warmer, more plastic layer termed the *asthenosphere*. The lithospheric layer slides over the plastic asthenosphere, which contains shear zones or other zones of deformation that accommodate this movement. The lithospheric layer returns to the earth's interior by plunging downward (at a few centimeters per year) along *subduction zones*. The sites of the subduction (swallowing up of the lithosphere) are usually marked by deep-sea trenches (Fig. 10.4). The friction of the descending lithospheric layer as it drags down beneath the layer across the subduction zone creates heat, enough to melt some of the descending rocks, which then rise to surface as volcanoes. Most of the world's subduction zones are in the Pacific Ocean. The Aleutian Islands, Japan, the Philippines, and the Mariana Islands, for example, are volcanic island arc systems located behind subduction zones where the floor of the Pacific Ocean plunges back into the earth's interior (Fig. 10.3). The Andes Mountains of South America have a similar origin, lying as they do along the Peru-Chile trench. Since this subduction zone is next to a continent, the

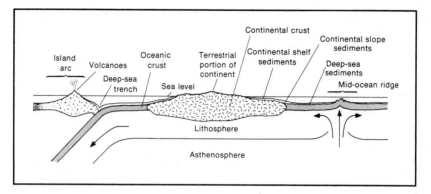

Fig. 10.4 Cross section of the upper mantle and crust of the earth to show a lithospheric plate, generated at a mid-ocean ridge and plunging back into the interior at a subduction zone sited at a deep-sea trench. A continent is embedded in the plate. Behind the deep-sea trench, volcanoes are formed by material from the crust of the descending plate, heated by friction as it drags down beneath the opposite plate.

arc system becomes a system of high coastal mountains rather than of islands. Incidentally, resistance to the descent of the lithospheric layer causes some of the earth's major earthquakes.

In effect, the seafloor segments—spreading from a ridge and then descending down a trench millions of years later—form large rigid plates. The surface of the earth is divided into about seven major plates (Fig. 10.3) and a number of smaller ones. The boundaries between plates are located at ridges or trenches, or at places where one plate can slip by the other. These latter places are marked by structures termed *transform faults,* which accommodate the differential plate movements that result when neighboring plates spread in different directions or at different rates. Movement along transform faults is also a major source of earthquakes; the famous San Andreas Fault of California is a transform fault, accommodating differential movement between the American and Pacific plates (Fig. 10.3).

Continental Drift and Plate Tectonics

Continents are part of the lithospheric layer and move with the lithospheric plate in which they are embedded. A continent may migrate across the earth's surface, moving away from a ridge as new seafloor appears behind it, and approaching a trench where seafloor in front of it is being consumed. Continents are so light, however, that they cannot be entirely subducted. When they collide with a trench the subduction zone may cease to function (in which case a new one may spring up elsewhere) or the zone may reverse its subduction direction and begin to swallow the plate behind it. If there is a continent on this second plate also, the two continents may eventually collide at the trench (Fig. 10.5) and become joined into a larger continent. This has happened a number of times in earth history. When continents collide, their margins are deformed and seafloor sediments trapped between them are uplifted into mountains. Sometimes pieces of the trench walls, including even parts of the earth's upper mantle, are sliced off by the leading edge of continents and caught up in the deformation, to be exposed in mountain ranges. The supercontinent Pangaea was formed by the coalescence of several continents that collided one after another over a period of time exceeding 200 million years and culminating about 225 million years ago when Asia was added.

Thus continental masses were generated as the earth's layers were differentiated billions of years ago, and for the last billion years at least, these masses have moved about the earth's surface because of seafloor spreading processes, growing by collisions and shrinking by fragmentation in ever-changing patterns. Their movement is termed *continental drift*. The processes of seafloor spreading and consumption and the deformations that accompany collisions of continents, arcs, and ridges are referred to as *plate tectonics* (tectonics is the science of crustal deformation).

Consequences of Continental Movement

Consider the environmental consequences of seafloor spreading and continental drift. A very obvious effect is that migration barriers and routes are altered so that the biogeographic pattern is altered. Faunas inhabiting separate continents (including the marine forms on the shelves) may be brought together to intermingle as the continents collide. When continents split up, their faunas may become isolated and evolve independently thereafter, barring occasional migrations across filter or sweepstakes routes. For shallow-water marine animals, island arcs form migratory corridors, and as the arcs change their geographic locations they can initiate or interrupt the flow of migration between distant regions according to the circumstances. We can imagine a situation in which all the continents are joined so that there are no oceanic barriers for the terrestrial species and no land or deep-sea barriers for the marine species on the continental shelves. There would be few prov-

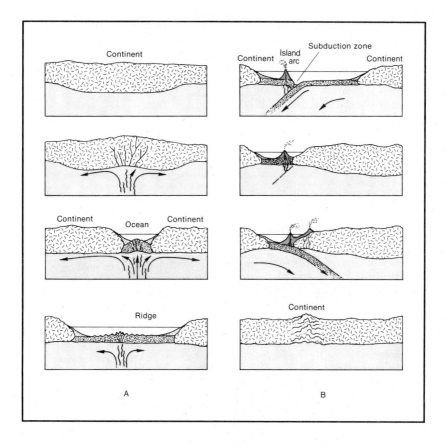

Fig. 10.5 Two scenarios illustrating the fragmentation and joining together of continents. *A,* a continent (dashes) is embedded in a lithospheric plate. Heating from below bulges the crust and rifting ensues. Emplacement of liquid rock in the fractures separates the continental parts. Continued spreading creates new lithosphere between the continental fragments, which therefore drift apart. *B,* two continents are separated by a subduction zone, so they approach each other as seafloor is consumed. One continent collides with the subduction zone, which flips under the continent and then proceeds to consume the remaining seafloor fronting the opposite continent. This creates Andes-type mountains on the continental margin behind the subduction zone. Finally, the continents collide over the subduction zone, which decays. A single continent has been created, sutured at the site of the subduction zone. The suture is marked by belts containing sediments from old seafloor and trench locales (now deformed and uplifted into continental margins), old island arc remnants, and Andes-type mountain remnants. (After D. H. Tarling and M. P. Tarling, 1971.)

inces in the world relative to today, and therefore many fewer species could be accommodated (Fig. 10.6).

Another obvious consequence of continental movement is that climates would change as the continents change latitudes. An equatorial continent is largely tropical, but if it drifts poleward it becomes temperate and finally arctic in climate. The fauna would have to become continuously adapted to the changing conditions. Just as important, continents affect the world climate by their po-

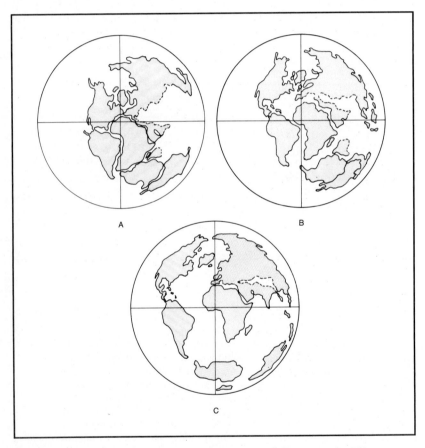

Fig. 10.6 Continental positions at three times in the past. *A*, 220 million years ago. *B*, 100 million years ago. *C*, 20 million years ago. The supercontinent in *A* is called Pangaea. It included all the world's major land masses. Rifting then occurred between the Americas and Africa, and between Antarctica and South America and Africa. The Atlantic Ocean had appeared by 100 million years ago. By 20 million years ago India and Australia had also broken away from Antarctica and the world began to look familiar. (Maps from Smith and Briden, 1977.)

sitions. If the continents change their geographic pattern, the climatic pattern itself is affected, and this would also require adaptive changes in the biota.

One important climatic effect of continental position concerns the transport of heat toward the poles. Because the angle of the earth's surface slopes increasingly away from the sun at higher latitudes, less heat is received per area of surface in higher latitudes; above 37 degrees of latitude more heat is radiated to space than is received from the sun. To keep the higher latitudes warm requires a poleward transfer of heat from low latitudes where it is in surplus. The heat is carried by winds and ocean currents.

If the continents are placed so as to inhibit the poleward heat flow—if they cut off ocean currents, or if mountainous areas restrict poleward air flow—then the poles will cool. If the continents are then positioned so that heat flow is relatively uninhibited, the poles will warm. Today the poles are quite cool. The reasons are most obvious in the sea. The Arctic Ocean is cut off from warm waters by the surrounding landmasses and by cold currents that circulate in high latitudes. Antarctica is surrounded by the Circumantarctic Current, which flows in an easterly direction, radiating heat into space and blocking the southerly transport of warm water (Fig. 10.7). In the past, warm water from low latitudes spread nearer the poles so that ocean temperatures were more similar than now between low and high latitudes. At those times there

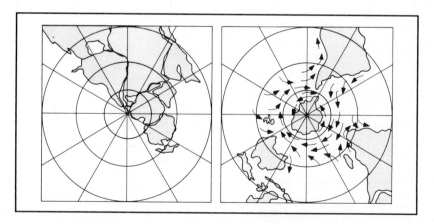

Fig. 10.7 The southern end of the world 220 million years ago, A, and today, B. Today a wide easterly current flows around Antarctica (arrows), blocking the poleward transport of heat in the warm south-flowing currents that drift down the eastern sides of the southern continents. We cannot yet infer the current patterns of 220 million years ago, but similar impediments to the poleward transport of warm water were lacking. (Maps from Smith and Briden, 1977.)

was less latitudinal climatic zonation and consequently fewer provinces. The most poleward conditions of extremely cool water found today did not exist.

The degree of spatial heterogeneity is also affected by continental geographies. If a large continent is formed out of two small ones, for example, its interior will be more remote from the sea than the interiors of the smaller continents were (Fig. 10.8). This situation creates entirely new conditions. The ocean has a moderating influence on climate because it takes a relatively large amount of heat to change water temperature. Continental rocks heat up or cool off much more rapidly for a given amount of heat input. The atmosphere is heated chiefly from below, by reradiation of the sun's heat from the land or sea surface. The air over the ocean thus tends to be of more constant temperature than the air over the land. The farther removed a region is from the influence of marine air, the more extreme the local temperatures are apt to be; continental interiors have notoriously hot summers and cool winters (Fig. 10.9). Thus the creation of larger continents enhances the continentality of the interiors and can create entirely new environments, for which evolution may then develop a new complex of communities. On the other hand, there is a tendency

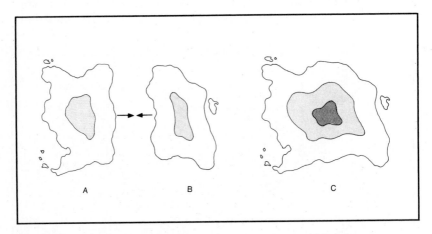

Fig. 10.8 The effects of continental suturing on spatial heterogeneity of the terrestrial environment. *A* and *B,* two continents that are converging; their interiors are remote from the oceans and have continental conditions (lightly shaded areas). *C,* the two continents have joined to create a single large continent, the interior of which is still more remote than was the case with either small continent, creating even more extreme continental conditions (darker shaded area).

for the terrestrial environments to become more homogeneous when continents break up into smaller masses.

Sea Level Variations

Changes in the level of the sea relative to the land surface can lead to important environmental changes that affect the course of evolution. Sea level variations arise from several causes. We have

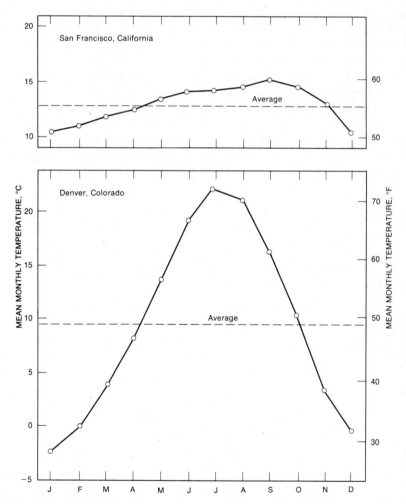

Fig. 10.9 Moderating effect of an ocean on terrestrial climates. San Francisco is cut off from the continental interior by the Sierra Nevada and other mountain ranges and has an equable, maritime climate; Denver has a continental climate. (After Gross, 1977.)

mentioned that glaciers lock up water in ice, resulting in lower sea levels; glaciations have caused sea level swings of more than a hundred meters. Other sea-level changes are related to changes in the volume of the ocean basins caused by plate-tectonic events. At times, the pace of plate tectonic activity seems faster than at others; the ridges are more active than average, with greater heat flow into them as the molten rock supply rises. The ridges therefore tend to bulge up more than average. Ridges are very large features —200 to 300 kilometers at their crests, rising halfway from the abyssal ocean floor to the ocean surfaces, a distance of about two kilometers. When they bulge up, the average depth of the seafloor becomes appreciably lower. Since the ocean basins are full, the oceans must spill out over the continents when the volume of the basins is decreased. The areas of continental shelves therefore grow appreciably and the exposed terrestrial environments shrink.

There are other sorts of ocean-basin volume changes as well. Increasing the number of trenches raises the volume, while the flow of sediments from the continents into the ocean lowers the volume, and so on. Sea levels in the past have stood hundreds of

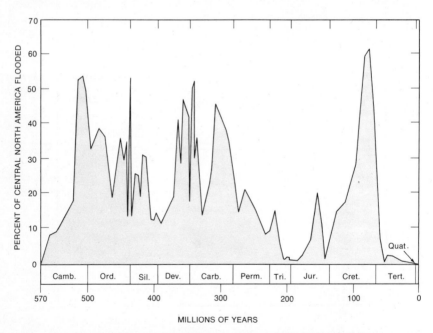

Fig. 10.10 The generalized history of flooding of North America by shallow seas. The estimates of maximum flooding are probably conservative. Even so, over half of the North American platform is shown as submerged at times. (After Wise, 1974.)

meters higher than today, drowning more than half of the world's land at times. It is very difficult to separate evidence of worldwide sea level changes in the past from local changes due to warping of continental margins, so we do not have as yet a convincing history of worldwide sea level positions. Figure 10.10 shows conservative estimates of sea-level fluctuations over the North American continent during the last 570 million years. Note that sea level today is much lower than average.

The environmental effects of sea level changes are pervasive, partly owing to the moderating effects of sea water on climate. When sea level is high the smaller continents are more accessible to marine-influenced air masses, and there is simply less continental area to be heated or cooled. The world as a whole becomes more equable and seasonal climatic changes become less extreme; the world climate becomes more *maritime*. When sea level is low, the climate is more heterogeneous in space and more extreme in time—more *continental*. The accompanying changes in area of environments must have important effects as well. At times of high sea levels the broad shallow shelves should support a greater variety of marine species than at times of low sea levels. For the terrestrial biota, high sea levels would mean less habitat space and, usually, lower diversity than low sea levels.

Biologically Based Changes in the Physical Environment

Organisms are affected by the environment, and in turn they affect the environment, both generally and locally. General environmental changes that have been accomplished by organisms include a change in the atmosphere from a primitive *anoxic* ("lacking oxygen") one that was neutral or slightly reducing in character to one that is chemically dominated by free oxygen. Most of the oxygen has been released by plants during photosynthesis. Since oxygen is so active chemically this change has had a profound effect on earth chemistry. Compounds accumulating at the earth's surface are usually oxidized today. This certainly applies to organic compounds, which are quickly oxidized unless buried rapidly or otherwise removed to an anoxic site.

On a primitive earth, however, with little or no free oxygen, organic compounds could become rather highly concentrated in the sea (see Chapter 11), and thus the stage could be set for the origin of life. Life could probably not form anew under present conditions. On the other hand, animals require oxygen for respiration. They could not exist on the primitive earth until the concentration of free oxygen became sufficiently high to support them. Even nitrogen, which forms such a high percentage of our present atmosphere, may owe its presence there primarily to organisms (bacteria). At any rate the composition of the atmosphere and the biochemical and chemical processes that go forward at the earth's surface today are due to a large degree to the activity of organisms.

As an example of organisms affecting the environment on a local scale, take the burrowing of marine worms in the muddy bottoms of shallow seas. Large numbers of marine animals can live on or within bottom muds. Some of these species feed chiefly by filtering minute organisms and bits of organic debris that are floating in seawater. These species usually live on top of the seabed (*epifaunal*) or burrow shallowly (*infaunal*) but retain a connection with the seawater by pumping it into their burrows or extending tentacles or some such organ up into the water column. Still other mud-dwelling species feed by eating the deposits through which they burrow. A deposit feeder cannot usually remain in the same place, where it will soon use up the organic contents of the sediment, but must burrow much of the time to find new, organic-rich deposits. In many bottom muds, the bottom is constantly stirred up and turbated by the burrowing activity of the deposit feeders. This changes the character of the mud, which becomes rather liquified and soupy and is then unsuitable for habitation by many suspension feeders, for they or their young suffocate in the soft muds.

Biological Environmental Changes

Species are components of the environment, and their presence can change the adaptations required of other species. Coral reefs, formed by frame-building organisms such as corals and buttressed by calcareous algae and other encrusting organisms, are a case in point. The reef communities are quite diverse, and this is in part due to the specialization of one species upon others. It is obvious with food relations. If first-level predators tend to specialize on one or a few herbivores, then more species will be present on this predator level. The second predator level will then have the opportunity to be relatively prey-specialized, and since there are more kinds of first-level predators there can be many second-level predators, and so forth. On reefs there are many trophic specialists. Other types of interactions are also found; sea anemones form the habitats for anemone fish; numerous fish and invertebrates burrow within or under the corals and large mollusks; parasites are numerous—the opportunities for food and living space presented by the animals themselves are put to good use. The effects of diversity are thus multiplied, for the organisms themselves increase the spatial heterogeneity.

Biological Responses to Environmental Change: the Fossil Record

It can be inferred from the theory of natural selection that rates and intensities of environmental change should affect the rates of evolution. Many past environmental changes can be interpreted from geological evidence. Biological responses to the changes appear in the associated fossil record. The fossil record is notoriously incomplete, however, for only a small fraction of organisms have left traces or remains of their existence. Usually these are mineralized skeletal remains such as the bones of dinosaurs or the skeletons of ancient corals. Sediments that were deposited in shallow marine waters are particularly rich in fossil material.

However, organisms that lack hard durable parts, like jellyfish and worms, are fossilized only under truly exceptional circumstances (Fig. 10.11). Perhaps only one in three or four species has an easily preserved skeleton. Even for the species that have very durable hard parts, becoming a fossil is no easy task. Commonly the sediments in which skeletons come to rest are eroded away during later periods of earth history, with destruction of their fossil contents. This is the fate of the bulk of fossiliferous shelf deposits. In the sediments that do remain, fossils are often dissolved by percolating groundwater. Sometimes sediments are caught up in crustal deformations and are heated, recrystallized, invaded by hot fluids, or otherwise altered so that fossils are destroyed. Many fossils that are preserved are inaccessible, buried beneath thick covers of younger rocks or deep under ocean floors. It is likely that very few animal species, of all that have ever lived, are actually found as fossils—we cannot say just how many; one in a hundred is a reasonable guess. The fossil record of plants may be even worse, although spores and pollen are commonly preserved and give us a rough idea of the nature of plant life in the past.

An additional difficulty in using fossils is that our methods of dating are not exact. Radiometric age determinations are subject to errors of a percent or two even under the best of conditions and can often be further off. In dealing with rocks of great age, such errors can be significant; 2 percent of 300 million years is 6 million years, which can be a long time in terms of evolutionary events. Such dating problems do not make too much difference so long as we restrict our interest to a local region, where the relative age of fossils can be determined by the laws of superposition: the stratigraphically lower rocks (that is, rocks that were deposited earlier) contain the older fossils, and higher rocks which were deposited on top of the lower contain younger fossils. The rocks containing one set of fossils can often be traced or mapped from one region to another to determine whether they are older or younger than rocks with a different set of fossils. Once the sequence of fossils has been worked out for one area, the fossils themselves can be used as indices to the ages of rocks elsewhere. This sort of fossil *correlation* can be more accurate than radiometric dating. In some well-studied cases, rocks can be dated to within half-million-year intervals. However when correlations are attempted between distant areas, such as on separate continents, they are often accurate to only two million years or so.

At a single fossil locality, the amount of time represented by a few centimeters of rock may range from hours to thousands of years. The fossils usually represent many generations and thus a fossil association may include, in addition to species that were present the whole time, some that were present for only a short while but left their skeletons behind, those that happened in by chance and never actually bred or occupied the habitat, and those that have been transported from other sites by stream or current action. Sometimes the remains of two or more communities are thoroughly intermingled. When we reconstruct communities we must take these factors into account as much as possible.

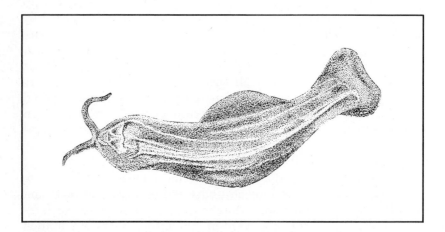

Fig. 10.11. The Burgess shale, in British Columbia, has yielded a remarkable collection of rarely fossilized soft-bodied Cambrian marine animals, such as this specimen, *Amiskwia,* that even reveals internal structures.

Still other problems arise when we try to reconstruct ancient provinces. Since we cannot correlate with perfect accuracy, the distributional patterns that we find may represent geographic distributions of the fossils over hundreds of thousands or even millions of years, during which time much migration may have occurred. Another difficulty is that the places where fossil-bearing rocks occur today may be very far from the place where they were deposited and where the fossilized organisms lived. Some fossil beds now thousands of kilometers apart across the Atlantic were actually deposited in close proximity within the same province, but have been separated by continental drift. On the other hand there are fossils in Scotland that were deposited in North America when the continents were separated before their last collision; they do not compare at all with other contemporaneous Scottish fossils with which they are now closely associated geographically. To make meaningful maps of ancient biotic provinces we must draw maps with the continents and continental fragments returned to the geographic positions they held when the provincial biotas lived.

Despite the incompleteness of the fossil record and the difficulties in correlation, fossils provide our best evidence for evolutionary rates and modes and the circumstances under which they have occurred. They also permit us to trace the history of various taxa and of ecosystems. Fortunately, the fossil record can be superb at some localities, where fossils are preserved in living positions and associations. From these remarkable localities we learn to interpret the ordinary ones, and by extrapolation we can work out histories.

CLOSER LOOK 10.1   Continental Patterns and Species Diversity

A strong control on the species richness of the world is exercised by the geography of continents and ocean basins and by the distribution of migration routes and barriers. As an example, take the models of continental geographies illustrated in the accompanying figures, which are made simpler than any real geographies so that the principles are clear.

In the first figure, all the continental masses are gathered together into a single continent which is circular in outline and centered on a pole. If the sea extends over a strip of the continent to form a narrow shelf around the perimeter, then the shallow-water marine fauna will not face any geographic barriers but can extend entirely around the shelf. Since the

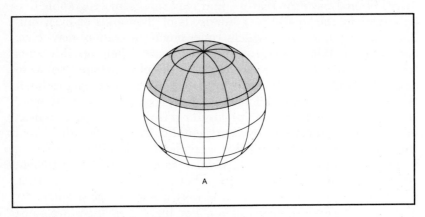

entire shelf is at about the same latitude, the marine climate will be about the same everywhere on the shelf and only a single marine province will exist. There may have been times in the past (the early Triassic period and times during the Jurassic period, for example) when provinciality was practically this low. There is only one set of communities in such a situation, and the total number of marine species is relatively low.

In the second figure, the continental masses are gathered into a belt that encircles the world from pole to pole, dividing the oceans in two. There will now be two sets of marine provinces, one in each ocean, and owing to the barrier of land they will contain totally different species. The world's marine fauna will be at least doubled from the previous model. In addition, the shelves of each ocean extend from the equator to

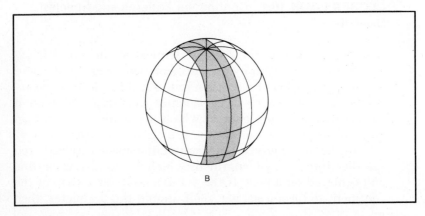

the poles, ranging through whatever climatic differences exist latitudinally. If the polar waters are significantly cooler than equatorial waters they will possess a cold-adapted fauna of their own. In this event there will be eight provinces present—two equatorial provinces in each ocean, separated by a province at each pole, for a total of four to an ocean or eight total. The latitudinal barriers are climatic, and their effect is multiplied by the longitudinal barriers, which are geographic (land or deep-sea). Since each province contains a large percentage of endemic (native) species, the total species richness of the marine shelf grows as new provinces are added. This

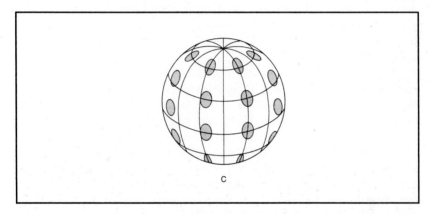

c

sort of pattern is present today—predominantly north-south shelves with chains of latitudinal provinces along each, separated by land or deep-sea barriers.

Finally, in the last figure, the continental masses are broken into 32 small continents and dispersed regularly over the earth's surface. If the shelf of each continent contains a large endemic element, then the total number of species would be very much greater than in either of the other figures. This situation has never occurred so far as we know. At times during the Paleozoic there were a few very large scattered continents, rather than many small ones.

Other model geographies that yield interesting provincial patterns include a belt continent that runs east-west and is centered on the equator, and a single circular continent centered on the equator rather than a pole. It is interesting also to speculate on the terrestrial diversities that we would expect to be associated with each of the models.

Extinction and Pseudoextinction

Many strange or bizarre animals of the past are revealed by the fossil record: dinosaurs, ammonites, trilobites—all these are long extinct. When a fossil species or other taxon is never found in rocks younger than a certain age, then it is assumed that it became extinct at the time it disappeared from the fossil record. Since the record is so spotty we do not catch the very last years of most taxa, but the approximation of disappearance to extinction is probably sufficiently good so that we can calculate extinction rates and trends from fossil data on disappearances. Disappearances of lineages can be due not to extinction but to evolution, when a lineage changes rapidly into a different descendant form. This is termed *pseudoextinction,* and probably underlies at least 10 to 15 percent of disappearances. For most purposes we can lump pseudoextinctions with regular extinctions.

Why are extinctions so common? Natural selection acts to improve or maintain adaptation to prevailing conditions. If environmental change is very slow, then selection may be able to maintain a high level of adaptation. However, change cannot ordinarily be for the better. Adaptations are not selected for future conditions, but for present ones. Organisms are bundles of complexly interwoven adaptations, legacies of the past conditions that the organisms have faced. In the general case, then, environmental change is environmental deterioration so far as organisms are concerned. For slight changes there would ordinarily be only a few species that failed to maintain their adaptation. For large or fast changes, many species might be unable to respond appropriately.

Deterministic and Probabilistic Factors

Since change seems to be incessant, we expect extinctions to occur continuously, rising in intensity during periods of greater and more rapid change. The species present at any time cannot predict the future; natural selection has no foresight. When an environmental change does occur, it often does so at random with respect to the adaptations of all the species. Which species become extinct and which succeed may be a matter of chance, as Leigh Van Valen has pointed out. Of course the species that endure do so because they are better adapted to the changes than the species that are extinguished. In this sense, the extinctions are *deterministic* (controlled by selective processes). However, which of the species are going to be best adapted to the next change cannot be predicted; this is a *probabilistic* matter (a matter of chance).

Seen in this light, extinction is part and parcel of the process of evolution; it is not "failure." In some ways it indicates success for selection. Evolution has successfully equipped a species with the ability to cope with a given environmental situation. When the situation changes, the species becomes extinct; if the environment had changed in some different ways the species could have persisted, while other species, successful in the first instance, would have succumbed in the second. It is not adaptation that is probabilistic, it is environmental change. Nevertheless, this does give the process of extinction a strong probabilistic aspect.

There are exceptions to such an unselective model of extinction. One exception occurs because some species are especially adapted to be generalists. They survive by being broadly adapted to a wide range of conditions. When environmental change occurs, these forms are more likely than the average species to be able to tolerate the new conditions.

Another kind of exception may occur when the environmental changes are not random, but when a later change is influenced by and follows in some way from an earlier change, so that the direction of environmental change follows some sort of trend. Plate tectonic processes might well provide such correlated sequences of environmental change—for example, to warmer and warmer climates as a continent drifts toward the equator, or to drier climates inland as mountains rise along a coastline. The species that evolve to best accommodate the early changes in such a trend might often be equipped to accommodate further changes. Pseudoextinctions might be particularly common in such situations.

Diversity Regulation and Extinction

Extinctions can be classed into two broad categories according to their relations to diversity regulation. Obviously, diversity is lowered by extinction. In some cases, extinction may be required precisely because the capacity of the environment to support species has declined. Joining two continents that had previously been separated, or lowering the temperature differences between poles and equator, causes extinction by lowering provinciality. Increasing the temporal variability of the environment lowers diversity within communities, according to some theories of diversity regulation (see Chapter 9). The resulting extinctions would be *diversity dependent*. In other cases, species become extinct simply because the environment has changed so as to exceed some of their tolerances or failed to meet some of their requirements. These are *diversity-independent* extinctions; their severity depends on the number of species that happen to find the new conditions lethal. There is nothing to prevent the evolutionary processes from developing new species adapted to the new conditions and replacing those that have become extinct. Here is another way in which extinction can be regarded as an important part of the evolutionary process: it opens up areas of the environment—potential adaptive zones—and thus provides the creative side of evolution with new or renewed opportunities.

Besides the usual range of extinction rates that appear to be due to average sorts of environmental change, called *background extinction,* there are a number of past times when severe waves of extinction, called *mass extinctions,* have carried off significant proportions of the living species. The distribution of extinction rates for various groups of marine invertebrates is indicated in Figure 10.12. The rates are determined by tabulating the number of taxa that disappear from the record during successive epochs of geological time. Figure 10.13 depicts extinctions for all shallow-water marine invertebrate taxa that are regularly represented as fossils. The extinction pattern of each group is unique in some respects, yet there are times when nearly every group suffers heavy extinction to produce a very high total. Presumably these times are when the environment was changing most rapidly, or when it underwent some key change that lowered the world's capacity to support species.

To interpret the actual causes of extinction of any given species or higher taxon is one of the most difficult tasks in history of life studies. This is partly because the proximate cause of extinction may be quite different from the ultimate cause. Consider one event that can lead to extinction, the coming together of two small continents to produce a larger continent, reducing the provinces from two to one. Before the collision there were two sets of communities, one in each province; afterwards there is only one set, in a larger province. Suppose that the new larger province can support 20 percent more species than either of the old provinces. Then if there were 2,500 species in each small province, they would become not 5,000 but 3,000 species in the large province; 2,000 species must become extinct.

How are all these extinctions actually accomplished? The actual change in physical conditions accompanying the collision may be lethal for some species. The remaining species will presumably be eliminated by biological interactions, by predators or parasites or diseases, or by competitors, introduced when the two biotas came into contact. One or several of these new sources of mor-

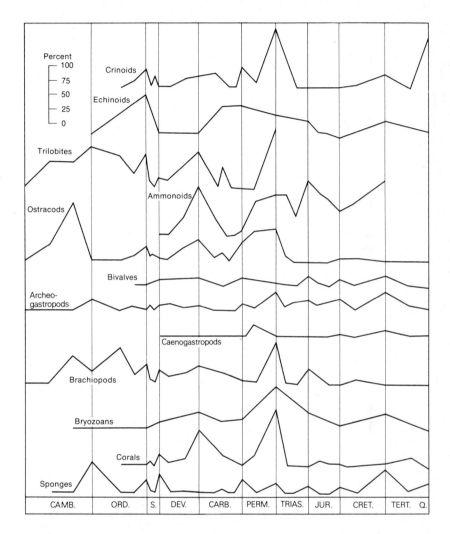

Fig. 10.12 Levels of extinction for the major groups of fossil invertebrates through the last 570 million years, expressed in percentage of families present. Note that at certain times, such as near the Permian-Triassic boundary, large numbers of groups suffer contemporaneous waves of extinction. (After Newell, 1967.)

tality, added to the load of mortality already being carried by the species, will prove so great for many species that their birthrates cannot be raised in time to meet them and extinction ensues. Usually the proximal causes of extinction are so complex or subtle as to defy discovery from only the data of the fossil record.

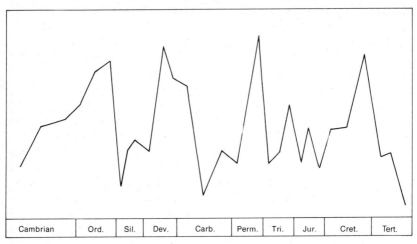

| Cambrian | Ord. | Sil. | Dev. | Carb. | Perm. | Tri. | Jur. | Cret. | Tert. |

Fig. 10.13 Levels of extinction of families of all major groups of fossil invertebrates from shallow marine seas through the last 570 million years. Extinction is assumed when the family disappears from the fossil record. (After Valentine, 1973.)

Extinction processes work on the species level, yet higher taxa —families, even classes and phyla—have also become extinct. For some small higher taxa, extinction may occur at least partly by chance—all the species happen by chance to be eliminated, perhaps due to different causes. For a richly diverse taxon this is unlikely. Extinction in these cases is probably due to adaptive traits shared by all (or most) members of the taxon. For example, if a taxon originates and diversifies in a stable environment and then the environment becomes very unstable, all species may be lost.

Competition over long stretches of time may often be involved also. For example, during middle and late Paleozoic time, fish become more diverse while large aquatic arthropods called eurypterids become fewer in species (Fig. 10.14). It is plausible that eurypterids were gradually replaced by fish with similar modes of life. This need not have been due to actual biological competition, when two species compete for a common resource. Rather, as a eurypterid species became extinct from whatever cause, its place might have been filled by a lineage of fish rather than by a lineage of eurypterids. The adaptive ensemble of eurypterids was probably inferior or at least less evolutionarily flexible than that of fish, and while fish might not have evolved into eurypterid niches while these niches were occupied, they might have taken them over when niches became available through extinctions.

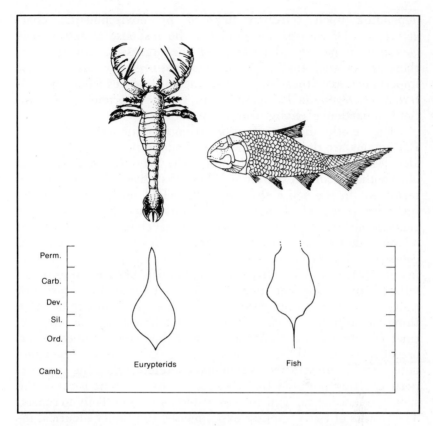

Fig. 10.14 Major trends of eurypterids and fish during the Paleozoic era. As the fish expanded, the eurypterids declined. It is plausible that fish replaced eurypterid species as they became extinct, until finally the entire group disappeared in Permian time. (Modified from Simpson, 1953.)

Diversification

In previous chapters many processes of diversification have been described: isolation of gene pools leading to new species; the development of new modes of life, which can be looked on as the invasion of new adaptive zones, leading to species that display adaptive novelties; and adaptive radiations into other adaptive zones or subzones that propagate the novelty so that the invading lineage becomes the trunk lineage of a new clade that may represent a genus, family, or even a higher taxon. According to these notions, the development of a new taxon is highly deterministic, depending upon selection imposed by environmental characteristics. Of course, the nature of the response to selection is governed in part by the nature of the organism. Nevertheless, there is evidence that environmental requirements are of major importance, as witness the convergence of lineages from very different phylogenetic backgrounds in similar morphological solutions to particular environmental opportunities.

On the other hand, extinctions seem to occur rather randomly. Extinctions are determined by loss of adaptation, but which species lose adaptation depends upon the kind of environmental change that occurs. It therefore follows that diversifications are determinant in the sense that they are called forth from the evolutionary potential of organisms by environmental change, but random in the sense that the environmental change is independent of the potentials, or chiefly so. Some studies by D. M. Raup and colleagues illustrate this point well. Using a computer, they experimented with a random diversification model. The computer began with a trunk lineage at time zero. At each successive interval of time, times *1, 2, 3,* and so on, the lineage could do one of three things: continue as before, split into two lineages, or become extinct. If splitting occurred, then each of the daughter lineages did one of these three things at the following time interval. The alternate events were selected at random; therefore, when the same experiment was repeated the outcomes were different because the same sequence of random choices would be very unlikely to repeat. The results of two computer experiments, run under identical instructions as to rates of splitting or terminations, are illustrated in Figure 10.15.

Each experiment produced a series of branches that were considered clades when they reached a certain size. The shapes of the clades reflect the diversification and extinction patterns formed by the randomized choices. Some of these clades have diversity histories very much like fossil groups. Clade A-16 displays an early burst of diversity but then gradually declines to extinction, much like the arthropod class Trilobita. Clade B-3 displays a gradual increase in diversity terminated by a relatively abrupt wave of extinction, much like the dinosaurs. Clade B-4 suffers a massive extinction and recovers to become highly diverse a second time; ammonites had a similar history, actually recovering repeatedly. Thus diversification under random control can mimic many of the patterns that we find in the fossil record, supporting the notion that environmental change is largely random with respect to adaptation.

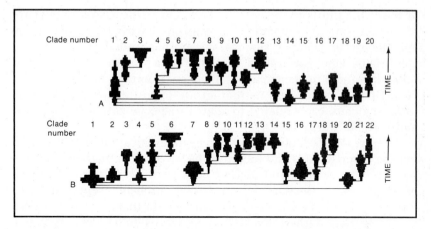

Fig. 10.15 Clades produced by random diversification and extinction models on a computer. The width of the clade indicates its diversity. The computer run was terminated arbitrarily, leading to the flat tops on clades that were diverse at that time. Some of the diversity patterns closely mimic actual fossil clades. (After Raup, Gould, Schopf, and Simberloff, 1973.)

Some patterns that occur in the fossil record have not shown up in these sorts of simulation models and are unlikely to because their occurrence by chance would be highly unlikely. These include mass extinctions in which many clades disappear at the same time, mass diversifications when many appear at the same time (except for the very first diversification, such as occurs in run A), and the persistence of clades at very low diversities.

When interpreting fossil patterns of diversification, clades that become more diverse or that endure longest are usually considered more successful. This is true by definition, but often such success is taken to imply superiority as well, and that is quite another matter. Superiority, like adaptation, must be related to a particular set of environmental conditions, and it can change from one taxon to another with environmental change. Since environmental change has a large random component, so does evolutionary success.

Biotic Turnover: Quality and Quantity

The pace and extent of environmental change can be regarded as a primary regulator of transspecific evolution. Extinctions are caused by environmental changes that exceed species' tolerances or lower the limits to species numbers. Diversifications are permitted when the environment is not saturated with species or when one lineage or clade is superior to another under the prevailing conditions. Undersaturation occurs after a wave of extinction or following a rise in the capacity of the environment to support species. Diversification and extinction are opposite sides of the evolutionary coin, each related to adaptive specialization in a patchy and variable world.

Figure 10.16 depicts the rates of diversification of families of marine invertebrates and includes the waves of extinction for comparison. The earliest and highest peak of diversification, the apex of which is in the Ordovician period, probably represents the filling up of marine habitats by animal phyla that appeared during the Cambrian. Other diversification waves follow closely upon mass extinctions. There is a continuous background of faunal turn-

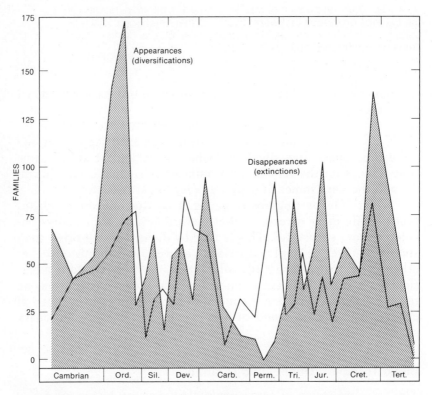

Fig. 10.16 Levels of diversification, indicated by their earliest fossil appearance, of families of shallow-water marine invertebrates. Extinction levels from Figure 10.12 are included for comparison. (After Valentine, 1973.)

over. The apparent coupling of diversification with extinction reminded Leigh Van Valen of a remark made by the Red Queen to Alice in *Through the Looking Glass:* "Now here, you see, it takes all the running you can do, to keep in the same place." It is as if evolution is hard-pressed to replace extinct species, and when it has done so it has merely returned the biosphere to a preextinction equilibrium to be upset by additional extinctions as fast as it is achieved. This is not literally the evolutionary case but it does tell part of the story.

The quality of the diversifications that follow mass extinctions is revealing. Figure 10.17 shows the times of appearance of orders and classes of marine invertebrates with durable skeletons, again with the peaks of family extinction included for reference. After the first great wave of diversification, higher taxonomic categories tend to appear following large waves of extinction (a class last appeared about 330 million years ago, and the last order to appear was some 50 million years ago). This suggests that the rise of higher taxa is particularly dependent upon the presence of prospective, unoccupied adaptive zones of considerable scope. Probably such wide adaptive zones could only be created by extinction once an initial diversification had "filled up" most of the marine habitats.

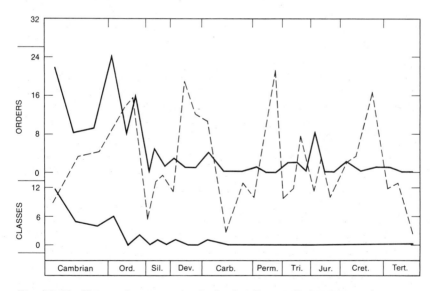

Fig. 10.17 Times of appearance in the fossil record of orders and classes of shallow-water marine invertebrates. Extinction levels from Figure 10.12 are included for comparison. (After Valentine, 1973.)

Figure 10.18 indicates the numbers of families, orders, and classes of marine invertebrates found at any one time during Phanerozoic time (the last 570 million years comprising the Paleozoic, Mesozoic, and Cenozoic eras). There were actually many more classes 400 million years ago than there are today and somewhat more orders. There are, however, many more families today than formerly. The general trend is for the higher categories to become relatively less diverse and the lower relatively more diverse. This is because of the pattern of extinction and replacement. Ordinarily, a class is not swept away at the height of its diversity but only after it dwindles away. Presumably the extinct species are being replaced, when permitted, by species from the surviving classes, which do not become modified sufficiently to warrant the erection of a new class; they are placed in new families or orders only, since they inherit the body plans of their own class.

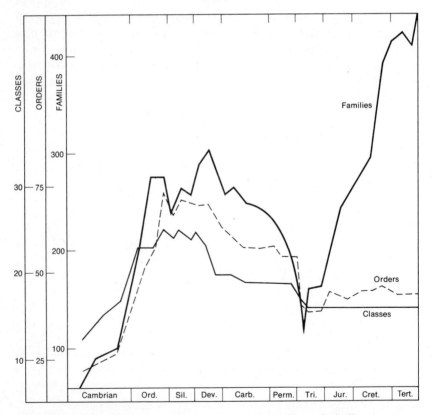

Fig. 10.18 Diversity of classes, orders, and families of shallow-water marine invertebrates during geological time. (After Valentine, 1973.)

Furthermore, it is likely that several classes may possess lineages than can exploit opportunities opened up by mass extinctions, so that the replacements for a given class must often be drawn from several surviving classes rather than just one. Thus as a phylum, class, or order becomes extinct it is usually replaced, not by another taxon on the same level, but by a lower taxon. The marine fauna has become less and less diverse as regards truly different adaptive types with distinctive body plans, but more diverse as regards different families with different minor morphological variations.

Extinction and Diversification

The long-term trend in species diversity has evidently not been one of steady increase, however. Because the fossil record and the taxonomy are poor at the species level, the numbers of described fossil species for any given time in the past may bear little relation to actual diversity trends. They may instead simply reflect the amount of fossiliferous sedimentary rocks of that age that happens to be preserved and exposed. Species diversity trends of the past must be inferred from other evidence. Genera have a better fossil record than species, since many species and therefore many more individuals make up the average genus, giving it a better chance of becoming fossilized. Families should have a better fossil record still, but at higher taxonomic levels we get progressively further from the species level where the processes of extinction and diversification chiefly operate. Several interpretations of species diversity trends have been proposed, but all must be regarded as tentative at present. We need not become embroiled in details of the arguments here.

Instead, consider the low diversity indicated at about 225 million years ago for all the higher taxa depicted in Figure 10.18. It is as certain as it is possible to be when dealing with biological events of such antiquity that the depressed diversity levels of that time are real. Note the lack of diversification for many millions of years before this low point (Fig. 10.16). This is not an artifact of preservation, since there are ample fossiliferous sediments known from this interval. The mass extinction wave near 225 million years ago is not an artifact either; most of the families that disappear do not have descendants that appear in later ages. The rediversification proceeded by and large from the few known survivors. Thus with little diversification and massive extinction, the number of families in the seas was greatly reduced. Since there are many species in an average family, the extinction of species must have been enormous, and species diversity must have reached its lowest point since the diversification of the early Paleozoic first "filled" marine habitats with invertebrates.

The massive marine extinctions correspond to the final assembly of the world's main continental masses into a single giant continent, Pangaea. As noted in Chapter 9, this reduced shelf provinciality by eliminating geographic barriers to dispersal. Marine climates seem to have been milder than today in high latitudes, at least insofar as temperature is concerned. Poleward-flowing currents would not have been blocked off from high latitudes the way modern currents are. Thus latitudinal provinciality was low. So far as we can tell there were only one or two marine provinces following the extinctions, although there may have been about six before. Additionally, sea level fell (Fig. 10.8), perhaps because the continents were joined and deep-sea ridges subsided, with a consequent increase in ocean basin volume. At any rate, there would have been less shelf area and less diversity of shelf habitats than formerly.

One result was that the coral and sponge communities of the Permian, which formed richly diverse reeflike associations, disappeared. The Permian communities were exposed as sea-level fell. Perhaps the muddy slopes that appeared under the shoaling water could not be colonized by the corals and other forms. Nearly all the Paleozoic coral species became extinct, as did vast arrays of associated shallow-water species. Finally, it is likely that the shelf environments became less stable, perhaps more seasonal, as the continent grew and emerged from its oceanic cover. Probably, all these effects would serve to lower the capacity of the shallow marine environment to support species. A more potent combination of causes could scarcely be found. Indeed this appears as the greatest crisis in history for the life of the marine realm.

Subsequently, Pangaea broke up into the present continents (plus, for a time, India, which separated from Africa but later joined Asia), and the diversity-repressive factors were ameliorated. As the continents spread apart, sea levels rose markedly (though they are rather low today), local faunas evolved in the separating continents, the climate became progressively more zoned and latitudinal provinciality increased, and in some regions maritime conditions replaced continental ones. One effect has been that marine species diversity has increased by several times at least.

The opportunities provided by the rising diversity capacity have been exploited by evolution to produce some new higher taxa. For example, modern corals (which comprise an order) were evolved from some Paleozoic allies of corals to fill the adaptive zones vacated by Paleozoic corals.

QUESTIONS FOR DISCUSSION

1. If seafloor has appeared at the Mid-Atlantic Ridge at the average rate of six cm per year at the ridge, how long ago did the Atlantic open? What were conditions like before then?

2. What will become of North America if seafloor spreading continues as at present?

3. What are the major environmental effects of continental drift?

4. If evolution is largely mediated by natural selection, why do extinctions and diversifications have some of the properties of randomness?

5. How could you arrange the continents to get the maximum marine diversity? The maximum terrestrial diversity?

Recommended for Additional Reading

Hallam, A. 1973. *A revolution in the earth sciences: from continental drift to plate tectonics.* Oxford: Clarendon Press.

Wilson, J. T. 1976. *Continents adrift and continents aground.* San Francisco: W. H. Freeman.

Windley, B. F. 1977. *The evolving continents.* London: John Wiley.

Three introductions to plate tectonic theory and its consequences. Hallam is the most basic, with Wilson and Windley each more advanced.

Raup, E. M., and Stanley, S. M. 1978. *Principles of paleontology.* 2nd ed. San Francisco: W. H. Freeman.

An excellent general introduction to the study of fossils.

11

The Major Evolutionary Pathways

In order to function in organisms, genetic material must possess three properties: it must contain information for the development and operation of an organism; it must replicate this information in descendants; and it must change the information in order that descendants may adapt to new conditions.

Nucleic acid genes possess these very properties. However, they cannot capitalize upon these properties without the aid of enzymes. In order that their information be transcribed, in order that protein products be synthesized, and in order that they be replicated, genes require the operation of enzymes. At least 30 enzyme systems are necessary to operate the genetic machinery of a simple cell stripped to the bare essentials. It seems clear that no such complex system of interacting molecules could have originated at a single step. Even the simplest imaginable nucleic acid-enzyme system must have evolved from much simpler forerunners. There must have been simpler ancestors to the simplest organisms known, and their genetic systems must have operated differently.

The Origin of Life

The ultimate question in phylogeny, then, is what was the very first organism which formed the trunk stem of all subsequent living beings? And the ultimate question in evolution is, how did the first organism arise? In other words, how did life originate? This problem has been under heavy investigation in recent years and the outlines of an answer are beginning to appear. The subject can be conveniently broken down into three major problems: (1) whether all the complex organic compounds required for life can be synthesized by nonliving processes; (2) whether such compounds, once formed, can be concentrated in sites where the reactions appropriate to the initiation of life can go forward; (3) and whether a process that improved and elaborated these early protobiological reactions can have occurred, so that their evolution into complex living cells could proceed.

Prebiotic Compounds

Simple organic compounds are regularly found in nature, independent of any biological activities. They are *abiotically* produced. Complex organic compounds (macromolecules such as proteins) are known to be forming only *biotically* today. Earth conditions today are so different from those of three and a half to four and a half billion years ago, when life evidently originated, that these facts may not be relevant. What we need to know is whether complex organic molecules of the sort that are required for life to begin can have formed under primitive earth conditions in the absence of life—under *prebiotic* conditions, as they are called.

An important experiment designed to test this question was performed in 1953 by S. J. Miller in collaboration with Harold Urey. Miller constructed a simple apparatus designed to imitate some aspects of the primitive earth environment. At that time, the primitive earth atmosphere was believed to have somewhat resembled the present atmosphere of Jupiter. Miller accordingly experimented with a mixture of methane, hydrogen, and ammonia, major components of the Jovian atmosphere. These gases were placed in a large flask and circulated through tubing by driving them with steam from a flask of boiling water (Fig. 11.1). The gas mixture was subjected to electrical sparks, a source of energy that is prevalent today (as lightning) and that was doubtless available in prebiotic times. Condensate from the steam and gas mixture was trapped at the bottom of the apparatus where nonvolatile products of any reactions could accumulate. The apparatus was

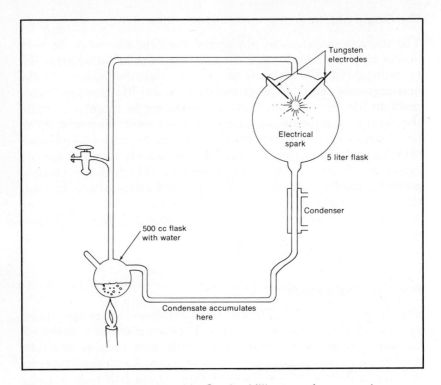

Fig. 11.1.   The apparatus used by Stanley Miller to perform a semi-nal experiment in prebiotic synthesis. A mixture of methane, hy-drogen, and ammonia gases (with water vapor) circulated through the tubing. Several biologically important organic compounds, in-cluding amino acids, were formed.

sealed and permitted to operate for a week. When the resulting condensate was analyzed, several organic compounds were found, including four amino acids. About 15 percent of the carbon in the system (in the methane) had become incorporated into new or-ganic compounds. Here was a demonstration that the prebiotic synthesis of the building blocks of proteins was an easy matter.

It no longer seems likely that the early earth atmosphere was Jovian, although free oxygen must have been essentially lacking. However, a great number of other gas mixtures have now been used in prebiotic experiments and many of them have provided ex-cellent yields of complex organic compounds. Indeed, 18 of the 20 amino acids found in proteins have been synthesized prebiotically, together with purines and pyramidines (constituents of nucleic acids), sugars, and other molecules that form building blocks for the large biological macromolecules. Amino acid chains have also been formed prebiotically into compounds called *proteinoids*. The prebiotic synthesis of nucleotides has not yet been accomplished, and this remains a problem. Nevertheless the ease of synthesis of the compounds most critically required as building blocks in biologically important large molecules has come as a surprise. There are many thousands of organic compounds, yet the very ones employed by biosynthetic processes are among the easiest to form prebiotically. Some of these key compounds are even known to occur naturally in meteorites and in interstellar dust clouds. It seems that abiotic reactions under quite a range of prebiotic and abiotic conditions can lead to the appearance of compounds required when life originated.

Supply and Concentration of Prebiotic Compounds

The yields of organic material during prebiotic syntheses are often large, so that when extrapolated to natural conditions they suggest that great accumulations could have occurred on the primitive earth. Based on the amounts of natural electrical discharges today it has been estimated that a three-foot layer of organic material over the entire surface of the earth could accumulate in only 100,000 years. Even if natural prebiotic syntheses were at much lower rates than this, there should have been quite a large amount of organic chemicals dissolved in prebiotic waters. L. E. Orgel estimates that there should have been on the order of a gram per liter, or about one-third the strength of a chicken bouillon. This primitive organic solution is commonly referred to as the *prebiotic soup* (or Haldane's soup, after the biologist who first suggested it).

Even at its near-bouillon strength, the prebiotic soup was too thin for the synthesis of biological macromolecules. Some type of concentration mechanism must have operated. Several possibilities have been considered—evaporation of the soup in shallow embayments or lakes, partial freezing of the soup into ice (which would concentrate organic materials in the unfrozen fraction), and the spontaneous formation of colloidal droplets of organic material, inside which prebiotic syntheses might go on, have all been suggested. Perhaps organic materials were absorbed onto the surfaces of minerals. Most minerals have well-defined crystal structures which, however, include imperfections that can give rise to net electrical charges on the mineral surfaces. Clay minerals, many of which are weathering products of the most common rocks and are abundant on the seafloor, are known to adsorb organic substances, preferentially concentrating the large chainlike molecules. Prebiotic lake or sea bottoms may have been sticky with adsorbed layers of organic molecules. Today such accumulations would be quickly destroyed by oxidation, but under some oxygen-free prebiotic conditions they might have been preserved. A further possibility, explored by Sidney W. Fox, is that proteinoids became aggregated into microspheres with catalytic properties—which they do spontaneously under certain conditions.

The Origin of Natural Selection

The creation and concentration of active molecules could not lead to living organisms unless there were some mechanism that preserved and duplicated the desirable molecules in preference to less useful ones. The number of possible molecules is enormous. Recall that each protein is constructed of a unique sequence of 20 kinds of amino acids. A moderate-sized protein may have 150 amino acids; the possible combinations of the 20 kinds in such a molecule are 10^{195}. This is a *very* large number; if the amino acid sequences in a prebiotic protein were generated by random processes, we could never hope to produce any given protein of this length on earth. It is estimated that there are about 10^{22} stars in the known universe. If each of these stars had a planet on which prebiotic syntheses were producing one billion different 150-amino acid proteins per second for five billion years, about 7.8×10^{25} combinations of amino acids would be produced—billions upon billions short of the total possible combinations. The likelihood of discovering a particular 150-amino acid combination by chance would still be minuscule. On earth we had only a billion years or less from the origin of the planet to the origin of life. We must conclude either that the particular proteins that exist on earth did not arise by random process, or that many amino acid sequences will lead to proteins having biologically interesting properties, so that the ones that happened to have appeared represent only one out of many billions of possible protein sets that could form the basis of living beings.

It is indeed possible that certain amino acid sequences were selected rather than created randomly. For example, natural templates may have existed—perhaps clay minerals—that controlled to some degree the ordering of the amino acids. Such a template would have some of the properties of a gene, in that the template structure would code for a particular protein or, perhaps more likely, a class of related proteins. A. G. Cairns-Smith has proposed a scenario in which the earliest genes were minute crystals that coded for proteins or protein classes because of their mineral structures. Sooner or later one of the proteins proved to have properties that preserved or increased the supply of its mineral template. Perhaps it catalyzed a chemical weathering process that produced the mineral. That particular crystal-protein combination would then be more fit than others and could become common and endure through time. Other proteins that further increased the fitness of this combination—by aiding in the capture of amino acids, or for binding the crystal-protein mass to the seafloor, or for whatever useful property—might become associated with it. At some point, nucleic acids were added to the evolving complex and their special replicating properties discovered; they then replaced the crystal genes. Eventually the colony of enzymes and other macromolecules was enclosed within a membrane and a cellular or protocellular organism had appeared.

This ingenious scenario is speculative but it does suggest that fitness can have arisen by some plausible prebiotic pathway. Another scenario envisions the early association of a nucleic acid with proteins, and the co-evolution of this primitive combination gradually to develop a more complex molecular behavior. The appearance of fitness is certainly the key to the spread of life; once fitness is established, natural selection may work to elaborate protoliving structures into more complex organisms.

Single-Celled Organisms: Prokaryotes

The earliest cell-like organisms are extinct; the most primitive living cells are prokaryotes, bacteria and cyanobacteria (usually called blue-green algae). They display great complexity but are nevertheless much more simply organized than the cells of all other organisms. Bacterialike forms appear in the fossil record in rocks about 3.5 billion years old and fossils that resemble blue-green algae appear in rocks of about 2.8 billion years old (Fig. 11.2).

The earliest life forms must have been heterotrophs, feeding upon the prebiotic soup. Probably the first actual cells were heterotrophs also. However, sources of energy other than the soup had to be developed, or life would have eventually depleted its food source. Autotrophy was evolved, probably many times. One bacterial group (Archaebacteria) is strictly anaerobic, that is, it cannot tolerate oxygen. Today these bacteria survive inside trees, in hot springs, and in other refugia where oxygen is absent; they take in carbon dioxide and hydrogen and give off methane. Other bacteria photosynthesize in anaerobic or weakly aerobic (oxygenic) conditions, and by methods slightly different from those of higher plants. The type of photosynthesis now dominant among auto-

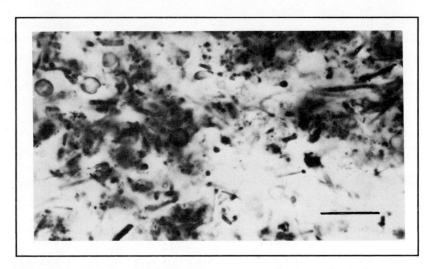

Fig. 11.2. An assemblage of the remains of primitive cells, both filamentous and unicellular blue-green algae, from the Gunflint Iron Formation, about two billion years in age. The bar is 20 microns long. (Photo courtesy of Dr. S. M. Awramik.)

Events in biosphere	Time (billions of years)	Events in planetary environment		
Extinction of dinosaurs	0.065			
Permian-Triassic extinctions	0.22		(Plate tectonics)	Free oxygen in atmosphere
First well-mineralized skeletons	0.57			
First animal fossils	0.70			
	1.0			
Possible early eukaryotes	1.4	Shallow marine sediments chiefly oxidized		
	1.91	Most banded iron formations	(Plate tectonics possible)	
	2.21			
(Prokaryotes becoming diverse)				Chiefly anoxic atmosphere
Probable blue-gree algae	2.8	Shallow marine sediments chiefly unoxidized	(Global tectonics unlike present)	
Oldest dated fossils (prokaryotic autotrophs)	3.4	Rocks chiefly granitic and gneissic, but sediments extensive		
	3.81	Oldest dated rocks (badly altered)		
	4.65	Origin of earth		

Fig. 11.3. Some of the principal events during the 4.65 billion year history of the earth. Life appears nearly as soon as rocks are found that could contain fossils.

trophic prokaryotes and universal among higher organisms probably arose before the advent of blue-green algae, which have chlorophyl molecules similar to those of higher plants. Some blue-greens can metabolize anaerobically but most are aerobic. Thus modern aerobic photosynthesis may have evolved by 2.8 billion years ago (Fig. 11.3).

Single-Celled Organisms: Eukaryotes

Since the sort of evolutionary advances that occurred in single-celled organisms are concerned chiefly with modifications in cellular machinery, they are not reflected in morphological changes that can be readily interpreted in terms of ecological significance, as with higher plants and animals. The record of these intracellular changes is largely equivocal. For this reason we are not sure when the single most important advance in evolutionary history actually occurred. This event was the development of the complex *eukaryotic* types of cells, without which multicellular organisms could not have evolved. It has been estimated that the evolutionary distance between prokaryotes and the advanced unicellular eukaryotes is twice the distance between animals and plants.

The most obvious differences between prokaryotic and eukaryotic cells is that the latter possess a nucleus, the dense, well-defined body that bears the chromosomes, and they undergo mitosis or meiosis during cell divisions. Eukaryotic cells also contain a host of metabolic machinery, termed organelles, that does not appear in prokaryotes. These include mitochondria (which perform respiratory and other functions) and sometimes chloroplasts (which contain chlorophyl and perform photosynthesis). The eukaryotic DNA is complexed (bound) with proteins in the chromosomes (prokaryotic DNA is not); and there is a convoluted intracellular membrane called the endoplasmic reticulum (lacking in prokaryotes). Finally, the ribosomes themselves, all alike in eukaryotes, are different in prokaryotes, for which two types are known; one type occurs in the anaerobic Archaebacteria, another type in all others.

Since all eukaryotes have basically similar cellular machinery it is usually assumed that they evolved only once. Lynn Margulis has developed a model of eukaryotic evolution that attributes the origin of many organelles to a process of symbiosis (Fig. 11.4). In this view, prokaryotic radiations produced a wide array of adaptive types, all prokaryotic in organization, that included heterotrophic ingesting forms. These ingestors took in a sequence of other cell types that became symbionts and eventually were integrated with the ingestor so as to become an inseparable part of the organism. Mitochondria were perhaps originally small aerobic bacteria, chloroplasts were perhaps originally blue-green algae, and so on. It may be that the large chromosomal complements of eukaryotes were built up from the DNA of ingested cells. Thus the remarkable jump in cell complexity achieved by eukaryotes can be explained as resulting from the amalgamation of separately

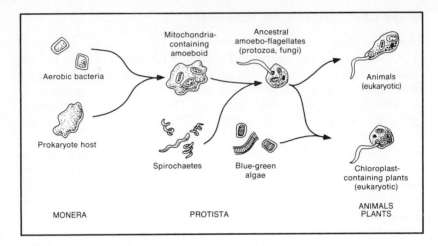

Fig. 11.4. The symbiotic theory of the origin of the eukaryotic cell from symbiotic prokaryotes. An ingestor prokaryote accepts as a symbiont an aerobic bacteria to give rise to an amoeboid form. Some of these become associated with prokaryotic spirochaetes to give rise to a flagellated protozoan ancestor. Some of these accept blue-green algal symbionts to become eukaryotic protistant ancestors of the higher plants. (After Margulis, 1970.)

evolved entities. Of course, evolution has subsequently modified the organelles and hosts alike in various ways, to accommodate each other and to harmonize with external conditions. This hypothesis has been debated and may yet prove untenable, but it seems likely that the chloroplasts at least have originated from blue-green algal cells.

However it happened, it is likely that the eukaryotic grade had appeared before 0.9 billion years ago, and possibly as early as 1.5 billion years ago. One of the more interesting evolutionary potentials of this grade was revealed with the evolution of sex. Even before sex developed, eukaryotes had probably been diploid (had two chromosome sets) by suppressing a reduction division during reproduction. The advantage of the diploid condition was presumably that its possessors displayed hybrid vigor. Once established, further advantages were uncovered—the ability of a dominant allele to mask the ill effects of a deleterious recessive mutant, for example. Reduction divisions to produce haploid gametes could then be employed as part of a sexual cycle. Thus gene pools could be formed, with all their adaptive versatility.

George C. Williams has suggested that when sexuality first arose, it alternated with asexual reproduction in the same populations; alternation of sexual with asexual generations is common in protozoa, cnidarians (jellyfish and their allies), and in many plants today. Since the asexual generations would possess the same genotypes as their parents, they would be most advantageous when the parents were well-adapted and when conditions were stable. If conditions changed, however, entire populations of asexual organisms could be extinguished. Sexual generations, on the other hand, produce a great variety of recombinants, many of which will be less fit than the parents (usually 50 percent), but when conditions change there is more likely to be some well-adapted recombinant that can perpetuate the lineage.

Primitive Multicellular Organisms and the Kingdoms of Organisms

As yet, fossils of truly primitive multicellular organisms have not been discovered. They arose after the eukaryotic grade had appeared and after sex had evolved, but probably well before the first confirmed traces of animals, near 700 million years ago. Actually, multicellularity was achieved independently numerous times; G. Ledyard Stebbins has put the minimum number at 17. This indicates that it was a great advantage, and perhaps it occurred in some lineages soon after the perfection of the processes of eukaryotic cell division.

The most obvious advantages of multicellularity arise from the modular repetition of cellular machinery. This permits size increases, shape variations and increased reproductive potential, and promotes homeostasis (the ability of an organism to maintain stability in its internal environment). Because individual cells can be replaced, multicellularity greatly enhances longevity. It also permits the differentiation of function among the cells. Probably the particular advantages that were promoted by selection varied from case to case as various lineages explored the multicellular grade.

Two major evolutionary paths from unicellularity to multicellularity have been suggested (Fig. 11.5). In one, cells first aggregate to form colonies to exploit size or shape advantages. The efficiency of the colonial association is increased by bringing the cells under a central control, increasing the advantages of intracellular exchange of cell products and regulating metabolic patterns so as to benefit the entire aggregation. The numbers and differentiations of cell masses may then be specified in the regulatory portion of the genome, which accordingly becomes more elaborate. The other path begins with a complex, multinucleate unicellular form such as is represented by some of the ciliate pro-

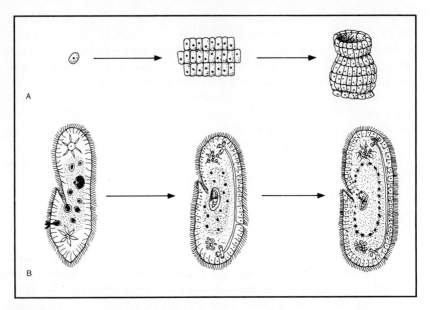

Fig. 11.5. Two possible pathways from unicellularity to multi-cellularity. *A*, a unicellular species becomes colonial and then is increasingly integrated to produce a multicellular organism. *B*, a unicellular species with multiple nuclei (a ciliate) evolves into a multicellular form by partitioning some of the nuclei into individual cells. (*B* after Hadzi, 1963.)

tozoa today (Fig. 11.5). If such a form were to become cellularized so that each nucleus and surrounding parts of the cell became enclosed in a membrane, a multicellular organism would result. The ensuing advantages would be the same as in the colonial pathway. A general difficulty with this idea is that multinucleate unicells usually exhibit highly specialized behavior, which would have to be deprogrammed when multicellularity appeared.

The highly polyphyletic nature of multicellular organisms makes it difficult to classify them at the highest taxonomic level, the kingdom. This is especially true for plants, many different lineages of which have achieved multicellularity independently. The concept of a plant as an autotrophic multicellular organism is nevertheless most useful. As a practical matter, then, kingdoms have been defined so as to distinguish organisms of different major grades and, when appropriate, to distinguish those with major differences in energy sources. Figure 11.6 illustrates a scheme, modified from one proposed by Robert H. Whittaker, that has gained some acceptance. Five kingdoms are recognized. The most

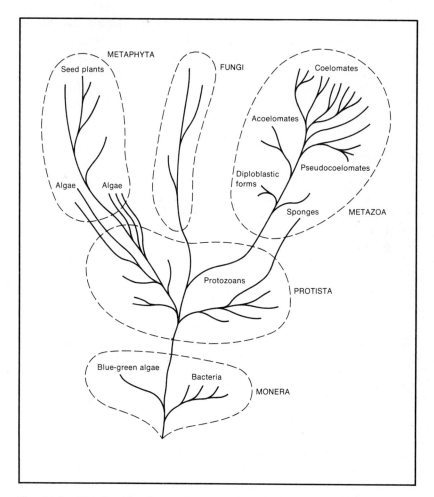

Fig. 11.6. The five kingdoms of organisms: unicellular prokaryotes (Monera), unicellular eukaryotes (Protista), and arising from various protistan branches, multicellular plants (Metaphyta), fungi, and animals (Metazoa). (After Whittaker, 1969.)

primitive kingdom, the Monera, consists of the unicellular pro-karyotic bacteria and blue-green algae. The second kingdom, Protista, includes the unicellular eukaryotes that arose from the Monera, perhaps only once. The Monera and Protista each include both autotrophs and heterotrophs, and indeed the distinction between these types of organisms is of little value in unicells. Different families of the same order, or even different genera of the same family, may be autotrophic and heterotrophic.

From the Protista, multicellular forms developed, and these are classed into three kingdoms according to their energy sources: Metaphyta (plants), which photosynthesize; Fungi (such as mushrooms), heterotrophs that absorb their food but are otherwise rather plantlike; and Metazoa (animals). Each of these three kingdoms is polyphyletic, but there is not special advantage to be gained by breaking them up into, say, 17 kingdoms.

Lower Metazoan Grades and Body Plans

It is likely that living animals represent only two independent lines of descent from protistan ancestors; one is represented by a single phylum, the Porifera (sponges), and the other by all the remaining animals. Sponges are therefore placed in a separate division, the subkingdom Parazoa, and other animals are in the subkingdom Metazoa (or sometimes Eumetazoa). Since there is no direct fossil evidence of the pathways by which the Metazoa arose and radiated into phyla, we must use inferences derived from studies of the functional significance of primitive characters of living phyla and of the fossils that we do have, and also from the comparative developmental morphology of living groups. These data do not define a unique solution, however, and a wide variety of hypotheses have been suggested to account for metazoan origins. The differences between them revolve around the basic structures of body plans of living phyla and the identification of primitive features.

The chief grades of organization of animals are depicted in Figures 11.7 and 11.9. Complex animal body plans are composed of a hierarchy of structures, of which the cell forms the basic building block (see Chapter 8). The simpler organisms are single cells, and the simpler multicellular organisms are at the tissue level of construction. These are the sponges, which have differentiated cells but lack organs. Their body walls (Fig. 11.7, A) contain outer and inner cell layers. Sponges are all aquatic; they pump water through intracellular pores in their body walls into a central cavity, ingesting minute food items intracellularly. The water then passes out through a top opening (Fig. 11.7, A). The next most complex animals are cnidarians, such as corals and jellyfish. They have two tissue layers, termed ectoderm and endoderm, which line the exterior and the gut respectively (Fig. 11.7, B, C). This condition is termed *diploblastic* (two layered). These animals have organs such as gonads differentiated from the tissue layers.

A still more complex grade is represented by the phylum Platyhelminthes, the flatworms, which have three well-defined tissue layers: ectoderm, endoderm, and mesoderm, an internal tissue

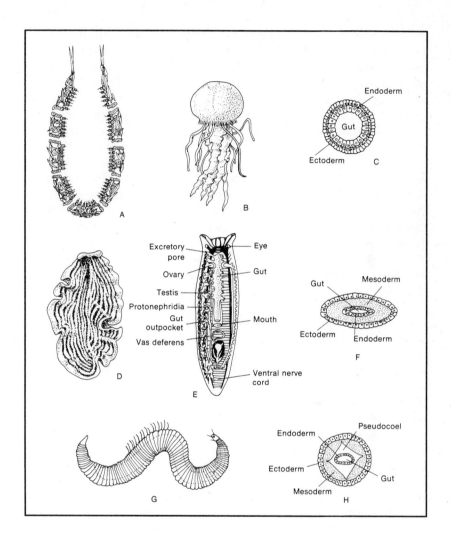

Fig. 11.7. The grades of organization of lower invertebrates. *A,* cross section of a simple sponge, a vase-shaped animal with minute incurrent pores and an excurrent opening at the top of the body. Several cell types are present. *B-C,* a cnidarian, a diploblastic animal such as the jellyfish *(B)* with a two-layered body wall *(C)* containing numerous cell types and having some differentiated organs. *D-F,* a flatworm *(D),* a triploblastic animal lacking a body cavity *(E,* a simplified transverse section of a flatworm) and with a serial repetition of internal organs *(F). G-H,* a nematode *(G),* a pseudocoelomate with a jelly-filled body space lying between the gut endoderm and the mesoderm lining the body wall *(H,* a diagrammatic transverse section of a pseudocoelomate). Note that the body is highly annulated (with numerous encircling rings) but not internally divided.

layer. They reach a triploblastic (three layered) condition (Fig. 11.7, *D-F*). Flatworms have complex systems of reproductive, excretory, and nervous systems organized in a characteristic body plan. They lack distinct respiratory and circulatory systems, breathing through the body surface. This accounts for their flattened shape, which minimizes the distance that oxygen must diffuse from the surface to the most interior cells. Flatworms feed through a subcentral mouth that leads into an elongate gut; intracellular digestion occurs in the endodermal gut lining. In order to nourish tissues throughout the body, the gut is provided with a series of lateral outpocketings, particularly in the larger species. Excretory, reproductive, and other organs occupy the spaces between the gut outpockets. This results in a seriated architecture, a more-or-less regular repetition of organs down the length of the body.

Another phylum at this grade, the Nemerteans, are wormlike animals less flattened than flatworms that contain a core of tissue that acts as a hydrostatic skeleton to antagonize muscles. When body wall muscles contract in one region so as to squeeze in the body wall, the rather incompressible tissues bulge out elsewhere. If the contracted muscles then relax and the muscles at the bulges contract instead, tissues can bulge into the former contractions to restore the body to its original shape. By coordinated muscular contraction, bulges can be made to flow down the length of the worm; this is *peristalsis*. Burrowing into soft sediments can be accomplished by peristalsis. The bulges are pushed laterally into the sediment and the worm can flow forward as the bulges migrate backward over the body. Flatworms cannot burrow by displacing sediment, but nemertine worms can.

There is a group of minute triploblastic worms of several phyla, the relationships of which are yet to be worked out, which have a fluid-filled body cavity developed between the gut and body wall (Fig. 11.7, *G, H*). During the ontogeny of metazoa there is an early stage in which the dividing cells form a single-walled sphere, the *blastula*, with a hollow center (the *blastocoel*). The hollow blastula is then invaginated to form a double-walled sphere, the *gastrula* (Fig. 11.8). It is the hollow of this new sphere that becomes the digestive cavity, and its lining becomes endoderm; the lining on the outside of the sphere becomes ectoderm.

In this group of phyla, the blastocoel cavity persists into the adult body, where it is termed a pseudocoel, forming a jelly- or fluid-filled space between the endoderm of the gut and the body wall. The pseudocoel can act as a hydrostatic skeleton to antagonize muscles for hydrostatic locomotion or for some other func-

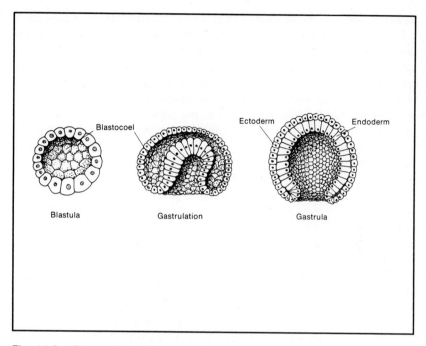

Fig. 11.8. The early developmental stages of metazoa include the formation of a hollow blastula which invaginates to form a double-walled gastrula. The external layer of the gastrula becomes ectoderm, the internal layer, endoderm. This sequence is highly modified in some lineages.

tion. Some minute animals with pseudocoels have evolved long proboscislike probes by which they burrow or capture food; the proboscis is operated by intrinsic muscles and the hydrostatic skeleton of the pseudocoel. Many of the minute pseudocoelomates are parasitic; many species of the phylum Nematoda cause diseases, including elephantiasis in humans. This grade of organization is termed *pseudocoelomate*.

Higher Metazoan Grades and Body Plans

The highest major grade of metazoan architecture is the *coelomate* grade, to which belong the arthropods (such as insects), the chordates (including humans), and numerous other phyla. In coelomates, the blastocoel is occluded during development, but a different internal cavity appears within mesodermal tissues. This is the *coelom* or "true" body cavity. It is filled with fluid and lined with mesoderm and communicates to the exterior only through special ducts (*coelomoducts*) that carry reproductive or excretory products. The coelom may, then, house such organ systems as gonads and kidneys, and provide space for such pulsating organs as the heart. Robert B. Clark has extensively examined the functional aspects of coelomic body plans. He concludes that in primitive wormlike coelomates the coelom functioned as a hydrostatic skeleton, just as did the tissue and pseudocoel skeletons of other grades. Most coelomates have circulatory systems and so possess still another system of fluid-filled spaces that contain the blood (the *haemocoel*).

There are several major kinds of coelomate-grade body plans. The phyla that share similar coelomic plans usually resemble each other in other ways as well and have probably descended from common ancestors. The ancestors of some plans seem to be extinct but others may be represented among living phyla.

Perhaps the most familiar coelomate body plan is exemplified by the bilaterally symmetrical annelid worms (including the common earthworm) and the arthropods (including insects). In earthworms, the body is divided by a large number of transverse partitions or *septa* into numerous separate *segments* or *meres* (Fig. 11.9, A-C). In what is evidently the more primitive condition, each segment is equipped with a paired set of organs, usually including pairs of excretory organs, coelomoducts, and nerve ganglia. This architecture is termed *metamerous*.

The septa prevent transfer of coelomic fluid between segments; blood vessels and other organs that pass between segments are surrounded by sphincter muscles to ensure against fluid leakage where they penetrate the septa. Every few segments are tied together by short longitudinal muscles in the body wall. Without the septa, effects of muscular contractions associated with peristaltic locomotion would be transferred the length of the coelom, but in annelids these effects are damped down and localized within a few neighboring segments. Thus some terminal segments can be freed from the pressures of locomotion. More than one peristaltic wave can be present at the same time. This greatly enhances burrowing efficiency. Those worms that burrow most continuously, chiefly deposit feeders and predators, are the ones with the more complete segmentation. In annelids that burrow only occasionally, as is the case with many suspension feeders, many of the septa become obsolete to create large coelomic chambers. The regularized repetition of the internal organs is retained, however.

Arthropods have developed a hardened exoskeleton of organic material that in some groups is further strengthened by mineralization. This skeleton provides support for a muscular system that operates a number of jointed appendages. The septation seen in annelids is lacking and the coelom itself greatly reduced. The burrowing functions of the hydrostatic skeleton are no longer required. Body turgor is provided by the haemocoel or blood system rather than by coelomic fluid. However, the metameric origin of the group is still recorded by the regularized repetition of paired internal organs.

A second coelomic architecture, termed *oligomerous,* is displayed by worms with one or two transverse septa that divide the coelom into two or three regions. (Fig. 11.9, *D, E*). It is likely that the primitive oligomerous forms were relatively sedentary burrowers with tentacles that were used for food gathering and respiration, somewhat resembling the living phylum Phoronida. Phoronids have coelomic trunks and tentacles that are separated by a septum. The trunk coelom is used in burrowing while the tentacular coelom functions to manipulate the tentacles. Two other living phyla (Brachiopoda, Bryozoa) resemble phoronids in coelomic plan but live upon the seafloor; they have calcareous exoskeletons.

There is another group of oligomerous phyla that have an unusual developmental pattern and are sometimes considered to represent a major branch of the coelomates equivalent to all other groups together. This group includes the chordates and hence human beings. A major developmental difference is that in most of the other animal groups, the opening into the gastrula hollow eventually develops into the location of the mouth (*protostomous* condition) while in this group, the Deuterostomia, the mouth develops at a new site (*deuterostomous* condition). Some of the simpler living deuterostomes have tentacles that resemble those of the phoronids. Some of these deuterostomes are primarily suspension feeders and have evolved slits behind the mouth region to enable water, from which food items are being trapped, to escape the pharynx without regurgitation through the mouth. One deuterostome phylum, the Echinodermata, elaborated a tentacle system into an extensive hydrostatic water-vascular system, as it is called. Starfish, for example, have water-vascular canals that radiate from a central ring to run down each ray. From these canals, short tentaclelike projections arise (tube feet) that are used for locomotion, capturing prey, and sometimes for respiration.

A third coelomic plan is the *amerous* one, which consists of a simple unregionated and unsegmented coelom that extends most of the body length. Amerous worms such as the Sipunculida (Fig. 11.9, *F*) possess a proboscis that can be extended for burrowing or

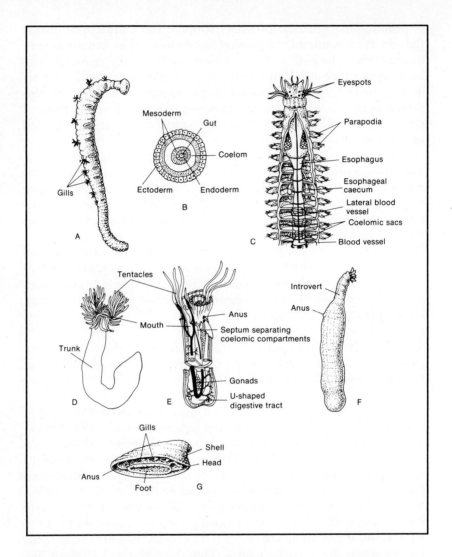

Fig. 11.9. The grades of organization of higher invertebrates. *A-C*, annelid worms (*A*, a marine example) which are coelomates (*B*, diagrammatic transverse section to show body cavity) and internally segmented (*C*, anterior end of annelid showing compartmentalization of the coelom). *D, E*, a phoronid (*D*), an oligomerous coelomate with a coelomic compartment in the trunk for burrowing and a second in the tentacular region (*E*). *F*, a sipunculid, an amerous coelomate with a proboscis (or introvert); there is no internal coelomic compartmentalization at all. *G*, a primitive mollusk, tilted to show serial arrangement of gills; some other organs are also repeated in series.

to feed. The proboscis is operated hydrostatically by coelomic fluids; it is probably an advantage to have a large undivided coelom in this case.

Finally there is a further coelomic type called *pseudometamerous* and exhibited by the phylum Mollusca (clams, snails, octopuses, and so on). Primitive members of this phylum have unsegmented or unregionated coeloms that are virtually restricted to ducts and to spaces around the heart. They also have seriated organs (muscles, gills, excretory organs, reproductive organs) that recall this condition in flatworms, from which they probably evolved (Fig. 11.9, G). Some advanced mollusks have more extensive coelomic spaces, but there is not a large coelomic space surrounding the gut, as in all the other coelomic body plans. The molluscan coelom may have evolved independently of those others, though this is not yet certain. The primitive living mollusks have exoskeletons and were probably epifaunal in habit when they evolved, browsing on algae or ingesting detritus on the seafloor.

The Fossil Record of Early Metazoans

The wide array of invertebrate metazoan body plans implies an array of opportunities for multicellular organisms. Despite the fact that the major metazoan grades can be arranged in a sequence of increasing complexity, it is not certain that this represents an evolutionary sequence.

The early fossil record of metazoans helps in some respects but does not provide us with decisive answers to phylogenetic questions. The fossils that are most likely to be preserved are trace fossils, such as burrows and trails, which affect the texture of sediments. When sediments are lithified, these traces are commonly preserved and even amplified. The trace fossils provide evidence of animal activities. Modern marine sediments are highly disturbed by burrowing and crawling organisms. In a like manner many ancient sedimentary rocks are crowded with trails and burrows (Fig. 11.10).

The next most easily preserved sorts of fossils are durable skeletons, usually composed of calcite or of some other carbonate mineral. Some organic skeletons also preserve well under certain conditions. Finally, soft-bodied forms that lack durable skeletons are only preserved in abundance at a few precious fossil localities, as impressions in the sediments or as thin mineral films. The first fossil appearance of a taxon depends to a large extent upon

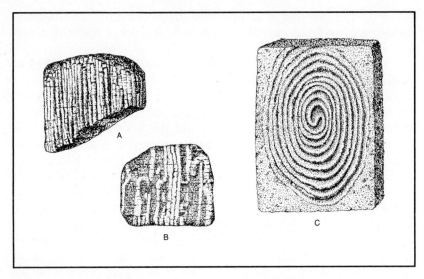

Fig. 11.10. Two kinds of trace fossils; *A,B, Skolithos*, a vertical burrow type known from the late Precambrian onwards. *C, Spirorhaphe*, probably a grazing trail, known from the Cretaceous and early Cenozoic.

whether it leaves a trace that is definitive (few taxa do) or has a durable skeleton; otherwise, it is largely a matter of luck.

The earliest animal fossils are not closely associated with rocks that can be dated radiometrically, so we are not sure of the exact sequence of occurrence. The earliest of all appear to be traces, both horizontal and vertical burrows, that may date from around 700 million years. These are most interesting fossils because the ability to burrow implies a relatively advanced, or at least complex, metazoan grade; it implies the possession of an efficient hydrostatic skeleton. While some diploblastic animals manage to burrow by using water-filled guts as hydrostatic skeletons, the long horizontal burrows of the late Precambrian suggest a more active animal, perhaps a pseudocoelomate but more likely (since the burrows are not minute) a true coelomate.

The next younger animal fossils that are known are, surprisingly, soft-bodied forms. An assemblage of chiefly diploblastic animals has been recovered from late Precambrian rocks between about 680 and 580 million years old. The animals are referred to as the Ediacaran fauna, from rocks in southern Australia from where the fauna is best known. The clearly identifiable fossils are all cnidarians, diploblastic in grade (Fig. 11.11, *A*). A number of the fossils are quite enigmatic and cannot be assigned with certainty to living phyla (Fig. 11.11, *B*). A few appear to be worms and may

Fig. 11.11. Late Precambrian soft-bodied animals of the Edi-acaran assemblage. *A, Cyclomedusa,* a jellyfish; *B, Marywadea,* which resembles some annelids (although the internal structure is unknown). *C, Tribrachidium,* a triradiate form of unknown affinities. (After Glaessner and Wade, 1966, courtesy of M. F. Glaessner.)

even be annelids or protoannelids, although their relationships are still in doubt (Fig. 11.11,C). The Ediacaran fauna is now known from a number of localities on several continents (Australia, North America, Africa, Europe), which is quite remarkable considering the difficulties of preserving such organisms.

Mineralized skeletons finally appeared around 580 million years ago. The first ones, described from Russia, are minute and of uncertain affinities; probably they are supporting or protecting structures of larger organisms. Near 570 million years ago, at the beginning of the Cambrian period, living phyla with mineralized skeletons make their first appearance. By the close of the Cambrian, near 500 million years ago, all but one of the mineralized living phyla are known (Fig. 11.12). With few exceptions the primitive members of these phyla lived epifaunally (upon the seafloor) rather than infaunally (within the sediments), perhaps in the case of trilobites digging shallow pits, but not occupying fully infaunal burrows. One not uncommon group of fossils, the Archaeocyatha, is known only from Cambrian rocks and may well represent an extinct phylum, perhaps at about the same grade of organization as sponges.

One other notable fossil occurrence bears heavily on the origin of animal phyla. A diverse assemblage that includes many soft-bodied animals has been recovered from a middle Cambrian for-

Eon	Proterzoic	Phanerozoic										
Era		Paleozoic							Mesozoic			Ceno.
Period	Ediacaran	Camb.	Ord.	Sil.	Dev.	Carb.	Perm.	Tri.	Jur.	Cret.	Tert.	Q.
Age (10⁶ yr)	700	570	500	440	395	345	280	225	190	135	65	1.8

Approximate first appearance of a phylum

- Cnidaria
? • Annelida
- Porifera
- Mollusca
- Brachiopoda
- Arthropoda
- Pogonophora
- Echinodermata
  - Priapulida
  - Hemichordata
- Chordata
  - Bryozoa
  ? • Urochordata
- Phoronida
- Sipunculida
  - Chaetognatha
  - Nemaroda
  - Echiuroida
  - Nemertina

Ctenophora •
Rotifera •
Acanthocephala •
Gastrotricha •
Kinorhyncha •
Entoprocta •
Platyhelminthes •

Fig. 11.12. The times of appearance in the fossil record of living phyla.

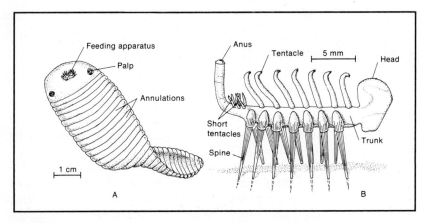

Fig. 11.13. Some soft-bodied animals from the middle Cambrian Burgess shale of Canada. *A, Odontogriphus; B, Hallucigenia.* In company with some other Burgess shale forms, they cannot be assigned to living phyla, for each has a body plan as distinctive as a separate phylum. (After S. Conway Morris, 1976, 1977.)

mation in British Columbia called the Burgess shale, roughly 540 million years in age. We are extremely fortunate to have the Burgess fauna; it is near enough to the early Cambrian appearance of mineralized skeletons so that it supplements them nicely, and it can be contrasted with the Ediacaran assemblage of late Precambrian times. The Burgess is teeming with fossil types, 20 percent of which cannot be assigned to living phyla (Fig. 11.13). Some of the remaining 80 percent are phyla that have not been found in older rocks (the Priapulida, usually considered to be pseudocoelomates; the Hemichordata, a deuterostome phylum; and the Chordata). S. Conway Morris has carefully restored several of the fossils that cannot be assigned to living phyla; their anatomical organizations have been well enough preserved to establish that they belong to previously unknown phyla that are not ancestral to living groups. Rather, they represent additional phyla that became extinct at some as yet unknown time.

Clues to Early Metazoans From Living Models

From the fossil record a model of the events that underlie the appearance of metazoan phyla can be suggested. As with all such historical models, it is subject to change as new evidence is brought to light. Some time before 700 million years ago, multicellular organisms that were ancestral to living Eumetazoans were evolved. There are some famous speculations as to the form of the earliest metazoans. Ernst Haeckel, a leading late nineteenth century proponent of the idea that ontogeny recapitulates phylogeny was forced by this notion to hypothesize that the earliest metazoa had the form of the earliest ontogenetic stages.

The early hypothetical ancestor that had the form of a blastula was called a *blastaea*. The posterior cells of the blastaea were supposed to have become specialized for feeding and invagination of the blastaea (now preserved in ontogeny as the invagination that produces the gastrula) produced a hypothetical animal called a *gastraea*. The cavity of the gastraea was a primitive gut lined with feeding cells that became endoderm. These ideas are no longer taken very seriously. E. Metschnikoff, a contemporary of Haeckel, suggested a different primitive metazoan, one that was solid. Digestive cells arose in the external cell layer and then migrated into the interior, where digestion proceeded intracellularly. An organism of this sort would closely resemble the larvae of coelenterates, called *planula* larvae. This hypothesis still commands some support.

We have previously mentioned the idea, championed especially by J. Hadzi, that the earliest metazoans were flatworms that arose by cellularization of a multinucleate protozoan. If this is correct, then the diploblastic phyla have evolved a simpler architecture than that of their ancestors. We cannot rule out this idea completely at the present time. In general, however, the more advanced we consider the original metazoan to have been, the more difficulties we encounter in developing a historical model to explain the subsequent evolutionary pathways.

It is clear from the examples above that a great variety of early metazoan forms can be imagined. Most investigators have sought clues from different living models—advanced protozoans, early developmental stages, early larvae, and so forth. This is not unreasonable, but nevertheless there is no guarantee that any living model at all resembles the earliest metazoan. Effective developmental or larval stages are adapted to their own requirements and need not reflect phylogeny.

Functional Clues to Early Metazoans

If we attempt to construct an early metazoan animal from, say, a colony of protozoan cells, we can imagine quite a variety of plausible forms, depending upon the environment. On the seafloor in shallow water, domelike shapes can be particularly advantageous because they provide for considerable surface area and at the same time rise a bit into the water column to create turbulence in any ambient current, thus increasing the volume of water that actually bathes the organism. This increases food and oxygen supplies. In very quiet waters, cylindrical shapes would provide access to a larger part of the water column. Elongate filaments or tentacles would have the same effect. Forms that became pelagic (floating) might be relatively globular or have radial symmetry. Detritus feeders that keep on the bottom would find advantage in bilateral, elongate forms.

In any of these hypothetical forms, cellular differentiation would be advantageous. For benthic detritus feeders, for example, food would be most easily available to cells in contact with the substrate, and it is reasonable that these cells would become specialized for digestion. Internalization of these cells into a cavity would increase their numbers and help stabilize the digestive process. The upper cell layers would then provide support and protection. Locomotion might fall to ciliated cells peripheral to the digestive region. Well-nourished cells near the gut lining might specialize for reproduction. The morphological pattern of cell differentiation would be different in suspension feeders (digestion might be dorsal within a cup-shaped invagination) or in floating forms (digestion might be posterior or in some pocket or localized region so placed as to best trap food particles).

We can imagine indefinite varieties of simple metazoa, but without historical evidence, and without a compelling theory to limit the possibilities, we cannot choose the more likely metazoan ancestor from among the possibilities. On the other hand, the origin of selective pressures to develop the array of differentiated functions that typify metazoa presents no conceptual difficulties. By 700 million years ago, it appears that floating and bottom-dwelling diploblastic forms were fairly common, and that burrowing forms had arisen as well.

Evolution of Coelomate Epifauna

The early metazoans lacked protective skeletons and could find both protection and a poorly exploited food resource in shallow layers of marine bottom sediments. Burrowing efficiently, however, required the development of a behavioral repertory not available to acoelomates, and thus the coelom was developed, probably from preexisting body spaces, within mesodermal tissue. The appearance of the coelomic space during ontogeny varies among phyla, suggesting that it may have arisen by more than one pathway and may be polyphyletic. However, the developmental stages might easily have diverged from a common ancestor; we simply have no evidence as yet that can be trusted implicitly. At any rate, once developed, the coelomic worms proved highly successful and exploited several distinctive modes of infaunal life, feeding on suspended material (oligomerous forms), surface detritus (amerous forms), buried detritus (metamerous forms), and probably living prey. The pseudometamerous lineages may have returned to the surface to feed on bottom detritus, or they may have developed as primitive surface-dwellers. Obviously, it is not known which coelomate body plan was the more primitive.

Near 570 million years ago a distinctive coelomate epifauna evolved, and it is likely that it descended primarily from the infaunal worm stocks of late Precambrian times (Fig. 11.14). Many of these new lineages were equipped with durable skeletons. In most cases the coadaptation of the skeletons and the soft-part anatomy is so close that it is very likely that they evolved together. That is, the origin of these phyla was not through a soft-bodied form developing a mineralized skeleton to suit its body plan; but of a new body plan arising that included a mineralized skeleton as one of its elements. The ancestors of the skeletonized phyla were different phyla.

Preston Cloud has presented an example that makes this point clearly. The articulate brachiopods are a bottom-dwelling oligomerous group that appeared during the early Cambrian (Fig. 11.15). They have paired shells. The body consists of a trunk containing viscera and digestive track, prolonged posteriorly into a

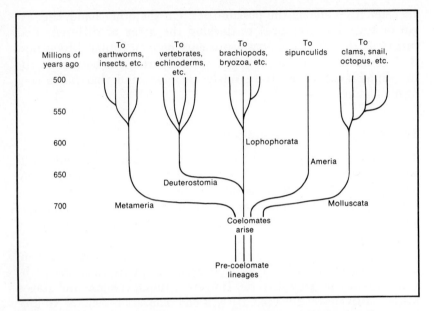

Fig. 11.14.   Radiations of the coelomates suggested by the fossil record. The coelom first appeared near 700 million years ago; subsequently there was a radiation into a number of distinctive coelomic architectures (Metameria, Deuterostomia, and so on), chiefly used by burrowing worms. Near 570 million years ago a wave of diversification occurred among several separate coelomate stocks to produce many of the modern phyla, which lived chiefly on the seafloor. The mollusks may have evolved their coelomlike body spaces separately. (After Valentine, 1973.)

fleshy stalk (the pedicle) that attaches the animal to the seafloor; and of a tentaculated crown, or *lophophore,* for feeding. There are a number of muscles that open, close, and adjust the shell parts and manipulate the pedicle. The shell and lophophore are arranged so as to provide an efficient circulation of seawater over the lophophore. If the shell were not present, many of the muscles would not be required or could not function; the organisms would become a saclike form with tentacles, much like a phoronid worm. The brachiopod mode of life, skeleton, and body plan are interdependent—a naked brachiopod could not exist. Since brachiopods first appear in lower Cambrian rocks, it is likely that they first evolved during that epoch. Similar arguments lead to the conclusion that most or all of the durably skeletonized phyla originated only shortly before their remains appear in the fossil record; the early Cambrian thus marks a major radiation of coelomate phyla.

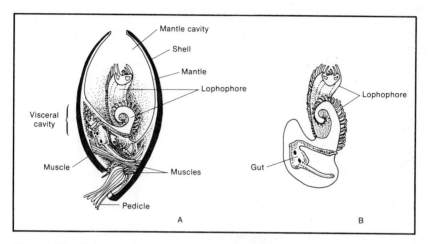

Fig. 11.15. Gross anatomy of an articulate brachiopod. *A,* longitudinal section to show visceral coelomic cavity with gut, glands, and other organs, and a lophophore containing another coelomic cavity, all enclosed within a mantle-and-shell and provided with pedicle and musculature. *B,* saclike polyp with tentacles resulting if shell is removed. Probably the soft-bodied brachiopod ancestor resembled a phoronid (Fig. 11.9, *D, E*). (After A. Williams, 1956.)

The general advantage of rigid skeletons as supporting structures is that they do not deform. Unlike hydrostatic skeletons, no energy is required to control their shapes. Thus although they require some energy to secrete, they save energy thereafter and are efficient evolutionary devices. When soft-bodied worms burrow, their body walls are partly supported by the confining burrow walls. When these worms crawl on the seafloor, however, they must expend energy in controlling their body shapes, and furthermore the peristaltic locomotory waves, which work so well when they are burrowing and are surrounded on all sides by sediments, would only be useful along the narrow strip of body in contact with the seafloor. Rigid skeletons can greatly reduce energy expenditures for epifaunal locomotion. Furthermore for sessile forms (attached or restricted to one locality) the skeleton is employed to enhance feeding capabilities.

The brachiopods furnish one example. The protective function of the burrow can be replaced in the epifauna by rigid skeletons. These considerations may explain the association of the mineralization of invertebrate skeletons with the expansion of metazoan coelomates into epifaunal communities. New body plans, involving rigid skeletons in many cases, were required to cope with conditions above the seafloor. The new epifaunal phyla in-herited the basic coelomic plans of their wormlike ancestors but modified them for life in their new adaptive zone. Each phylum that developed a rigid skeleton did so independently.

Coelomate Radiation

Since the coelom probably appeared at least 100 million years be-fore the Cambrian, it is a puzzle why coelomates so long delayed their major radiation. Many paleobiologists suspect that oxygen levels may have been too low to sustain the energy required for ac-tive epifaunal life. Another possibility is that primary productivity became stable enough at that time to support relatively specialized animals in epifaunal communities, perhaps due in turn to a sta-bilizing trend in the environment at large.

Most of the animal phyla for which we have good records un-derwent relatively early adaptive radiations that produced a num-ber of distinctive classes. Presumably each class was originally evolved for life in a different adaptive subzone. Most of the early classes are now extinct. The echinoderms provide a particularly impressive example (Fig. 11.16). At least 14 classes appeared in the Cambrian and 12 more in the Ordovician, when five became extinct. Only five classes of echinoderms are present today, one of which did not arise until the early Mesozoic.

Early extinction of many of the products of an adaptive radi-ation is a common pattern at every taxonomic level. The usual ex-planation is that the early radiations produce some taxa that are less efficient than the rest and they are eventually eliminated by competition. This may be the case at times. Another possibility is that the short-lived taxa are not necessarily less efficient in any ab-solute sense, but that the environment happened to change so as to put them at a special disadvantage. If the environmental history had been different, some of them might have diversified and pros-pered while others were eliminated. In any event many of these early taxa contain few species, so far as we can tell, so that ex-tinctions need not have been very massive in order to carry off all species of these small clades.

The First Vertebrates

The middle Cambrian Burgess shale fauna includes a wormlike form with the characteristically chevron-shaped muscle blocks (*myotomes*) that are adapted for swimming in amphioxus (prim-itive living chordates) and fish (Fig. 11.17). This wormlike organ-

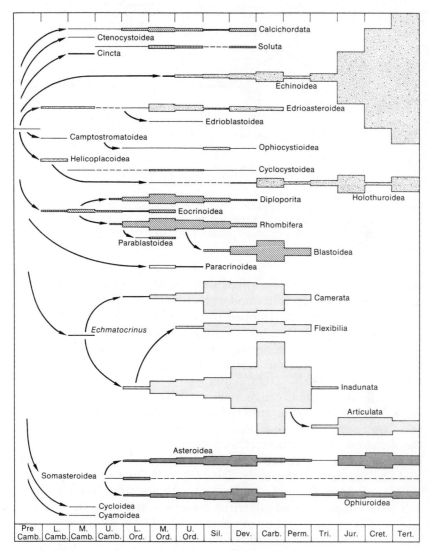

Fig. 11.16. The evolutionary history of the phylum Echinodermata; each clade is a separate class. The width of each clade reflects the numbers of genera known as fossils at that time. The patterns (Stippled, crosshatched, and so on) identify classes belonging to the same subphyla. (After Paul, 1977.)

ism is the earliest known fossil chordate. Armored dermal (skin) plates of fish have been found in late Cambrian rocks. Speculation on the descent of chordates is centered on three main possibilities. One is that they evolved by neoteny from the swimming larvae of early deuterostome invertebrates with bottom-dwelling adult

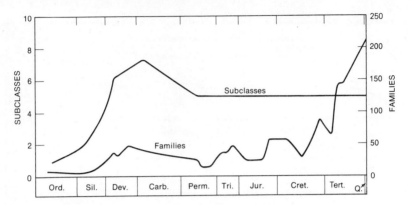

Fig. 11.17. The diversity of subclasses and families of fish. Fish arose during the Ordovician but did not become very diverse until the Devonian, when jaws evolved. (Data from Romer, 1966, and K. S. Thomson, 1977.)

stages. Reproduction was shifted into the larval stage, perhaps in response to heavy mortality and/or selection for a prolonged larval life, and the adult form dropped from the life cycle, creating a novel adaptive type. A second possibility is that chordates evolved from an extinct echinodermlike stock, the so-called *carpoids,* which have flattened calcareous skeletons with taillike structures. A third is that they arose from deuterostome worms.

These are strikingly different ideas. However they arose, early vertebrates do resemble armored deuterostome worms. The characteristic axial skeleton of the vertebrates is another example of the development of a mineralized skeleton to enhance efficiency, this time for a swimming rather than a bottom-dwelling mode of life. One lineage of the early fish may still survive as lampreys. Lacking jaws, with small mouths, and with unpaired fins, the early fish were suspension and detritus feeders, continuing the trophic modes of their ancestral invertebrates (Fig. 11.18, *A*). They underwent only a modest diversification so far as we can tell, exploring their limited adaptive potentials until the appearance of jaws and paired fins in the Devonian. This greatly widened the choice of food items and improved swimming balance. The jawed fish diversified dramatically, while the jawless forms declined. The fish pattern from here on is reminiscent of that for invertebrate clades. The major taxa reached an early peak, in the Devonian and early Carboniferous, and then declined as extinctions took their toll. The extinct groups were replaced by lower taxa. Fish were well established in fresh waters during the mid-Paleozoic, and indeed much of their diversity then is based on freshwater groups.

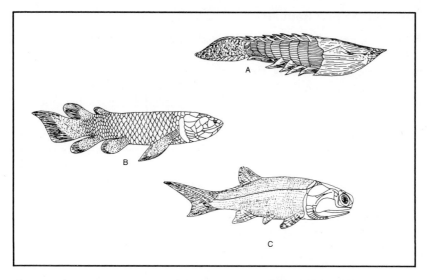

Fig. 11.18.   Early fish. *A, Anglaspis,* a jawless fish (ostracoderm).
B, Holoptychus, a rhipidistian lobe-fin. *C, Moythomasia,* a primitive
ray-fin. (*A* from Romer, 1969; *B* and *C* from Moy-Thomas and Miles,
1971.)

Relatively early in the history of jawed fishes, two major types
diversified, ray-fins and lobe-fins. The ray-fin fish (Fig. 11.18, C)
gave rise to most modern fishes, while the lobe-fins (Fig. 11.18, *B*)
produced the first land vertebrates, the amphibians. Land plants
had become established by the Silurian. At first they seem to have
been closely tied to moist habitats, but they soon spread into drier
regions. Although we do not have a good fossil record, it is clear
that several invertebrate groups had spread into fresh water by
this time. Some of these followed the plants onto land. Arthropods
seem to have been among the more adventurous, and given the
growing food resources of the spreading terrestrial plant associations they soon diversified into taxa adapted primarily to terrestrial life, chief among these being the insects. With terrestrial
food sources available, a group of primitive land vertebrates developed from perhaps only a single lineage of lobe-finned fishes belonging to a group called *rhipidistians* (Fig. 11.18,*B*). Limbs
developed from the bony supports of the fins. The rhipidistians are
extinct, but modern lobe-fins include the lungfishes, some of which
live in ponds and streams that become seasonally dessicated. It is
sometimes suggested that the evolution of limbs by rhipidistians
was originally to cope with dry periods, perhaps to seek ponds
where water remained.

Amphibians and Reptiles

The early amphibian groups diversified extensively into predatory and herbivorous types living in aquatic, semi-aquatic, and terrestrial habitats during the early Carboniferous (Fig. 11.19). In the absence of competition, they completely dominated the terrestrial ecosystems. The more fully terrestrial lineages became increasingly independent of water as adults (though still tied to water to complete their reproductive cycles). Perhaps a number of the early amphibians approached the terrestrial hardiness of reptiles. It was evidently the development of the reptilian egg, quite protected against dessication, that permitted full-fledged reptiles to emerge from one of the more primitive of amphibian lineages (the anthracosaurs).

Freed from ties to standing water bodies by an ability to lay eggs in a terrestrial environment, reptiles diversified rapidly. Some stocks remained in, or returned to, aquatic habitats, but others invaded forests and radiated into a variety of predatory and herbivorous lineages. The pelycosaurs (Fig. 11.19) appeared in the late Carboniferous and were particularly successful, both as predators and competitors. Amphibians were rather overwhelmed by the appearance of these reptiles and their diversity declined dramatically (Fig. 11.19). In fact, the ancient amphibian taxa that had dominated prereptilian land environments became extinct before the end of Triassic time. Modern amphibians are not connected by fossil intermediates to those early amphibians, but probably radiated from a group of small-bodied forms that originated in the late Paleozoic, living in cryptic habitats and utilizing resources not regularly exploited by their larger cousins. They thus escaped much of the interaction with larger reptiles and survived to form the rootstocks for living frogs, toads, salamanders, and others. These living groups are thus very different in structure and probably in biology from the larger amphibians that first spread across the continents.

By the beginning of the late Permian the decline of the amphibians was well advanced and the larger terrestrial fauna was composed of pelycosaurs and one of their offshoot clades, the therapsids, which had split off in the early Permian (Fig. 11.19). The therapsids flourished, diversifying until they came to dominate a wide range of terrestrial habitats, eclipsing the pelycosaurs. During the Triassic, continuing reptilian radiations produced the dinosaurs, which probably arose from an offshoot of the lineage that produced the pelycosaurs. The dinosaurs included a wide range of sizes, but the general impression that they were pretty large is correct; the smallest was about the size of a rooster. They first appear

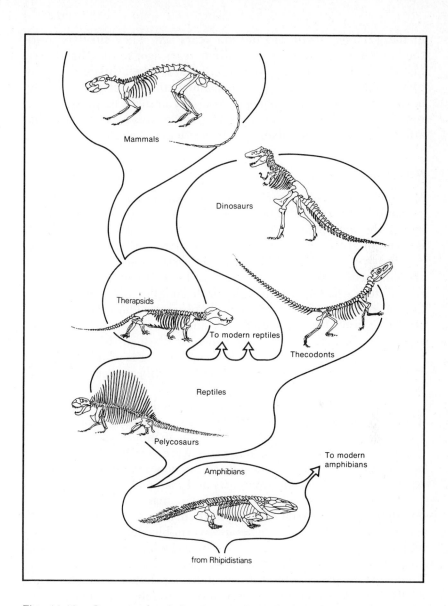

Fig. 11.19. Some tetrapod (land vertebrate) types. Ancient amphibians (here represented by *Ichthyostega,* the oldest known, from the late Devonian) gave rise to reptiles, which diversified into several major lineages. One, the pelycosaurs, gave rise in turn to therapsids, from which mammals evolved. Another gave rise to thecodonts, a group ancestral to dinosaurs. The ancient amphibians became extinct except for one group of small forms that survive as the modern representatives of the class. (After Jarvik, 1955; Romer, 1966; Gregory and Camp, 1918; Ewer, 1965; Osborn, 1924; and Wood, 1962.)

in the Triassic, about 215 million years ago. They survived until the late Cretaceous, about 65 million years ago, so they had a long history stretching over 150 million years. Their history is complex; Bakker has recognized four major faunal developments ("dynasties") in which dinosaurs played a prominent or predominant role. The dynasties are separated by widespread extinctions that affected the larger herbivores with particular severity.

Mammals

Mammals also began during the Triassic period, evolving from predatory therapsids. The changeover from the reptilian therapsids through mammallike therapsids to therapsidlike mammals is so gradual that paleontologists have great difficulty in drawing a definitive boundary. Many characteristic mammalian features are not easily inferred from fossils. These include warm-bloodedness, hair, the facial skin and muscle complex that permits suckling of young, and a respiratory diaphragm. These features and attributes probably arose gradually and perhaps at different rates and times. Somewhat arbitrary rules for distinguishing mammals and reptiles have been proposed, based on details of jaw and tooth fossils, which fortunately are the most readily preserved skeletal parts.

By about 200 million years ago, mammals had clearly arrived. They thus shared the terrestrial environment with dinosaurs for 135 million years or so. The Mesozoic mammals are all rather small; most of their adult body sizes were smaller than dinosaur hatchlings. Just as mammals and insects now inhabit two different body-size spheres, so did dinosaurs and mammals.

Since the passing of the dinosaurs, mammals have come to dominate the terrestrial habitats in their turn. Their success is often attributed to their *endothermy* (warm-bloodedness) and to the high activity levels that this can support, which may also be associated with a higher level of intelligence and general alertness. The question arises, why did mammals not diversify much earlier to displace the dinosaurs? There were a number of opportunities during the faunal turnovers that separated the dinosaur dynasties. In fact, mammals seem relatively or even entirely unaffected by those extinction events.

A model of dinosaur biology, worked out by R. T. Bakker, J. H. Ostrum, and others, provides an attractive explanation if it can be confirmed. They believe that, by and large, dinosaurs as well as mammals were endothermic. Early clues that this might be so were furnished by reconsideration of dinosaur postures and gaits.

Reptiles have a rather sprawling posture, and dinosaurs are commonly represented this way. However, it is likely that by late Triassic time dinosaurs had developed erect postures and were highly active. (Mammals also inherited a sprawling posture from reptiles and were not yet erect by this time.) Other studies seeking to test the idea of dinosaur endothermy have provided supporting evidence; dinosaur bones, for example, display canallike spaces that can be interpreted as an indication of warm-bloodedness by comparison with modern reptile and mammal bone morphology and physiology.

Another line of argument involves the dinosaurian ecosystems. *Ectotherms* (cold-blooded animals) have much lower energy requirements than do endotherms, which metabolize at a higher level and normally maintain an internal body temperature above that of the environment. When food webs dominated by ectotherms are compared with those dominated by endotherms, it turns out that the prey-predator biomass ratio is much higher in the former than the latter, since less food is burned by ectotherms. Bakker has compared the fossil evidence of such predator-prey ratios of dinosaurian faunas and finds that they agree more closely with an endothermic than ectothermic interpretation. The evidence is not decisive, however, and the question is much debated. Certainly the very large dinosaurs were so massive that the heat generated by their metabolic activities would have served to raise their body temperatures even in the absence of special thermoregulatory mechanisms.

If dinosaurs were endotherms, then the putative mammalian advantages vanish. The dinosaurs surviving the extinction waves tended to be the smaller ones, but still larger than the mammals and able to hold their own and to furnish stocks from which large animal niches were eventually filled. The mammals had to wait their turn, until dinosaurs were completely extinct, before they could become dominant. What actually caused the final dinosaur extinction remains uncertain. Bakker has suggested that extinctions which terminated the tetrapod dynasties were caused by diversity-dependent factors, chiefly a lowering of habitat diversity associated with the episodic appearance of topographically monotonous continents. The evidence is not strong enough to test this idea at present. The species that survived the Cretaceous-Cenozoic extinction were all small, below 10 kilograms in body weight; mammals were smaller still.

Even while they played their subsidiary role to dinosaurs, mammals were far from stagnant. The most advanced group, the subclass of placental mammals, arose by the middle Cretaceous and diversified into a number of clades now recognized as orders. These include the order Primates, to which *Homo sapiens* belongs; it appeared before the dinosaurs became extinct. The mammals did not respond immediately to the dinosaur extinction. Perhaps the diversity capacity remained low for a while. Or perhaps it simply took time for evolutionary processes to accomplish the expansion that followed. The eventual diversification does appear spectacular when charted (Fig. 11.20). Some of the new orders, particularly those that typified Paleocene faunas, became extinct in the Eocene or shortly thereafter, and a few other orders have also disappeared; in a pattern we have seen before, they were not replaced by other orders.

Peak mammalian evolutionary activity on the level of species and genera has been within the relatively recent past, say the last two million years; a great wave of diversification in the early part of this time was then succeeded by a great wave of extinction. P. D. Gingerich has suggested that the diversification was caused by increasing spatial heterogeneity, climatic rather than topographic, associated with a climatic diversity brought about through continental glaciations. The wave of extinction has been attributed by Paul Martin and others to the activity of early humans. This is

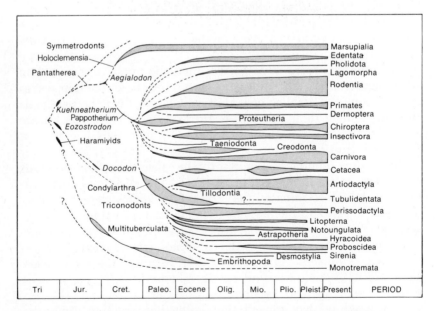

Fig. 11.20. Rise and diversification of mammalian orders. Note their explosive appearance in the late Cretaceous and the Paleocene. (After Gingerich 1977, with data from Romer.)

another much-debated point. The genus *Homo* suffered a decline in diversity across this time span (see Chapter 12).

Plant Evolution

Although we come to plants last, they have in many ways had evolutionary precedence. Plants, or at least autotrophs, must precede animals in any given environment (once the prebiotic soup has gone) to provide the primary productivity necessary to support the ecosystems. Because of the fundamental position that plants thereby hold, it has often been suggested that the changes we observe in the fossil record of animal groups, as for example some waves of diversification and extinction, merely reflect changes in the flora or protistan assemblages on which the animals depended. It is therefore of special interest to contrast the evolutionary patterns of plants with those of animals, particularly their timing and intensities.

There is little evidence on the evolutionary patterns of marine autotrophs. Modern protistan autotrophs, which account for 85 percent of the production in today's oceans, do not appear until the Mesozoic. The organisms that produced the main oceanic biomass during earlier times are extinct and their biological properties are uncertain. The earliest autotrophic invaders of the land are unknown.

Vascular land plants appear in the Silurian (Fig. 11.21). The earliest consists of simple horizontal runners from which vertical shoots arise. These diversified into horsetails, club mosses, and ferns to form a primitive plant grade, the pteridophytes. All these forms reproduce via spores that require moisture for germination. They dominated the land flora through the Devonian, Carboniferous, and into the Permian, producing major forests. Of the surviving stocks at this grade, only the ferns remain diverse.

During the Devonian, the gymnosperm ("naked seed") grade arose from primitive members of the pteridophytes. A key adaptation developed in the trunk stock of this grade was the evolution of ovules that mature to form seeds, and of wind-transported male gametes or pollen grains that fertilize the ovules. This development freed the reproductive system from dependence on water.

Seed-bearing plants diversified into several major groups during the Carboniferous, of which the seed ferns (Pteridospermales) were most important in the floras of the times. In the Permian the conifers and their allies began an expansion that led them to dominance in the Mesozoic. This shift is spoken of as the Paleophytic-Mesophytic transition ("ancient plant-middle plant transition"). It

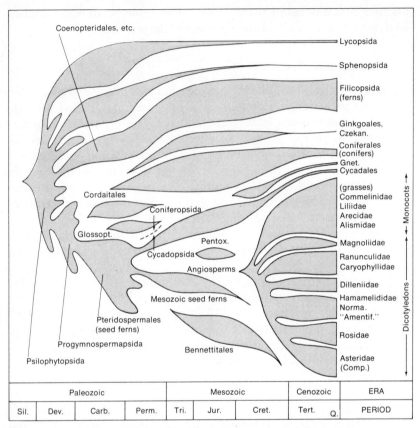

Fig. 11.21. The rise and diversification of vascular land plants. Note the change from seed ferns and lycopsids to conifers well within the Permian, and the great expansion of the angiosperms (modern flowering plants) in the mid-Cretaceous. (After Doyle, 1977.)

was most pronounced in low latitudes and marks the passing of a more humid flora, typified by the coal swamps, into one adapted to drier climatic conditions. Conifers continue today as important constituents of some temperate and boreal forests.

Modern floral associations are dominated by plants of the angiosperm ("vessel seed") grade—the flowering plants of the modern world, of which about 250,000 species exist. They are first known in the mid-Cretaceous and seem to have begun as weedy, shrubby opportunists. They radiated with startling suddenness to dominate terrestrial floras during the late Cretaceous. Their success is attributed to the evolution of flowers and the associated insect pollinating routines, and other refinements in reproduction. The usefulness of their reproductive advances for rapid-growing, opportunistic shrubs was translated into a general advantage over the more slowly growing gymnosperms.

The major replacements of land plant types do not correlate well with those of tetrapods. The changes from amphibian to reptile dominance and from dinosaur to mammal dominance were significantly preceeded, not accompanied by the more important floristic changes. The several tetrapod dynasties cannot be correlated with any obvious events in floral evolution. Since land plants and tetrapods are members of the same ecosystems, it is not surprising to find some correlations in the pulses of their extinction and diversification patterns, but it is difficult to sustain a case for the regulation of animal evolution by floristic changes on present evidence. Whether evolutionary patterns of the marine phytoplankton significantly influenced the patterns of marine invertebrate history remains moot.

QUESTIONS FOR DISCUSSION

1. Why do you think the easiest organic compounds to synthesize prebiotically are among the key compounds in living matter?

2. Why have we not yet discovered a fossil record of animallike unicellular ancestors of animals?

3. What evolutionary mechanisms are probably involved in the "sudden" appearance of modern animal phyla in the fossil record?

4. Can you propose a scenario to explain the extinction of dinosaurs?

5. If there is life on other worlds, are there ways in which we could predict that it would resemble earthly life? That it would differ from earthly life?

Recommended for Additional Reading

Cairns-Smith, A. G. 1971. *The life puzzle.* Edinburgh: Oliver and Boyd.

Orgel, L. E. 1973. *The origins of life: molecules and natural selection.* New York: John Wiley.

Two fine accounts of the problems associated with life origins. The Orgel book should be read first.

Hallam, A., ed. 1977. *Patterns of evolution as illustrated by the fossil record.* Amsterdam: Elsevier Scientific Pub. Co.

Laporte, L., ed. 1973. *Evolution and the fossil record.* San Francisco: W. H. Freeman.

Valentine, J. W. 1973. *Evolutionary paleoecology of the marine biosphere.* Englewood Cliffs, N.J.: Prentice-Hall.

These three works range from basic (Laporte) to advanced (Hallam); the latter is a particularly outstanding collection of essays on the evolutionary patterns of individual plant and animal groups.

12

The Evolution of Mankind

In *The Origin of Species* (1859) Darwin avoided extending his theory of evolution to the origin of man, although near the end of the book he remarked that evolutionary studies would throw "much light . . . on the origin of man and his history." Darwin refrained from discussing the evolution of mankind in order to avoid providing additional grounds for attacks against his theory. He directly faced the question of the evolutionary origins of the human species a few years later in his *The Descent of Man* (1871). The idea that man was a descendant of nonhuman animals was contrary to established philosophical and religious views. Not surprisingly, the publication of *The Descent of Man* provoked a flurry of attacks against the theory of evolution.

Man and the Apes

The evolutionary origin of living organisms, including humans, is today beyond reasonable doubt. We do not know all the details of the process, although we know the evolutionary history of mankind better than that of most other living species. Anybody taking the effort to become familiar with the evidence cannot doubt that our ancestors of millions of years ago were not human. To be sure, there still exist people who deny the reality of evolution, and in particular human evolution, but these either are ignorant of the evidence or have so prejudged the matter that no evidence is meaningful to them. The negation of evolution is often based on religious grounds, such as a belief in the literal truth of the Bible. Enlightened theologians are, however, convinced that the evolution of life is not incompatible with Christianity, Judaism, and other religions. But be that as it may, the incontrovertible evidence for biological evolution stands.

Table 12.1
TAXONOMIC CLASSIFICATION OF THE LIVING APES AND MAN

Family	Genus	Species
Hylobatidae	*Hylobates* (gibbon)	*H. agilis*
		H. concolor
		H. lar
		H. pileatus
	Symphalangus (siamang)	*S. syndactylus*
Pongidae	*Pongo* (orangutan)	*P. pygmaeus*
	Pan (chimpanzee)	*P. troglodytes*
		P. paniscus (pygmy chimpanzee)
	Gorilla (gorilla)	*G. gorilla*
Hominidae	*Homo* (man)	*H. sapiens*

Our closest living relatives are the apes, which are classified in the families Pongidae (pongids) and Hylobatidae (gibbons and siamangs). There are five ape genera; three of them Asiatic, the other two African. The Asiatic apes are the relatively small gibbon (*Hylobates*) and siamang (*Symphalangus*), which are the only two genera of Hylobatidae, and the larger orangutan (*Pongo*), which belongs to the Pongidae. The Pongidae also include the two genera of African apes, the chimpanzee (*Pan*) and the gorilla (*Gorilla*). The human species, *Homo sapiens,* is the only living species in the family Hominidae (hominids). The Hominidae, the Pongidae, and the Hylobatidae are included in the superfamily Hominoidea (hominoids). Table 12.1 summarizes the classification of living hominoids; their appearance and some relevant characteristics are shown in Figure 12.1.

The phylogeny of the hominoids is known through the study of the fossil record, as well as through the comparative study of living organisms. Since the 1960s, molecular techniques such as protein sequencing, immunology, and gel electrophoresis have

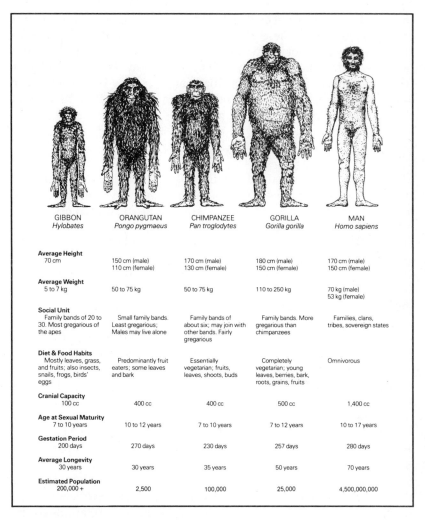

	GIBBON *Hylobates*	ORANGUTAN *Pongo pygmaeus*	CHIMPANZEE *Pan troglodytes*	GORILLA *Gorilla gorilla*	MAN *Homo sapiens*
Average Height 70 cm		150 cm (male) 110 cm (female)	170 cm (male) 130 cm (female)	180 cm (male) 150 cm (female)	170 cm (male) 150 cm (female)
Average Weight 5 to 7 kg		50 to 75 kg	50 to 75 kg	110 to 250 kg	70 kg (male) 53 kg (female)
Social Unit Family bands of 20 to 30. Most gregarious of the apes		Small family bands. Least gregarious; Males may live alone	Family bands of about six; may join with other bands. Fairly gregarious	Family bands. More gregarious than chimpanzees	Families, clans, tribes, sovereign states
Diet & Food Habits Mostly leaves, grass, and fruits; also insects, snails, frogs, birds' eggs		Predominantly fruit eaters; some leaves and bark	Essentially vegetarian; fruits, leaves, shoots, buds	Completely vegetarian; young leaves, berries, bark, roots, grains, fruits	Omnivorous
Cranial Capacity 100 cc		400 cc	400 cc	500 cc	1,400 cc
Age at Sexual Maturity 7 to 10 years		10 to 12 years	7 to 10 years	7 to 12 years	10 to 17 years
Gestation Period 200 days		270 days	230 days	257 days	280 days
Average Longevity 30 years		30 years	35 years	50 years	70 years
Estimated Population 200,000 +		2,500	100,000	25,000	4,500,000,000

Fig. 12.1. Resemblances and differences between man and pongids. Longevity for the pongids is based on animals in captivity; for humans, it is the average for American men.

given measures of genetic similarity between hominoids. Figure 12.2 represents a phylogeny based on this evidence as well as on morphological and paleontological data. There is first a split between a lineage leading to the gibbon and the siamang and a lineage leading to man and the great apes. The degrees of molecular similarity between man, gorilla, and chimpanzee are about equal, the differences between the values obtained being smaller in fact than probable differences due to experimental error. Thus the intersection between these three lineages is covered in black indicating that we do not know for sure the precise order in which the lineages split from each other. The orangutan is more different from man, chimpanzee, and gorilla than these are from each other but less than the gibbon and the siamang are, and thus its lineage has an intermediate branching point.

The Fossil History of Mankind

Skeptic contemporaries of Darwin spoke of the "missing link" between man and the apes—if apes and man have evolved from common ancestors and if evolution is a gradual process, then organisms must have existed in the past intermediate between man and the apes. Where are the fossil remains of such links? asked the anti-evolutionists. The "missing link" is not missing anymore; not one but many "links"—organisms intermediate between the living apes and humans—have been found as fossils and are available in the collections of museums and universities.

Already in 1848 a human skull had been found in Gibraltar with some apelike features. A similar one was discovered in 1856 in Germany and was called the Man of Neanderthal (meaning "Neander Valley," after the place near Düsseldorf where this second skull was found). Numerous Neanderthal skulls and skeletons have since been found in a territory stretching from Spain through central and southern Europe to Palestine and central Asia. Some saw Neanderthal Man as the missing link. Neanderthal Man, however, is not a middle link between living man and his apelike ancestors but is much closer to modern man. The Neanderthals are fossil remains of human beings, very similar to ourselves although the configuration of the head was somewhat different, with a low, sloping forehead, heavy brow ridges, a receding chin, and a strong mandible (Fig. 12.3). The Neanderthals are now considered as a distinctive subspecies (*Homo sapiens neanderthalensis*) of our own species; modern humans are *Homo sapiens sapiens*.

More significant as a missing link were a skull cap and a thigh bone discovered in Java in 1891 and 1892 by a young Dutch doctor, Eugène Dubois. This fossil was named *Pithecanthropus,* meaning "ape-man." Many *Pithecanthropus* fossils were discovered during the 1920s in a cave at Choukoutien, near Peking. Still more have been discovered in various places in Asia, Africa, and Europe. These fossils are now classified as a different human species, *Homo erectus,* which lived between about 500,000 and more than one million years ago. Their cranial capacity is about 900–1,000 cc,

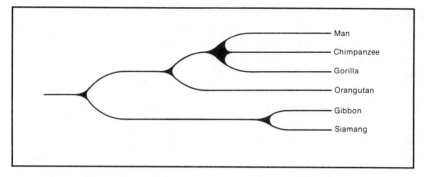

Fig. 12.2. Phylogeny of the hominoids based on all presently available evidence from paleontology, comparative anatomy, and molecular biology. The lineages leading to humans, gorillas, and chimpanzees must have branched off from one another at about the same time, but the precise order of branching is not known.

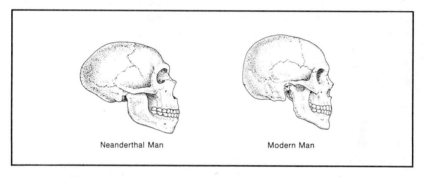

Fig. 12.3.   Skull profiles of Neanderthal and modern man. The Neanderthal skull shows a retreating forehead, heavy eyebrow ridges, an elongated brain case, a strong mandible, and a receding chin.

intermediate between that of gorillas (500 cc) and modern humans (1,400 cc). *Homo erectus,* however, walked erect, made stone tools, and is the first known user of fire (Fig. 12.4).

In 1924 Raymond Dart, a South African anthropologist, discovered in Taung, about 120 kilometers north of Kimberley, South Africa, a skull from a creature even more primitive than *Homo erectus,* which was named *Australopithecus* (for "southern ape") *africanus.* A second skull was discovered in 1936 at Sterkfontein, in the Transvaal, also in South Africa, and many others have later been found in various tropical and subtropical African localities. *Australopithecus africanus* individuals were bipedal (walked erect), were slightly more than one meter in height, and had a small brain of about 500 cc. Their head displayed an odd mixture of ape and human characteristics: a low forehead and a long, apelike face, but with teeth proportioned like those of a modern man (Fig. 12.4).

A second australopithecine, *Australopithecus robustus,* was discovered in 1938 in the Transvaal about one and a half kilometers away from the Sterkfontein location where an *A. africanus* had been discovered two years earlier. *Australopithecus robustus* individuals were larger and somewhat less manlike than *A. africanus,* had bigger teeth, and powerful jaws. *Australopithecus boisei,* an even more robust species, was later discovered in East Africa. The australopithecines lived in Africa between more than five million and about one million years ago.

An intermediate form between *Australopithecus africanus* and *Homo erectus* was discovered in 1961 in Olduvai Gorge, East Africa, by Jonathan Leakey, son of Louis and Mary Leakey and brother of Richard Leakey, all famous anthropologists who have made numerous important discoveries of man's evolution. This new finding was a 1.8 million-year-old skull from a creature named *Homo habilis,* who was bipedal, had teeth more manlike than those of *Australopithecus,* and a larger brain: about 700 cc, almost exactly

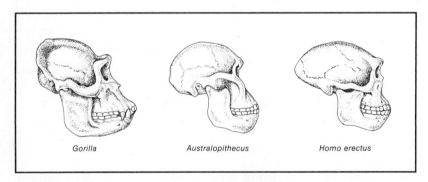

Gorilla   Australopithecus   Homo erectus

Fig. 12.4. Skull profiles of a gorilla, *Australopithecus,* and *Homo erectus.* The skull of a gorilla bears an overall resemblance to that of man but has a small brain capacity (about 500 cc). The thick bony ridge on top serves for attachment of the strong muscles needed to operate the heavy jaw with its large teeth. The brain capacity of *Australopithecus* is similar to that of the gorilla, but the skull is much more like that of modern man and the jaw lacks the large, sharp canine teeth of the gorilla. Compared to modern man, *H. erectus* has a small and flat braincase, heavy browridges, a robust jaw, and a pronounced ridge for muscle attachment at the rear of the skull. The brain case is large (900–1,000 cc), but still much smaller than that of *Homo sapiens* (1,400 cc).

midway between *A. africanus* and *H. erectus*. *Homo habilis* (meaning "handy man") made numerous stone tools and lived from more than two million years to about one million years ago (Fig. 12.5).

The rate of discovery of "missing links" between the apes and modern humans has been accelerating; more fossils of ape-men and early men have been unearthed since 1965 than in all previous times. Although many details remain unknown or are conjectural, the main stages of human evolution during the last six million years or so are fairly well established. In broadest outline those stages involved the transitions *Australopithecus africanus* → *Homo habilis* → *Homo erectus* → *Homo sapiens*.

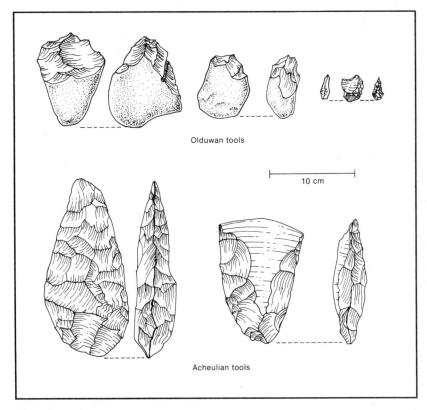

Olduwan tools

10 cm

Acheulian tools

Fig. 12.5. Stone tools made by primitive humans. *Top row:* Quartz and lava tools of the so-called Olduwan Industry, about 1.75 million years old. These tools were used by *Homo habilis. Bottom row:* Tools of the Acheulian Industry, which first appeared around 1.5 million years ago but were little changed until 125,000 years ago. Acheulian tools, made of quartz, limestone, and chert, were used by *Homo erectus*.

But the picture that has emerged in recent years is considerably more complex than this simple outline would suggest. Figure 12.6 shows in broad strokes the most likely course of events during the last six million years of human evolution. The ancestral lineage consists of *Australopithecus africanus*. Between three and four million years ago three lineages branched off *A. africanus*, which continued in existence. One lineage represented evolution toward larger brains and increasingly humanlike traits. This lineage evolved into *Homo habilis*, then into *Homo erectus*, and eventually into *Homo sapiens*. This is the only lineage that has continued to the present. The other two lineages represent evolution from frail australopithecine forms (*A. africanus*) towards the larger and coarser *Australopithecus robustus* and *A. boisei*. These two lineages probably represent a split of a common robust australopithecine stock rather than each having evolved independently from *A. africanus*.

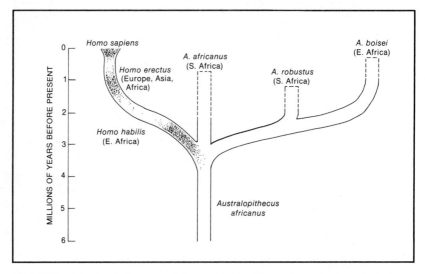

Fig. 12.6. Likely phylogeny of the hominids. The unbranched trunk may be an oversimplified representation of hominid evolution between six and four million years ago, but the scarcity of fossil remains during that period makes it impossible to know whether more than one lineage existed at the time. The *Homo* lineage branches from the *Australopithecus* lineage between three and four million years ago. *Homo habilis*, which evolved from *Australopithecus* at that time, evolved into *Homo erectus* about 1.5 million years ago. The appearance of *Homo sapiens* occurred around half a million years ago or even more recently.

Four different hominid lineages existed, therefore, in human evolution (the genus *Australopithecus* is included together with the genus *Homo* in the family Hominidae): three of them became extinct, while one eventually gave rise to modern man. There was, then, a time when four hominid species were simultaneously in existence—*Homo habilis, Australopithecus africanus, A. robustus,* and *A. boisei*—although not all sympatrically. *Australopithecus robustus* and *A. boisei* lived at the same time, at least during part of their existence, but in different regions, *A. robustus* in southern Africa and *A. boisei* in East Africa.

Common Ancestors to Humans and Apes

The oldest australopithecines appeared in the human lineage well after the human lineage had separated from the lineages going to the great apes. Who are the pre-*Australopithecus* ancestors in the human lineage, and who are the last common ancestors to man and the apes? At present these are unsettled questions, although there may be answers before very long if the current flurry of discoveries continues.

The last common ancestor to man and the great apes is likely to be *Dryopithecus,* which appeared somewhat more than 20 million years ago (Fig. 12.7). The dryopithecines are true apes and thus hominoids but not hominids. The dryopithecines were arboreal, fruit-eating quadrupeds, with brain size similar to that of modern apes, and not very different from these except that they had smaller cheek teeth and jaws. They may have originated in Africa, but they spread to Europe and Asia as far as India. By about 15 million years ago three, and perhaps more, lineages had split from the *Dryopithecus* stock. One lineage, a huge ground ape called *Gigantopithecus,* persisted in Asia for several million years but eventually became extinct. A second lineage led eventually to the orangutan. A third lineage was the ancestor to the chimpanzee, gorilla, and man. One branch of this lineage evolved into the two African apes. The other branch evolved into *Ramapithecus,* believed by many anthropologists to be the oldest ancestor of man which is not an ancestor of the modern apes.

Ramapithecus is represented by two species. *R. wickeri* from Africa (originally called *Kenyapithecus wickeri*) is represented by fossils about 14 million years old, and *R. punjabicus* from India by fossils about 12 million years old. Only a few *Ramapithecus* bone fragments, particularly teeth and jaw parts, have been found, but these seem to be intermediate between the earlier dryopithecines

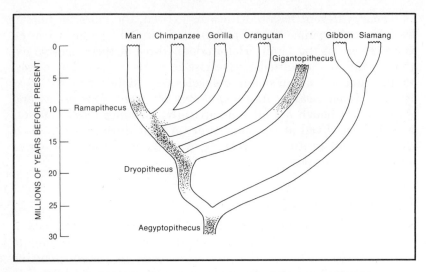

Fig. 12.7. Likely phylogeny of the hominoids. Many elements of the phylogeny are conjectural owing to the paucity of fossil remains.

and the later hominids, and thus they are placed by many anthropologists in the hominid lineage. The lineage going to man would, then, have diverged from the ape lineages about 15 million years ago. (We should note, however, that this is a controversial point. Some anthropologists think that the human and ape lineages may have diverged more recently than 15 million years ago. Others believe that any one of *Ramapithecus, Gigantopithecus,* and a related form known as *Sivapithecus,* may have been the ancestral hominid.) *Ramapithecus* was not a forest dweller properly; rather it lived in a mixed environment, at the boundaries between forests and savannas. Movement into a more open country may have affected its morphology and behavior and facilitated evolution towards bipedalism. Baboons, for example, tend to be bipedal when they feed in the open but not in the forests.

The last common ancestor to all living hominoids is thought to be *Aegyptopithecus,* an African ape that lived some 28 million years ago. The divergence of the *Hylobates-Symphalangus* lineage and the man-great ape lineage occurred about 25 million years ago.

Similarities Between Humans and Apes

In his classification of living beings, Linnaeus placed mankind together with the apes in the order of primates, thus acknowledging their morphological similarities. Thomas H. Huxley, Darwin's contemporary and staunchest defender, concluded in his *Man's Place in Nature* (1863) that "man differs less from the chimpanzee or the orang, than these do even from the monkeys, and that the difference between the brains of the chimpanzee and of man is almost insignificant, when compared with that between the chimpanzee brain and that of a lemur." The near complete correspondence of bone for bone between an ape and a human is apparent to anybody who has examined their skeletons in a museum (Fig. 12.8).

When molecular comparisons are made, the similarities between man and the apes are no less striking. Cytochrome *c,* a protein involved in oxygen transport, consists in vertebrates of 104 amino acids; all 104 amino acids are the same and appear in identical sequence in man and the great apes. The α and β hemoglobin chains, composed of 141 and 146 amino acids respectively, are identical in humans and chimpanzees, and differ from them by

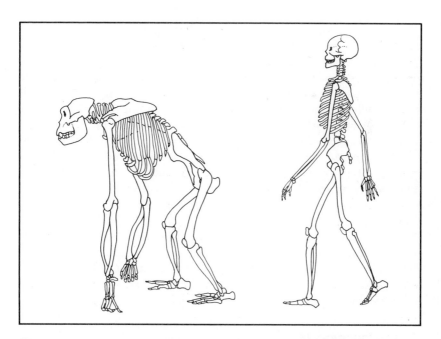

Fig. 12.8.   Skeletons of ape and man showing nearly complete correspondence of bone for bone. The skeletons are nevertheless quite different. In apes the arms are longer than the legs; in man the opposite is true. The long, straight pelvis of the ape provides support for quadrupedal locomotion; the short, broad pelvis of man curves backward, carrying spine and torso in bipedal position.

one amino acid each in the gorilla. M. C. King and A. C. Wilson have calculated that man and chimpanzee differ just by about one for every hundred amino acids in their proteins.

Table 12.2
GENETIC DISTANCE BETWEEN THE HOMINOIDS
Based on the study of 23 gene loci coding for enzymes and other proteins. The genetic distance, D, is calculated according to the method given in Closer Look 6.1 and estimates the number of allelic substitutions per locus that have occurred in the evolution of the organisms compared since their last common ancestor. (After Bruce and Ayala, 1979.)

	Chimpanzee	Gorilla	Orangutan	Gibbon	Siamang
Man	0.349	0.373	0.349	0.782	1.099
Chimpanzee		0.379	0.220	0.761	0.946
Gorilla			0.461	0.552	0.806
Orangutan				0.591	0.806
Gibbon					0.274

On the basis of electrophoretic data, the average genetic distance for all pairwise comparisons between man, gorilla, chimpanzee, and orangutan is $D = 0.355$ (Table 12.2), indicating that only about one allelic substitution for every three loci has occurred in the separate evolution of any two of those organisms. This amount of genetic differentiation is comparable to, or less than, that observed between closely related species of the same genus (see Chapter 6). Genetic differentiation is greater when comparisons are made between man or the great apes and the gibbon or siamang, but even these comparisons reflect no more genetic differences (average $D = 0.793$) than is observed in other groups of organisms between species of the same genus or between closely related genera.

The chromosomes of humans and chimpanzees are remarkably similar, chromosome for chromosome and band for band, except for a few chromosomal rearrangements, including a fusion of two chimpanzee chromosomes into one in humans, which thus have only 46 chromosomes while apes have 48.

The Uniqueness of Man

Humans and apes are similar because they are phylogenetic cousins; their evolutionary separation occurred in the (geologically) recent past and thus there has been no time for much differentiation. The similarities, however, should not obscure the differences. Some biological differences are of such great consequence that man may be considered a truly unique kind of organism, the most distinctive of the whole animal kingdom.

The molecular and genetic similarities between humans and apes are great, but examination of the proteins in a drop of blood or in a small amount of tissue makes possible unambiguous identification of the species to which the blood or tissue belong.

The morphological differences are even more conspicuous— one need not be a zoologist in order to be able to tell a human being from a chimpanzee or from a gorilla or orangutan. G. G. Simpson has listed 12 (among many) distinctive human anatomical traits:

1. Normal posture is upright.
2. Legs are longer than arms.
3. Toes are short, the first toe frequently longest and not divergent.
4. The vertebrae column has an S-curve.
5. The hands are prehensile, with a large and strongly opposable thumb.
6. Most of the body is bare or has only short, sparse, inconspicuous hair.
7. The joint for the neck is in the middle of the base of the skull.
8. The brain is uniquely large in proportion to the body and has a particularly large and complex cerebrum.
9. The face is short, almost vertical under the front of the brain.
10. The jaws are short, with a rounded dental arch.
11. The canine teeth are usually no larger than the premolars, and there are normally no gaps in front of or behind the canines.
12. The first lower premolar is like the second, and the structure of the teeth in general is somewhat distinctive.

The molecular, genetic, and morphological differences between humans and apes are valid as far as they go, but clearly they are not telling us why we perceive ourselves as unique. Indeed it is in the realm of behavior that the differences between humans and other animals are greatest. Man is a tool-using and tool-making animal; the development of technology has greatly contributed to the biological success of the species.

The ability to make tools is associated with bipedalism. Relieved of use in walking, the anterior extremities of man became specialized for precise manipulation of objects. But there is much more in tool-using and tool-making than simple dexterity to handle objects. The design and construction of tools depends on the ability of seeing them precisely as tools, that is, as instruments that serve a function. Humans can see the connection between means and the ends or goals they serve, between anticipated needs and the tools or instruments that might satisfy such needs. The ability to anticipate the future and to see the connection between means and ends depends on the existence of a large and complex brain. This may, then, be the most fundamental anatomical feature of man, in that it has made possible the development of technology and culture.

Anthropologists have raised the question whether bipedalism or an advanced brain came first. One view is that brain development provided the conditions that facilitated bipedalism; once the intellectual ability to plan for the future was present, there was a selective advantage in changes that freed the anterior extremities from walking and changed them into manipulating organs. The available evidence, however, leans towards the alternative explanation: bipedalism came first as our ancestors moved from the forest into more open environments. The anterior extremities were

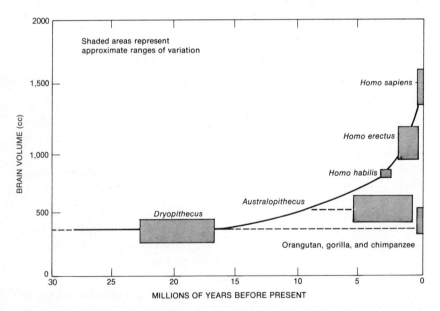

Fig. 12.9. Evolution of brain volume in the hominoids. The large and complex brain of man has made possible the development of technology and culture.

then available for object manipulation, which facilitated the development of the brain so as to make better use of such ability. But the two alternative explanations should be seen as complementary rather than as mutually exclusive. It seems likely that, for some time at least, there was a positive feedback between the two developments: incipient bipedalism favored brain development, which in turn favored further evolution of the anterior extremities for object handling, which favored further brain development, and so on.

A complex brain made possible another ability fundamental for the development of complex technologies, namely symbolic communication. A *symbol* is an object or act whose meaning is not self-evident, but rather is agreed upon by those who use it. In contrast, a *sign* is something whose meaning is apparent, without requiring social agreement. For example, smoke is a sign of fire, and crying is a sign of pain; but the word "smoke" is a symbol for smoke only because it has been so agreed, and the United States flag with its stars and stripes is a symbol of the United States because it has been chosen for that purpose. Animals, including humans, communicate by signs, while humans communicate by symbols; human languages are symbolic, while the so-called animal "languages" consist of signs rather than symbols. (It seems that chimpanzees, and perhaps other animals, can learn the meaning of symbols and use them in communication, but their ability in this respect is rather limited; in any case, animals do not create symbols nor use them in their ordinary communication, at least not in any substantial degree.)

Culture, the Human Domain

As G. G. Simpson has written, man "is another species of animal, but not just another animal. He is unique in peculiar and extraordinarily significant ways." The paramount distinctive attribute of mankind is culture, a term which encompasses all nonstrictly biological human activities and creations. Culture consists of the social structures, ways of doing things, religious and ethical traditions, language, scientific knowledge, art, technology, and in general of all the creations of the human mind. Although social organization exists in some animal groups, nothing closely resembling culture exists outside the human realm.

There are in mankind two kinds of heredity—the biological and the cultural, which may also be called organic and superorganic, or *endosomatic* and *exosomatic* systems of heredity ("internal to the body" and "external to the body," respectively). Biological inheritance in man is very much like that in any other sexually reproducing organism; it is based on the transmission of genetic information encoded in DNA from one generation to the next by means of the sex cells. Cultural inheritance, on the other hand, is based on transmission of information by a teaching-learning process, which is in principle independent of biological parentage. Culture is transmitted by instruction and learning, by example and imitation, through books, newspapers and radio, television and motion pictures, through works of art, and by any other means of communication. Culture is acquired by every person from parents, relatives and neighbors, and from the whole human environment.

Cultural inheritance makes possible for people what no other organism can accomplish—the cumulative transmission of experience from generation to generation. Animals can learn from experience, but they do not transmit their experiences, their "discoveries" (at least not to any large extent) to the following generations. Animals have individual memory, but they do not have a "social memory." Humans, on the other hand, have developed a culture because they can transmit cumulatively their experiences from generation to generation.

Cultural inheritance makes possible cultural evolution, that is, the evolution of knowledge, social structures, ethics, and all other components that make up human culture. Cultural inheritance makes possible a new mode of adaptation that is not available to nonhuman organisms—adaptation by means of culture. Organisms in general adapt to the environment by means of natural selection, by changing their genetic constitution to suit the demands of the environment. But humans, and humans alone, can also adapt by changing the environment to suit the needs of their genes (Fig. 12.10). In fact, for the last few millennia humans have been adapting the environments to their genes more often than their genes to the environments.

In order to extend its geographical habitat, or to survive in a changing environment, a population must become adapted through slow accumulation of genetic variants sorted out by natural selection, to the new climatic conditions, different sources of food, different competitors, and so on. The discovery of fire and the use of shelter and clothing allowed humans to spread from the warm tropical and subtropical regions of the Old World to the whole earth, except for the frozen wastes of Antarctica, without the anatomical development of fur or hair. Humans did not wait for genetic mutants promoting wing development; they have conquered the air in a somewhat more efficient and versatile way by building flying machines. Mankind travels the rivers and the seas

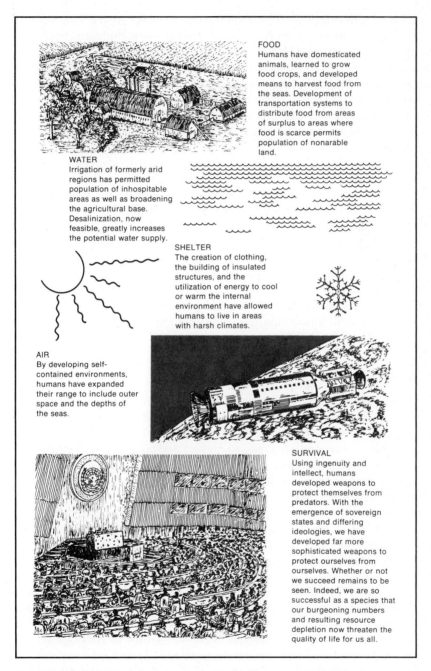

FOOD
Humans have domesticated animals, learned to grow food crops, and developed means to harvest food from the seas. Development of transportation systems to distribute food from areas of surplus to areas where food is scarce permits population of nonarable land.

WATER
Irrigation of formerly arid regions has permitted population of inhospitable areas as well as broadening the agricultural base. Desalinization, now feasible, greatly increases the potential water supply.

SHELTER
The creation of clothing, the building of insulated structures, and the utilization of energy to cool or warm the internal environment have allowed humans to live in areas with harsh climates.

AIR
By developing self-contained environments, humans have expanded their range to include outer space and the depths of the seas.

SURVIVAL
Using ingenuity and intellect, humans developed weapons to protect themselves from predators. With the emergence of sovereign states and differing ideologies, we have developed far more sophisticated weapons to protect ourselves from ourselves. Whether or not we succeed remains to be seen. Indeed, we are so successful as a species that our burgeoning numbers and resulting resource depletion now threaten the quality of life for us all.

Fig. 12.10.   Some ways in which humans have adapted their environment to suit the essential genetic needs of food, water, shelter, and survival.

without gills or fins. The exploration of outer space has started without waiting for mutations providing some humans with the ability to breathe with low oxygen pressures or to function in the absence of gravity; astronauts carry their own oxygen and specially equipped pressure suits. From their obscure beginnings in Africa, humans have become the most abundant species of mammal on earth. It was the appearance of culture, a superorganic form of adaptation, that made mankind the most successful living species.

Cultural adaptation has prevailed in mankind over biological adaptation because it is a more rapid mode of adaptation and because it can be directed. A favorable genetic mutation newly arisen in an individual can be transmitted to a sizeable part of the human species only through innumerable generations. However, a new scientific discovery or technical achievement can be transmitted to the whole of mankind, potentially at least, in less than one generation. Moreover, whenever a need arises, culture can directly pursue the appropriate changes to meet the challenge. Biological adaptation depends on the accidental availability of a favorable mutation, or of a combination of several mutations, at the time and place where the need arises.

Evolution and Ethics

People have moral values, i.e., they accept standards according to which their conduct is judged either right or wrong, good or evil. Systems of ethical values vary to some extent from individual to individual, and from society to society, although some values (for example, not to kill, not to steal, to honor one's parents) are widespread and perhaps universal. The universality of ethical values among humans raises the questions whether the moral sense is based on human nature itself, and whether natural selection may have indeed promoted the development of certain ethical precepts.

Philosophers, evolutionists, and others have written extensively about these questions, often arguing opposing points of view. According to some authors, ethical values are natural, i.e., determined by human nature, while other authors have claimed that ethical values are chosen by society so as to facilitate society's functioning or are derived from religious beliefs. One source of confusion in some writings has been the failure to distinguish between two related but separate questions: (1) Are humans ethical beings by their biological nature? (2) Does their biological nature prescribe specific ethical codes?

The first question is the more fundamental; it asks whether or not the biological nature of man is such that humans are necessarily inclined to accept ethical values, to identify certain actions as either right or wrong. Answers to this first question do not necessarily determine what the answer to the second question should be. Independent of whether or not humans are necessarily ethical, it remains to be determined whether particular moral prescriptions are in fact determined by the biological nature of man, or whether they are chosen by society, or by individuals. Even if we were to conclude that people cannot avoid having moral standards of conduct, it might be that the choice of moral standards would be arbitrary. The need for having moral values does not necessarily tell us what the moral values would be.

The answer to the first question raised must be affirmative: humans are genetically determined ethicizing beings. This is a consequence of the advanced intellectual development of humans which makes them unique within the living world. People are by genetic constitution capable of anticipating the consequences of their actions and of evaluating the actions as either right or wrong depending on what the consequences are. Moreover, people are able to choose between alternative courses of action. The ability of judging alternatives in ethical terms and the ability of choosing between these alternatives are jointly necessary for the existence of ethical behavior.

The capacity for ethicizing is a biological characteristic of the human species that was promoted by natural selection because it was adaptive. Human evolution is characterized by the adaptive development of the increased intellectual abilities that make humans unique among all animals. The ability of foreseeing the results of one's action and of evaluating alternative courses of action was adaptive in human evolution and was therefore favored by natural selection. The social organization of prehumans and early humans may also have directly favored genetic constitutions that predisposed them for accepting instruction and therefore authority. Humans are genetically predisposed for learning and, more specifically, for receiving moral standards from their parents and social milieu.

Humans are ethicizing beings, but they are not bound to specific moral systems by their biological nature. Thus the answer to the second question raised is in the negative, although some qualifications are necessary. The same intellectual abilities that make people able to choose between alternatives that are judged either right or wrong give them the ability to choose between alternative sets of moral precepts. Thus people are not necessarily bound to one moral code rather than another, and they may in fact change their moral codes. Moral codes are different in different individuals, and even more so in different societies or in the same social group at different times. This versatility is in fact adaptive; the complete fixation of a moral code would be maladaptive because the conditions of mankind change from place to place or from time to time, calling for different responses. As an example we may consider the biblical injunction, "Be fruitful, and multiply, and replenish the earth" (Genesis 1:28), which in the view of many people has now become replaced by a moral obligation to limit the growth of human populations.

Even though humans are not bound to specific moral codes and can change them, are there any ethical precepts that might be genetically conditioned? This is a difficult question, which can be only briefly treated here. We may distinguish two kinds of ethics: family and group ethics. *Family ethics* include the moral precepts governing the behavior among close relatives, such as parental care of children, cooperation between family members, and the like. Many family ethical precepts (e.g., "honor your parents") are genetically conditioned dispositions, favored by natural selection in our ancestors because the prescribed behaviors increase reproductive fitness. However, these predispositions can be willfully overcome, and thus humans are not unavoidably bound by family ethical precepts. This point deserves attention particularly with respect to parallel behaviors in nonhuman animals; care of the young, cooperation between relatives, and other altruistic behaviors exist in animals because they are favored by natural selection. Animals, however, are not subject to ethics because they lack the ability to discriminate between right and wrong and because they are not properly able to choose between alternative behaviors; they are, however, genetically bound to follow certain behavioral patterns that are similar to those prescribed by family ethics in humans.

Group ethics includes moral precepts governing behavior in human societies. These are products not of biological but of cultural evolution. Natural selection does not promote genetic predisposition to social ethical precepts because these confer no advantage and may be disadvantageous to individuals practicing them, although they may be indispensable to the maintenance of human societies. Some popular authors have written that human actions are driven by genetically determined aggressive tendencies or by territorial and other socially harmful "imperatives." Such imperatives do not exist, at least in the sense that our behavior would be necessarily determined by them. Whatever proclivities humans may have towards selfish or antisocial behavior can be counteracted by choice of appropriate ethical precepts promoting social cooperation. Human evolution is characterized by selection for plasticity of behavior rather than for genetically fixed egoistic or altruistic behaviors.

Evolution and Religion

Religion is a cultural universal of mankind. Every culture of the past that has left a historical record has had some system of religious views concerning the meaning (and proper conduct) of human life, and so does every known human society of the present Neanderthal Man practiced religious rituals, at least in the form of ceremonial burial of the dead (Fig. 12.11), and there are indications that Peking Man (*Homo erectus*) also may have practiced religious rituals (involving in this case extraction and ingestion of the brain). Religious systems may range from very primitive to very elaborate ones, such as are found in the Judeo-Christian tradition and in some Oriental religions. Moreover, different religions are not alike in their beliefs, and at some points they are incompatible; but their universal occurrence in all human cultures suggests a biologically determined predisposition toward religion. Can this predisposition be explained as a result of natural selection? Why are humans but no other animals religious?

As pointed out earlier, it is the human mind that sets our species apart from other animals. Except for young infants, every hu-

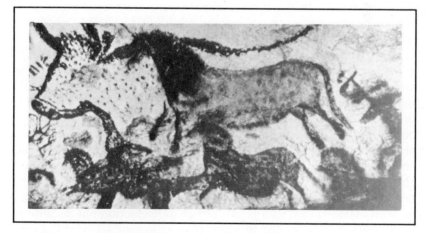

Fig. 12.11. Paleolithic cave paintings, Lascaux, France. Such paintings may have had religious significance. The bulls and horses have been interpreted as representing the male and female principle, an integral factor in an earth mother cult, evidence for which has been discovered at various sites in Europe and the Near East.

man being is conscious of himself as something different from other people and from the environment. Self-awareness is the most immediate and unquestionable of all realities. The French philosopher René Descartes (1596–1650) found that he could doubt the existence of the external world and every reality except his own existence: "I think, therefore I am." How do we know that other people have similar minds and self-awareness? We cannot enter into their consciousness, but they can communicate their self-awareness to us, and we can also infer this from their behavior.

Are other animals self-aware? We have no direct animal communication expressing their self-awareness or lack thereof, but we can judge from their behavior whether or not they possess self-awareness. The matter is far from settled; some people attribute to some animals at least the beginnings of mind. This is in part a matter of definition, but there is little doubt that human mind and self-awareness differ greatly from any rudiments of them that may be present in other animals. Pierre Teilhard de Chardin wrote: "Admittedly the animal knows. But it cannot know that it knows —this is quite certain." We could qualify his statement: The animal cannot know that it knows, at least not to the same degree that humans can. And the magnitude of the difference makes it a difference of kind; that is why nothing comparable to culture has been developed by any animal.

One consequence of self-awareness is death awareness; because we are aware of our own self, we are also aware of our transitoriness. Humans, at least beyond childhood, are aware of the inevitability of their death. One reliable sign of death awareness is ceremonial burial of the dead. This is universal in man and wholly absent in animals, which does indicate absence of death awareness and consequently of self-awareness in nonhuman animals. Animals ignore the death of individuals of their own species or treat dead individuals as rubbish (e.g., some social insects, such as ants and bees, that throw cadavers out of the nest together with other debris) or as food (e.g., termites which eat dead members of the colony as they do any other substances having nutritional value). Humans, at least since Neanderthal Man, ceremonially bury their dead, although burial rites are very diverse. Death awareness is genetically conditioned as a result of natural selection not because it is adaptive by itself, but because it is a by-product of mind and self-awareness, which are adaptive.

Humans are aware of the transitory character of their own existence and, as a consequence, develop anxiety about death. This anxiety is at least in part alleviated by religious beliefs that attribute the transience of life to supernatural powers and give meaning to one's own life even though this will end. Anxiety about death is further relieved in religions propounding the immortality of the soul, either through successive reincarnations or in the form of life beyond the death of the body.

The origin of religion cannot be attributed solely to anxiety about death. In primitive cultures at least, supernatural beliefs arise also in order to account for phenomena for which no natural explanation is known. But death awareness predisposes humans towards religious beliefs; and death awareness, as a correlate of self-awareness, is a genetically determined human attribute. In this sense at least, humans are genetically predisposed to religious beliefs and rituals.

QUESTIONS FOR DISCUSSION

1. Do you think that the various human groups or races are likely to become gradually more differentiated and eventually give origin to several human species? Why?

2. Discuss the relationship between erect posture, tool using, social behavior, and the intelligence increase that occurred in human evolution.

3. Enumerate the differences between humans and the other animals. Are all these differences in degree or do people have attributes qualitatively different from those found in animals?

4. What is meant by cultural evolution? Why is it a more rapid process of adaptation than biological evolution?

5. Does cultural evolution occur in organisms other than man? Why?

Recommended for Additional Reading

Dobzhansky, T. 1962. *Mankind evolving*. New Haven. Conn.: Yale University Press.

A classic, covering all aspects of human evolution. Moderately advanced.

Dobzhansky, T. 1967. *The biology of ultimate concern*. New York: New American Library.

A profound discussion of the biological bases of ethics and religion. Moderately advanced.

Isaac, G. L., and McCown, E. R., eds. 1976. *Human origins*. Menlo Park, Calif.: W. A. Benjamin.

The fossil history of man presented by multiple authors. Moderately advanced.

Moore, R., and the Editors of *Life*. 1976. *Evolution*. New York: Time, Inc.

An elementary, readable, and well-illustrated introduction to evolution. Human evolution is the single topic discussed most extensively.

Simpson, G. G. 1969. *Biology and man*. New York: Harcourt.

Insightful discussions of a variety of topics relating to human evolution including the biological nature of humans, race, language, and ethics. Moderately advanced.

13

Future Evolution

The evolution of mankind, at least for the last few millennia, has been marked much more by cultural change than by biological change. Culture is a more effective and versatile mode of adaptation than the biological mode. Yet, the superorganic has not annulled the organic: biological evolution continues in mankind, and it may be taking place at a faster pace than ever precisely because it is fueled by cultural evolution. Cultural and biological evolution are mutually interrelated. The existence and development of human culture are possible only so long as the genetic basis of human culture is maintained or improved; there can be no culture without human genotypes. At the same time, cultural evolution is doubtless the most important source of environmental change promoting the biological evolution of man.

Biological Evolution in Modern Mankind

There is no basis to the claim sometimes made in popular writings that the biological evolution of mankind has stopped. That mankind continues to evolve biologically can be shown because the necessary and sufficient conditions for biological evolution persist. These conditions are genetic variability and differential reproduction. There is a wealth of genetic variation in mankind. It was shown in Chapter 3 that, with the trivial exception of twins developed from a single fertilized egg, no two people who live now, lived in the past, or will live in the future, are likely to be genetically identical.

Does natural selection continue to occur in modern mankind? Natural selection is simply differential reproduction of alternative genetic variants. Therefore, natural selection will occur in mankind if the carriers of some genotypes are likely to leave more descendants than the carriers of other genotypes. Some writers have argued that due to the progress of medicine, hygiene, and nutrition most people now survive beyond reproductive age, and thus that natural selection is hardly or not at all operating in modern mankind. But this claim is based in a misconception. Natural selection consists of two main components: differential mortality and differential fertility; both persist in modern mankind, although the intensity of selection due to postnatal mortality has been somewhat attenuated.

Death may occur between conception and birth (prenatal) or after birth (postnatal). The proportion of prenatal deaths is not well known (death during the early weeks of embryonic development may go totally undetected), but it is known to be substantial. Such deaths are often due to deleterious genetic constitutions, and thus they have a beneficial selective effect in the population. The intensity of this form of selection has not changed substantially, although it has been slightly reduced with respect to a few genes such as those involved in Rh incompatibility.

Postnatal mortality has been considerably reduced in recent times, particularly in technologically advanced countries. For example, in the United States somewhat less than 50 percent of those born in 1840 survived to age 45, while it is estimated that more than 90 percent of those born in 1960 will survive to that age (Table 13.1). In other regions of the world, postnatal mortality remains quite high although there also it has generally decreased in recent decades. Postnatal mortality, particularly where it has been considerably reduced, is largely due to genetic defects, and thus it has a favorable selective effect in human populations. More than 2,000 genetic variants are known causing diseases and malformations in humans; such variants are kept at low frequencies due to natural selection.

It might seem at first that selection due to differential fertility has been considerably reduced as a consequence of the reduction in the average number of children per family taking place in many parts of the world during recent decades. However, this is not nec-

Table 13.1
PERCENT OF CAUCASIAN AMERICANS BORN BETWEEN 1840 AND
1960 SURVIVING TO AGE 15 AND TO AGE 45
(After Kirk, 1968.)

Year of Birth	Surviving to Age 15		Surviving to Age 45	
	Men	Women	Men	Women
1840	62.8	66.4	48.2	49.4
1880	71.5	73.1	58.3	61.1
1920	87.6	89.8	79.8	85.8
1960*	96.6	97.5	92.9	95.9

*The values for 1960 are projections.

essarily so. The intensity of fertility selection depends not on the *mean* number of children, but on the *variance* in the number of children. It is clear why this should be so. Assume that all people of reproductive age marry and that all have exactly the same number of children. In this case, there would not be fertility selection independent of whether couples all had very few or all had very many children. Assume, on the other hand, that the mean number of children per family is low, but some families have no children at all while others have many. In this case, there would be considerable opportunity for selection—the genotypes of parents producing many children would increase in frequency at the expense of those having few or none. Studies of human populations have shown that the opportunity for natural selection often increases as the mean number of children decreases (Table 13.2). Therefore, there is no

Table 13.2
MEAN NUMBER OF CHILDREN PER FAMILY AND OPPORTUNITY FOR
FERTILITY SELECTION IN VARIOUS HUMAN POPULATIONS

I_f is the "index of opportunity for selection due to fertility," which is calculated as the variance divided by the square of the mean number of children. The opportunity for selection usually increases as the mean number of children decreases. (After Crow, 1961.)

Human population	Mean number of children	I_f
Rural Quebec, Canada	9.9	0.20
Gold Coast, Africa	6.5	0.23
New South Wales, Australia (1898–1902)	6.2	0.42
United States, women born in 1839	5.5	0.23
United States, women born in 1871–1875	3.5	0.71
United States, women born in 1928	2.8	0.45
United States, women born in 1909	2.1	0.88
United States, Navajo Indians	2.1	1.57

evidence that natural selection due to fertility has decreased in modern human populations.

It may be that natural selection will decrease in intensity in the future, but it will not disappear altogether. So long as there is genetic variation and the carriers of some genotypes are more likely to reproduce than others, natural selection will continue operating in human populations. Cultural changes, such as the development of agriculture, migration from the country to the cities, environmental pollution, and many others, create new selective pressures. The pressures of city life are no doubt partly responsible for the high incidence of mental disorders in certain human societies. The point to bear in mind is that human environments are changing faster than ever owing precisely to the accelerating rate of cultural change; and environmental changes create new selective pressures, thus fueling biological evolution.

The Biological Future of Mankind

Where is human evolution going? Biological evolution is directed by natural selection, which is not a benevolent force guiding evolution toward some success. Natural selection is a process bringing about genetic changes that often appear purposeful because they are dictated by the requirements of the environment. The end result may, nevertheless, be extinction—more than 99.9 percent of all species that ever existed have become extinct. Natural selection has no purposes; humans alone have purposes and humans alone can introduce these purposes into their evolution. No species before mankind could select its evolutionary destiny; mankind possesses techniques to do so, and more powerful techniques for directed genetic change are becoming available. Because they are self-aware, humans cannot refrain from asking what lies ahead, and because they are ethical beings they must choose between alternative courses of action, some of which may appear as good, others as bad.

The argument has been advanced by some authors that the genetic endowment of mankind is rapidly deteriorating owing precisely to the improving conditions of life and to the increasing power of modern medicine. We saw in Chapter 4 (p. 129) that the frequency of deleterious mutations is determined by the opposing forces of mutation pressure and selection. The equilibrium frequency of deleterious dominant alleles is the rate of mutation divided by the selection coefficient (u/s), the equilibrium frequency of recessive alleles is the square root of that value ($\sqrt{u/s}$). The effect of health care is to decrease the value of s and thus to increase the frequency of deleterious alleles in human populations. Moreover, mutation rates may be increasing in modern societies due to exposure to high frequency radiations and to chemical mutagens; and if the rate of mutation increases, the equilibrium frequency increases proportionally.

As an example, let us consider retinoblastoma, a cancerous disease due to a dominant mutation. The unfortunate child carrying this gene develops a tumorous growth during infancy, which starts in one eye, rapidly extends to the other eye and then to the brain, causing death before puberty. Surgical treatment now makes it possible to save the life of the child if the condition is detected sufficiently early, although usually one eye at least is lost. The treated person can live a more-or-less normal life, marry, and procreate, but on the average one half of his children will inherit the retinoblastoma gene and will have to be treated. Before modern medicine, every mutation for retinoblastoma was eliminated from the population in the same generation in which it arose owing to the death of its carrier. With surgical treatment the mutant gene can be preserved and the new mutations arising every generation are added to those arisen in the past.

In the case of recessive conditions, such as phenylketonuria (PKU), the children of the cured person will not inherit the disease unless the other parent is heterozygous (or homozygous, of course) for the deleterious allele, but the two alleles that would have been eliminated from the population through the death of the homozygous phenylketonuric are passed on to the following generation.

CLOSER LOOK 13.1   Deleterious Genes and Incidence of Genetic Diseases

The mutation rate for retinoblastoma is approximately $u = 1 \times 10^{-5}$, which is a typical mutation rate for deleterious genes in humans. The gene for retinoblastoma is lethal and, therefore, $s = 1$. The equilibrium frequency of the gene for retinoblastoma, which is dominant, is under "natural" conditions

$$p = \frac{u}{s} = \frac{1 \times 10^{-5}}{1} = 1 \times 10^{-5}$$

or the same as the mutation rate. But retinoblastoma patients are heterozygotes and thus have a frequency of $2pq$ in the population. Since q is very nearly equal to one, the number of people suffering from retinoblastoma is $2pq \simeq 2p = 2 \times 10^{-5}$. In general, the incidence of a condition caused by a lethal dominant gene is approximately twice the mutation rate.

The equilibrium frequency of a deleterious gene is much higher in the case of a recessive gene ($q = \sqrt{u/s}$) than in the case of a dominant one ($p = u/s$). But the number of people suffering from a deleterious condition is of the same order of magnitude in the case of a recessive as in the case of a dominant gene, so long as the mutation rate is approximately the same for both. Consider a recessive lethal gene with a mutation rate, $u = 1 \times 10^{-5}$. The equilibrium frequency of the gene is

$$q = \sqrt{\frac{u}{s}} = \sqrt{\frac{1 \times 10^{-5}}{1}} \simeq 0.003$$

But the incidence of the disease will be (since, being a recessive condition, only homozygotes will suffer) $q^2 = 0.003^2 = 1 \times 10^{-5}$.

The incidence of a disease would increase relatively slowly even if all patients were cured and would reproduce as effectively as normal individuals. More precisely, in the case of lethal dominant genes the incidence of the disease would increase at twice the rate at which mutations occur. This is because the newly arisen mutants are added to the genes that were present in the cured patients and because the frequency of patients is approximately $2p$, or twice the frequency of the gene. In the case of recessive lethal conditions, the incidence of the disease increases much more slowly. The frequency of the defective gene increases by the mutation rate every generation, but the incidence of the disease increases every generation by the square of the mutation rate.

The proportion of individuals affected by any one serious hereditary infirmity is relatively small. For example, about two of every 100,000 newborns suffer from retinoblastoma. Moreover, even if a lethal condition were to be completely cured in every patient (which is never the case at present, since only a few have access to such cures), the incidence of the disease would increase only slowly.

The problem, however, is very serious when all hereditary diseases are considered in the aggregate. More than 2,000 gene variants causing diseases are known, and more than 5 percent of all people suffer at some point in their lives from serious physical or mental impairments determined by genes. Although the number of hereditary conditions that can be partially or totally cured is not very large, that number increases continuously, and thus the effects on the population cannot be ignored. Moreover, human beings cannot be treated as statistics. Consider one single dominant lethal condition, and assume that its frequency increases by two per 100,000 individuals in one generation; since the world population is about four billion people, the two per 100,000 increase would mean an additional 80,000 patients—the human misery and social cost caused by the increase would be staggering.

Positive and Negative Eugenics

Although the rate of genetic deterioration of mankind is not as large as some have claimed, there is little doubt that progress in health care entails increases in the frequency of deleterious genes in human populations. How can this process of genetic decay be stopped, or reversed?

Eugenics is the science and practice seeking to improve the genetic endowment of mankind. Two kinds of eugenics may be distinguished—positive and negative. *Negative eugenics* is concerned with avoiding the spread of undesirable genes, while *positive eugenics* seeks the multiplication of desirable ones. Eugenics is a matter fraught with sociopolitical and ethical implications. We shall, therefore, have to deal with these implications and will thus be moving out of scientific ground.

Methods proposed to improve the genetic endowment of mankind may be classified into four categories: genetic counseling, genetic surgery, germinal selection, and cloning. The first two are primarily methods of negative eugenics, the other two of positive eugenics.

Genetic counseling Increasingly practiced in the United States and other countries, genetic counseling informs prospective parents about the genetic nature of a given condition that may be known to exist in one of them or in their families, and about the chances of its transmission to their offspring. So advised, the prospective parents may choose not to have a child, or they may take their chances on a normal child. Genetic counseling can be supplemented with *amniocentesis:* a sample of the amniotic fluid surrounding the fetus inside the mother's womb is obtained and examined for chromosomal and other genetic abnormalities. The prospective mother can be informed whether or not the fetus car-

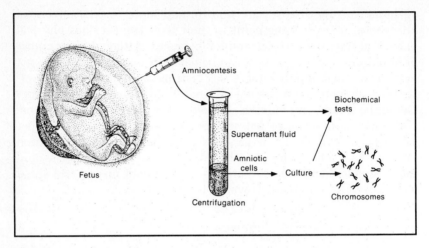

Fig. 13.1. Amniocentesis is practiced by extracting 15–20 ml of the amniotic fluid surrounding the fetus. Cells present in the amniotic fluid can be cultured in the laboratory and subjected to biochemical tests and chromosome analysis. Scores of deleterious conditions, including the very severe and common Down's syndrome, can be thus identified. A woman may choose to have an abortion rather than giving birth to a seriously defective child.

ries a certain genetic defect, and she may choose to have an abortion if such is the case.

The body politic could pursue a genetic program based on genetic counseling, amniocentesis, and abortion. Financial incentives, sterilization, and other coercive measures could be used to restrict carriers of unwanted genetic traits from procreating.

Genetic surgery Also called *genetic engineering* and *genetic therapy*, genetic surgery refers to the direct manipulation of the genetic material. Consider, for example, sickle-cell anemia, a condition caused by the substitution of a single nucleotide in the gene coding for the beta chain of hemoglobin (p. 90); the abnormal nucleotide could be replaced by the normal one, or the whole defective gene (or segment thereof containing the abnormal nucleotide) could be replaced by a normal one. Directed mutation and certain techniques known as *transformation* and *transduction* that are used with bacteria could be the methods to achieve the desired genetic change. The problem is that none of these methods can now successfully be applied to human beings, nor does successful application seem likely in the foreseeable future. The recently developed *recombinant DNA* techniques represent an advance in that direction, but their application to correcting genetic defects remains remote.

Germinal selection   This is a technique ardently proposed by the eminent geneticist and Nobel laureate H. J. Muller (1890-1967). The technique involves the extensive use of sperm and egg cells from individuals with desirable genetic constitutions through artificial fertilization; the frequency of the genetic variants possessed by such individuals would greatly increase in the population.

Muller's plan begins with the establishment of sperm banks for storing the seminal fluid of men of great achievement. This semen could be made available to any woman who would prefer to have a child fathered by a great man rather than by her husband or lover. Through artificial insemination, millions of women could be fertilized with the seminal fluid of a few eminent men. But Muller suggests going further: women produce some 500 eggs each through their lifetime; they can have only a few children because of the long nine-month pregnancies. Women of great excellence could be selected, their eggs flushed out and preserved under physiological conditions until requested by a prospective mother. A married couple could then select the genetic mother as well as the genetic father of their child: eggs fertilized in a test tube would be implanted in the prospective mother and allowed there to develop in the old-fashioned way.

More than a dozen commercial sperm banks already exist throughout the world. Several thousand cases of successful artificial insemination are estimated to occur per year in the United States alone; several hundred documented normal births have resulted from the use of semen obtained from sperm banks. Artificial insemination is often used by couples when the husband is infertile rather than for eugenic reasons, but eugenic goals are not necessarily precluded. (The British Academy of Sciences recommended to Parliament in 1975 that rock stars be prohibited from selling their semen to sperm banks. They feared that the popularity of rock stars could lead to thousands of offspring being produced from the sperm of a single star, which could result in inbreeding problems if some of the offspring intermarried, perhaps without knowing of their genetic relatedness.) There are not yet commerical banks for the storage of women's eggs.

Cloning   Also called *twinning*, cloning would ensure that a child be a true genetic copy of another individual. Cloning has been practiced with some success in frogs and toads by removing the nucleus from an unfertilized egg and replacing it with the nucleus of a somatic cell (which contains the same two full complements of genes and chromosomes as a fertilized egg). The egg with the replaced nucleus is then induced to develop; the resulting organism is genetically identical to the donor of the nucleus.

Cloning could produce a potentially unlimited number of people genetically as similar to each other and to the donor as identical twins are. Cloning has never yet successfully been accomplished with mammals, in spite of numerous attempts with mice and rats. A claim publicly announced in March, 1978, by a journalist that a baby boy had been successfully cloned from a rich man was a hoax.

Should Mankind Steer Its Own Evolution?

The ethical and sociopolitical implications of eugenics are enormous. Not all the methods mentioned in the previous section can presently be used as eugenic measures, but some could and others will become available in the future. We now raise the question whether such methods should or should not be applied to human populations. In so doing we leave the grounds of scientific discourse and enter the fields of ethics, sociology, and politics. We advance our views aware that not all readers will share them, but at least some of the important issues involved will become apparent. It should perhaps be pointed out that in developing our opinions in these matters we have been moved more by a concern for relieving individual suffering than by a desire for improving the gene pool of mankind, although this would be an additional benefit in some cases.

Genetic counseling is desirable, because it prepares prospective parents to make informed choices. We believe that parents should be advised not to have natural children whenever these have a high probability of having a very serious genetically determined defect—for example when a prospective parent has been cured of a dominant lethal condition, such as retinoblastoma, and has therefore, a one-half probability of a defective child; or when both parents are heterozygous for the same recessive lethal gene, in which case there is a one-quarter probability of having a child homozygous for the deleterious gene. Moreover, we believe that amniocentesis should be made available to parents, at least when the genetic risks are high. We also believe that abortion is morally and socially preferable to bringing into the world a severely handicapped child, such as an individual with Down's syndrome, although we realize that many people may disagree with us for religious or other reasons.

We maintain that, in general, known carriers of a severely deleterious dominant gene should be discouraged from reproducing, even when the gene is not lethal, although the lesser the deleterious effects of the gene the smaller is the moral and social ob-

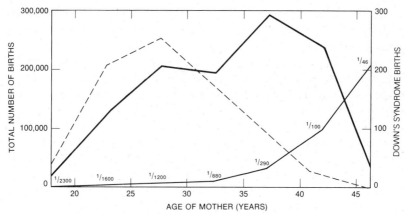

Fig. 13.2. Total number of births and of Down's syndrome births as a function of mother's age. The dashed line plots the total number of births; the thick solid line plots the number of babies born with Down's syndrome; the thin solid line gives the incidence of Down's syndrome. Although mothers more than 35 years old give birth to a small fraction of all children, they produce many of the babies with Down's syndrome because the incidence of the syndrome increases dramatically with age. Amniocentesis of the relatively small proportion of pregnant women over 35 years old would help identify a large fraction of all fetuses with Down's syndrome.

ligation of the carriers not to have children. However, there are factors, personal and otherwise, beyond the genetic considerations advanced here, that qualify the degree of moral responsibility of prospective parents in particular cases. An extensive treatment of the matter is impossible here, but what has been said should give some idea of the issues.

Governments could make it compulsory for people to follow the advice of genetic counselors. As a general policy this would be a flagrant violation of human rights, and thus we strongly oppose it. It should be further noted that as a method to improve the genetic endowment of human populations, compulsory genetic counseling is inefficient, particularly because its effects on changing genetic frequencies are extremely slow.

The use of genetic surgery to correct serious genetic defects appears to us ethically and socially unobjectionable. A person choosing to have a genetic defect corrected is making a decision as unobjectionable as a person who willingly chooses to have corrective surgery (such as the removal of a diseased kidney). Genetic surgery would be socially commendable if the defective gene were also corrected in the germinal cells of a person, because it would then benefit the progeny as well as the individual. The difficulty with genetic surgery is that, as previously noted, appropriate techniques usefully applicable to human beings do not yet exist.

The first point to make concerning the techniques of positive eugenics (germinal selection and cloning) is that the advocates of such Brave New World proposals generally ignore a fundamental genetics notion, namely that the phenotype of an individual is not determined exclusively by its genotype, but results from the interaction between the genotype and the environment. This is true of all organisms, but is most significant in the case of human beings because of the plasticity of their behavior and the decisive influence of the cultural components of the environment. The genotype does not determine the phenotype of an individual but only its "range of reaction." In different environments the same genotype may produce very different individuals. Thus the genotype of a great benefactor of mankind, of a great national leader, of a great scientist, or of a saint, might in a different set of environmental circumstances develop into a tyrant, a criminal, or a bum. This point has been cogently made by the Nobel laureate geneticist George W. Beadle: "Few of us would have advocated preferential multiplication of Hitler's genes. Yet who can say that in a different cultural context Hitler might not have been one of the truly great leaders of men, or that Einstein might not have been a political villain." In order to obtain another Einstein from Einstein's genotype, we would have to provide the latecomer with exactly the same environment and education, the same challenges and experiences, the same parents, teachers, and friends as those of the original Einstein, which is obviously impossible.

Positive eugenic measures raise a fundamental sociopolitical issue that cannot be satisfactorily resolved within a democratic society dedicated to the protection of civil liberties: Which one is the ideal genotype? What are the characteristics that should be multiplied? Who makes such decisions? Frequently, high intelligence is identified as a desirable characteristic; but artistic ability and a host of emotional and moral qualities are at least as important. John V. Tunney, former U.S. Senator from California, has asked: "How can we compare intelligence (even assuming it can be defined) with love?" We believe that few of the major problems facing the nations of this world could be solved with increased intellectual acuity, while much progress could be made if the individual and social morals were enhanced.

We believe that there is no way in which wise choices can be made as to the genetic characteristics to be multiplied, nor do we see how such decisions could be satisfactorily reached within the framework of a democratic society. But let us assume for a moment that agreement can be obtained about which individuals possess genetic characteristics to be multiplied. There are reasons to doubt that germinal selection would have significant overall favorable genetic effects on mankind for the following reasons among others: (1) As pointed out, the genotype does not unambiguously determine the phenotype. (2) The fitness of a genotype is determined by complex interactions between genes at different loci; but genetic recombination would occur in the formation of the sex cells of the selected individuals with unpredictable results, and the genomes received from the two genetic parents might not favorably interact. (3) Exclusive use of semen from a few men or of eggs from a few women would reduce the genetic diversity of mankind, a decidedly undesirable prospect. (4) It seems unlikely that a large fraction of women would choose to be fertilized with the semen of distinguished men rather than by their husbands; it seems even more unlikely that many women would choose to act as incubators for embryos altogether derived from other people's gametes.

As a means to change the genetic constitution of mankind, cloning the genotypes of chosen individuals would be more effective than any other technique. The possibilities of such a technique are stunning: the genotype of a rock star, a beautiful actress, a scientific genius, or a clever politician could be multiplied thousands or millions of times. It would also be possible to create a few genetic castes, each consisting of millions of identical individuals, dedicated to the service of a dominant elite. The production of even a single individual by cloning seems to us ethically repugnant; extensive human cloning would endanger the very survival of a democratic society. We believe that study groups consisting of biologists, physicians, sociologists, philosophers, legislators, and political and religious leaders should investigate this matter and provide advice and guidance to governments so that human cloning never comes to pass. Cooperation beyond national boundaries will be necessary; the future welfare, and even the survival, of mankind are at stake.

Mankind and the Evolution of the Organic World

Independently of whether or not it chooses to interfere in the process, the human species will continue evolving genetically in the future—unless, of course, it commits suicide through all-out atomic war. Some meddling will certainly occur in the form of negative eugenics; whether mankind will also choose to play a more active role in its own evolution through positive eugenics remains to be seen. If mankind chooses to do so, only future generations will know whether it was done wisely. We believe that the wisest course of action is to bypass the proposed methods of positive eugenics, while using negative eugenics as a means of decreasing the incidence of serious physical and mental diseases.

Human culture has influenced human evolution, but it has also had an enormous impact on the evolution of the living world; no other species ever had comparable effects. Mankind has expanded over the earth's surface. In the process many species have become extinct, while some weeds, mice, houseflies, and other organisms associated with humans followed them in their expansion. Some large animals have become extinct due to wholesale slaughter by man; but the extinction of many animal and plant species has been largely due to man's massive transformation of the environment (Fig. 13.3). Large areas are used for agriculture, cities, highways, and reservoirs. Mankind uses about 40 percent of the earth's land surface; the natural flora and fauna there have been replaced with agricultural crops, concrete, and other human products.

Fig. 13.3. Los Angeles on a clear day in the 1950s and 25 years later, visibility greatly impaired by smog, one of many undesired effects of massive transformation of the environment. By the early 1970s, Los Angeles County was discharging a daily average of 26,200,000 pounds

Species that have not become extinct have evolved to meet pervasive environmental challenges provoked by human activities. The disruptions of natural ecosystems have been large. Consider the effects on the flow of energy available to life. Agricultural ecosystems represent only 5 percent of the global photosynthetic activity although man has occupied 40 percent of the land. The effects of human activity have even reached into those regions where natural vegetation and animal life have not been destroyed, creating new environmental challenges for life by means of pollution, changes in the atmosphere due to the burning of fossil fuels, release of fertilizers and other chemicals into soils and streams, and the like.

The present and future evolutionary consequences of human activities are difficult to evaluate. Many species of insects have become resistant to DDT and other pesticides; scores of butterfly and moth species have coped with soot pollution of the environment by becoming dark-colored (industrial melanism) and thus less conspicuous to predators. Adaptations to specific environmental challenges of this kind will continue occurring throughout the living world. Adaptation to human-created ecological niches

of pollutants into the atmosphere, 7,500,000 more than in 1955. In the same period, carbon monoxide increased from 17,500,000 pounds a day to 18,200,000, while oxides of nitrogen increased from 1,100,000 pounds a day to 2,100,000.

may have occasionally produced new species. A possible example is the petroleum fly, *Psilopa petrolei*, whose larvae live in pools of crude oil in California oilfields and, as far as known, nowhere else. The process of organic evolution, including speciation, will go on in the future, promoted in part by human activities.

Mankind has affected organic evolution not only indirectly by changing the environment, but also directly through artificial selection. For about the last 10,000 years, mankind has cultivated plants and domesticated animals and has directed their evolution so as to satisfy human needs. The effects of artificial selection have been great, as witnessed for example by the diversity of dog or cattle breeds and by the great differences between cultivated grains or garden vegetables and their wild relatives (Fig. 13.4). Humans have produced new cultivated species, often by polyploidy, such as bread wheat (*Triticum aestivum,* hexaploid) and cultivated strawberries (e.g., a decaploid obtained from a cross between the octaploid *Fragaria chiloensis* and the diploid *F. vesca*).

In recent decades, agricultural practice and animal husbandry have increasingly relied on the use of a few highly productive strains or stocks. This has reduced the reserves of genetic vari-

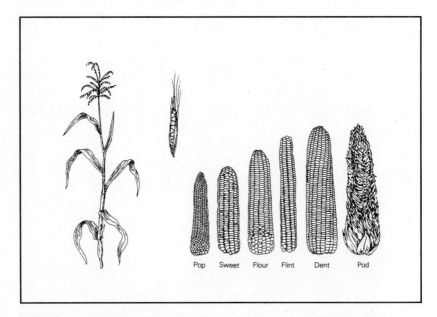

Pop   Sweet   Flour   Flint   Dent   Pod

Fig. 13.4.   Selective cultivation of corn over the centuries has resulted in much larger yields. Teosinte, *Euchaena mexicana,* the probable wild ancestor of corn, *left,* has small ears with few kernels. Modern corn varieties, *right,* include pop, sweet, flour, flint, dent, and pod corn.

ation, a potentially dangerous situation as evidenced, for example, by the epidemic of helminthosporiosis that in 1970 ravaged cornfields throughout the United States. The appearance of a new parasite or predator might completely destroy a single crop or breed, something considerably less likely if different genetic strains are cultivated. This trend is now being reversed in part through the initiative of various organizations such as the United Nations, the International Biological Program, and the National Academy of Sciences of the United States, which are fostering the creation of germplasm repositories to store seed varieties or to keep animal breeds, thus increasing the reserves of genetic variation.

Some techniques proposed for positive eugenics are potentially very useful in plant and animal breeding. Germinal selection is extensively practiced, for example, in cattle breeding, by fertilizing many females with semen from a few males. Cloning of domestic animals, whenever it can be successfully practiced, will allow the exact multiplication of genotypes with desired traits, genotypes that can then be bred under controlled conditions. The techniques of recombinant DNA would make possible introduction of useful genetic traits in food plants and in animal breeds. For example, research at the University of California in Davis and at other institutions is being directed toward the production of grains that carry genes (in microorganisms associated with the grain) that are able to obtain nitrogen from the atmosphere. Such grains would reduce the demand for nitrate fertilizer.

The effects of mankind on the evolution of the living world are likely to increase, rather than decrease, in the future. In particular, the galloping growth of the human population is placing increasing demands on the natural resources of the world. More and more areas with their natural flora and fauna remaining are being reclaimed for agriculture, recreation, and human habitation. Unless mankind's demographic explosion is soon contained, the consequences for the living world, including the human species itself, could be catastrophic.

QUESTIONS FOR DISCUSSION

1. Is the genetic endowment of mankind decaying rapidly? What arguments are used by those who believe that it is?

2. Do you think that some forms of negative eugenics should be used? If so, which ones and under what circumstances? What about the use of positive eugenics?

3. It has been reported that a child was conceived in a test tube using sperm from a man and an egg cell from his wife. The fertilized egg was then implanted in the mother, who eventually gave birth to a healthy child. Is this an instance of human cloning? Why?

4. During the last few hundred years the human population has soared. Do you think that the number of other species may have decreased as a consequence? Discuss the various ways in which the human population explosion affects the diversity of the living world.

5. Enumerate some of the ways in which the study of biological evolution has given you a better understanding of yourself and your place in nature.

Recommended for Additional Reading

Lerner, I. M., and Libby, W. J. 1976. *Heredity, evolution, and society.* 2nd ed. San Francisco: W. H. Freeman.

An elementary text covering many aspects of evolution, particularly as they relate to society.

Lipkin, M., Jr., and Rowley, P. T., eds. 1974. *Genetic responsibility.* New York and London: Plenum.

Milunsky, A., and Annas, G. J., eds. 1976. *Genetics and the law.* New York and London: Plenum.

Two multiauthored books dealing with social and legal questions associated with genetic engineering. Moderately advanced.

Rosenfeld, A. 1969. *The second genesis: the coming control of life.* Englewood Cliffs, N.J.: Prentice-Hall.

Genetic engineering presented in a highly readable and somewhat sensational fashion by a former science writer for *Life* magazine.

Volpe, E. P. 1975. *Man, nature, and society.* Dubuque, Iowa: Wm. C. Brown Co.

An introductory text emphasizing the aspects of evolution that particularly relate to society. Genetic engineering is discussed.

Appendix

Effects of Selection at a Single Gene Locus

The effects of selection on gene frequencies depend on the fitness of the various genotypes. The alternatives considered below are for gene loci with only two alleles, and therefore only three genotypes. The cases considered are: (1) selection against homozygous recessives; (2) selection against the genotypes with the dominant allele; (3) selection when there is no dominance; (4) selection against both homozygotes. The procedures used to calculate gene frequency changes are explained with greater detail in Case 1.

Case 1 Selection against homozygous recessives. It is assumed that homozygotes and heterozygotes carrying the dominant allele have fitness 1, but the homozygotes for the recessive allele have lower fitness, namely $1-s$. The model is the following:

$$
\begin{array}{cccc}
\text{Genotype} & AA & Aa & aa \\
\text{Fitness } (w) & 1 & 1 & 1-s
\end{array}
$$

The procedure is outlined in Table A.1. The frequencies of the two alleles at the beginning (zygote stage) of a certain generation are p and q, for A and a respectively. The genotypes are assumed to be in Hardy-Weinberg equilibrium and thus will have the frequencies given in the first row. Since there are only two alleles, p

$+ q = 1$, and therefore $p^2 + 2pq + q^2 = (p + q)^2 = 1$. The basic step is shown in the third row of the table: multiplication of the initial zygote frequencies (first row) by their fitnesses (second row), since these represent the relative rates of reproduction of the genotypes.

Table A.1
ALLELE FREQUENCY CHANGES AFTER ONE GENERATION OF SELEC-
TION AGAINST RECESSIVE HOMOZYGOTES

		Genotypes			Total	Frequency of a
		AA	Aa	aa		
1.	Initial zygote frequency	p^2	$2pq$	q^2	1	q
2.	Fitness, w	1	1	$1 - s$		
3.	Zygote proportions after selection	p^2	$2pq$	$q^2(1 - s)$	$1 - sq^2$	
4.	Zygote frequencies after selection	$\dfrac{p^2}{1 - sq^2}$	$\dfrac{2pq}{1 - sq^2}$	$\dfrac{q^2(1 - s)}{1 - sq^2}$	1	$\dfrac{q - sq^2}{1 - sq^2}$
5.	Change in allele frequency, Δq				$\dfrac{-spq^2}{1 - sq^2}$	$\dfrac{- sq^2}{1 - sq^2}$

However, when we sum the values given in the third row, we see that they do not add up to 1:

$$p^2 + 2pq + q^2 (1 - s) =$$
$$p^2 + 2pq + q^2 - sq^2 =$$
$$(p^2 + 2pq + q^2) - sq^2 =$$
$$1 - sq^2$$

In order to convert those values into frequencies, we divide each value by their total, as shown in the fourth row. These now add up to 1:

$$\frac{p^2}{1 - sq^2} + \frac{2pq}{1 - sq^2} + \frac{q^2(1 - s)}{1 - sq^2} =$$

$$\frac{p^2 + 2pq + q^2 - sq^2}{1 - sq^2} = \frac{1 - sq^2}{1 - sq^2} = 1$$

We calculate the frequency of the a allele after selection by adding up the frequency of the aa genotypes and half the frequency of the Aa genotypes (since only half the alleles of this genotype are a):

$$\frac{q^2(1 - s)}{1 - sq^2} + \frac{pq}{1 - sq^2} = \frac{pq + q^2 - sq^2}{1 - sq^2} =$$

$$\frac{q(p + q) - sq^2}{1 - sq^2} = \frac{q - sq^2}{1 - sq^2}$$

The gene frequency of the A allele after selection can be obtained by adding the frequency of the AA homozygotes and half the frequency of the Aa heterozygotes; or, alternatively, by subtracting the frequency of a after selection from one. Using the first procedure we obtain

$$\frac{p^2}{1 - sq^2} + \frac{pq}{1 - sq^2} = \frac{p^2 + pq}{1 - sq^2} =$$

$$\frac{p(p + q)}{1 - sq^2} = \frac{p}{1 - sq^2}$$

Finally, we are interested in the allele frequency change (fifth row). The initial frequency of the a allele was q, the frequency after selection is

$$\frac{q - sq^2}{1 - sq^2}$$

The change in allele frequency (represented as Δq) is obtained by subtracting the original frequency from the new frequency:

$$\Delta q = \frac{q - sq^2}{1 - sq^2} - q = \frac{q - sq^2 - q(1 - sq^2)}{1 - sq^2} =$$

$$\frac{q - sq^2 - q + sq^3}{1 - sq^2} = \frac{-sq^2 + sq^3}{1 - sq^2} = \frac{-sq^2(1 - q)}{1 - sq^2}$$

And, since $1 - q = p$, we obtain

$$\Delta q = \frac{-spq^2}{1 - sq^2}$$

The frequency of the *a* allele has decreased, as we would expect, since selection is acting against the homozygotes *aa*. What will be the ultimate outcome of selection? By definition, there will be no further change in allele frequencies when

$$\Delta q = \frac{-spq^2}{1 - sq^2} = 0$$

Δq will be zero when the numerator is zero, i.e., q will gradually decrease in value until it becomes zero. The ultimate outcome of this form of selection is the elimination of the *a* allele.

Another interesting question concerns the rate of selection. For a given value of s, the product pq^2 will become smaller and smaller as q changes from intermediate to lower and lower values. Although p increases as q decreases, the decrease in q^2 is greater than the increase in p because the square of numbers smaller than 1 is smaller than the number itself. Therefore, the rate of selection becomes extremely slow as q approaches zero.

Case 1a There is an interesting instance of Case 1, namely when the recessive homozygotes die before reproductive age or are sterile. An example is phenylketonuria, due to homozygosis for a recessive allele. Then, the fitness of the recessive homozygotes, *aa*, is $w = 0$, and $s = 1$; and the formulae given in Table A.1 become simpler. Substituting 1 for s, the frequency of *a* after selection becomes

$$\frac{q - sq^2}{1 - sq^2} = \frac{q - q^2}{1 - q^2} = \frac{q(1 - q)}{(1 - q)(1 + q)} = \frac{q}{1 + q}$$

and the change in allele frequency becomes

$$\Delta q = \frac{-spq^2}{1 - sq^2} = \frac{-pq^2}{1 - q^2} = \frac{-q^2(1 - q)}{(1 - q)(1 + q)} = \frac{-q^2}{1 + q}$$

Case 2 Selection against the dominant allele. The model is the following:

Genotype	AA	Aa	aa
Fitness (w)	$1 - s$	$1 - s$	1

Table A.2 summarizes the effects of selection. The sum of the zygote proportions after selection (third row) are obtained as follows:

$$p^2(1 - s) + 2pq(1 - s) + q^2 =$$
$$p^2 - sp^2 + 2pq - 2spq + q^2 =$$
$$p^2 + 2pq + q^2 - sp^2 - 2spq =$$
$$(p + q)^2 - s(p^2 + 2pq) =$$
$$1 - s(p^2 + 2pq + q^2 - q^2) =$$
$$1 - s(1 - q^2) = 1 - s + sq^2$$

Table A.2
ALLELE FREQUENCY CHANGES AFTER ONE GENERATION OF SELEC-
TION AGAINST GENOTYPES WITH THE DOMINANT ALLELE

	Genotypes			Total	Frequency of A
	AA	Aa	aa		
1. Initial zygote frequency	p^2	$2pq$	q^2	1	p
2. Fitness, w	$1 - s$	$1 - s$	1		
3. Zygote proportions after selection	$p^2(1 - s)$	$2pq(1 - s)$	q^2	$1 - s + sq^2$	
4. Zygote frequencies after selection	$\dfrac{p^2(1 - s)}{1 - s + sq^2}$	$\dfrac{2pq(1 - s)}{1 - s + sq^2}$	$\dfrac{q^2}{1 - s + sq^2}$	1	$\dfrac{p(1 - s)}{1 - s + sq^2}$
5. Change in allele frequency, Δp					$\dfrac{-spq^2}{1 - s + sq^2}$

The frequency of A after selection (fourth row) is the frequency of AA plus half the frequency of Aa, or

$$\frac{p^2(1 - s)}{1 - s + sq^2} + \frac{pq(1 - s)}{1 - s + sq^2} = \frac{(p^2 + pq)(1 - s)}{1 - s + sq^2} =$$

$$\frac{p(p + q)(1 - s)}{1 - s + sq^2} = \frac{p(1 - s)}{1 - s + sq^2}$$

The change in the frequency of allele A is

$$\Delta p = \frac{p(1 - s)}{1 - s + sq^2} - p = \frac{p(1 - s) - p(1 - s + sq^2)}{1 - s + sq^2} =$$

$$\frac{p - sp - p + sp - spq^2}{1 - s + sq^2} = \frac{-spq^2}{1 - s + sq^2}$$

The frequency of A will gradually decrease until it becomes zero. The rate of selection (when the frequency of the allele selected against is low) is greater in Case 2 than in Case 1 because p enters in the formula for Δp in simple form, not as a square. It is intuitively obvious that the rate of selection should be more efficient against a dominant allele than against a recessive one, because in the second case selection acts also against the heterozygotes rather than only against the homozygotes.

Case 3 Selection without dominance: it is assumed that selection is half as strong against the heterozygotes as against the disfavored homozygotes. The model is:

$$
\begin{array}{lccc}
\text{Genotype} & A_1A_1 & A_1A_2 & A_2A_2 \\
\text{Fitness } (w) & 1 & 1-s & 1-2s
\end{array}
$$

Table A.3 summarizes the effects of selection. The sum of the zygote proportions after selection is

$$
\begin{aligned}
&p^2 + 2pq(1-s) + q^2(1-2s) = \\
&p^2 + 2pq - 2spq + q^2 - 2sq^2 = \\
&p^2 + 2pq + q^2 - 2spq - 2sq^2 = \\
&(p+q)^2 - 2sq(p+q) = \\
&1 - 2sq
\end{aligned}
$$

The frequency of A_2 after selection is

$$
\frac{q^2(1-2s)}{1-2sq} + \frac{pq(1-s)}{1-2sq} = \frac{q^2 - 2sq^2 + pq - spq}{1-2sq} =
$$

$$
\frac{q^2 + pq - sq(2q+p)}{1-2sq} = \frac{q(q+p) - sq(p+q+q)}{1-2sq} =
$$

$$
\frac{q - sq(1+q)}{1-2sq} = \frac{q - sq - sq^2}{1-2sq}
$$

Table A.3
ALLELE FREQUENCY CHANGES AFTER ONE GENERATION OF SELEC-
TION WHEN THERE IS NO DOMINANCE

	Genotypes			Total	Frequency of A_2
	A_1A_1	A_1A_2	A_2A_2		
1. Initial zygote frequency	p^2	$2pq$	q^2	1	q
2. Fitness, w	1	$1-s$	$1-2s$		
3. Zygote proportions after selection	p^2	$2pq(1-s)$	$q^2(1-2s)$	$1-2sq$	
4. Zygote frequencies after selection	$\dfrac{p^2}{1-2sq}$	$\dfrac{2pq(1-s)}{1-2sq}$	$\dfrac{q^2(1-2s)}{1-2sq}$	1	$\dfrac{q-sq-sq^2}{1-2sq}$
5. Change in allele frequency, Δq					$\dfrac{-spq}{1-2sq}$

The change in the frequency of A_2 is

$$\Delta q = \frac{q - sq - sq^2}{1 - 2sq} - q = \frac{q - sq - sq^2 - q + 2sq^2}{1 - 2sq} =$$

$$\frac{-sq + sq^2}{1 - 2sq} = \frac{-sq(1 - q)}{1 - 2sq} = \frac{-spq}{1 - 2sq}$$

The frequency of A_2 will decrease until it becomes zero. The rate of selection will be approximately equal when p is small as when q is small since both enter into the numerator as simple, unsquared quantities (the rate will be slightly higher when q is large since the denominator will then be smaller). The rate of selection is greater, however, when both A_1 and A_2 have intermediate frequencies, since the product pq is greater when $p = q = 0.5$. Note that the product pq appears in the numerator of the expression for Δq in all the cases so far considered (as well as in the next one), and thus it is generally true that the efficiency of selection is greatest when both alleles have intermediate frequencies.

Case 4 Selection in favor of the heterozygote over both homo-zygotes, known as *overdominance,* or *heterosis.* The model is:

Genotype	AA	Aa	aa
Fitness (w)	$1-s$	1	$1-t$

The effects of selection are summarized in Table A.4. The sum of the zygote proportions after selection (third row) is obtained as follows

$$p^2(1 - s) + 2pq + q^2(1 - t) =$$
$$p^2 - sp^2 + 2pq + q^2 - tq^2 =$$
$$p^2 + 2pq + q^2 - sp^2 - tq^2 =$$
$$(p + q)^2 - sp^2 - tq^2 =$$
$$1 - sp^2 - tq^2$$

Table A.4
ALLELE FREQUENCY CHANGES AFTER ONE GENERATION OF SELEC-
TION WHEN THERE IS OVERDOMINANCE

	Genotypes			Total	Frequency of a
	AA	Aa	aa		
1. Initial zygote frequency	p^2	$2pq$	q^2	1	q
2. Fitness, w	$1 - s$	1	$1 - t$		
3. Zygote proportions after selection	$p^2(1 - s)$	$2pq$	$q^2(1 - t)$	$1 - sp^2 - tq^2$	
4. Zygote frequencies after selection	$\dfrac{p^2(1 - s)}{1 - sp^2 - tq^2}$	$\dfrac{2pq}{1 - sp^2 - tq^2}$	$\dfrac{q^2(1 - t)}{1 - sp^2 - tq^2}$	1	$\dfrac{q - tq^2}{1 - sp^2 - tq^2}$
5. Change in allele frequency, Δq					$\dfrac{pq(sp - tq)}{1 - sp^2 - tq^2}$

The frequency of a after selection (fourth row) is, as usual, obtained as the frequency of aa plus half the frequency of Aa, or

$$\frac{q^2(1 - t)}{1 - sp^2 - tq^2} + \frac{pq}{1 - sp^2 - tq^2} = \frac{q^2 - tq^2 + pq}{1 - sp^2 - tq^2} =$$

$$\frac{q^2 + pq - tq^2}{1 - sp^2 - tq^2} = \frac{q(q + p) - tq^2}{1 - sp^2 - tq^2} = \frac{q - tq^2}{1 - sp^2 - tq^2}$$

The frequency of A after selection is

$$\frac{p^2(1 - s)}{1 - sp^2 - tq^2} + \frac{pq}{1 - sp^2 - tq^2} = \frac{p^2 - sp^2 + pq}{1 - sp^2 - tq^2} =$$

$$\frac{p(p + q) - sp^2}{1 - sp^2 - tq^2} = \frac{p - sp^2}{1 - sp^2 - tq^2}$$

The change in the frequency of a due to one generation of selection (fifth row) is

$$\Delta q = \frac{q - tq^2}{1 - sp^2 - tq^2} - q = \frac{q - tq^2 - q + sp^2 q + tq^3}{1 - sp^2 - tq^2} =$$

$$\frac{sp^2 q - tq^2 + tq^3}{1 - sp^2 - tq^2} = \frac{sp^2 q - tq^2(1 - q)}{1 - sp^2 - tq^2} =$$

$$\frac{sp^2 q - tpq^2}{1 - sp^2 - tq^2} = \frac{pq(sp - tq)}{1 - sp^2 - tq^2}$$

Case 4 represents the most interesting case with respect to the outcome of selection. By definition, changes in allele frequencies will cease whenever $\Delta q = 0$, which requires that the numerator of the expression for Δq be zero. If at any given time neither p nor q is zero (i.e., both alleles exist in the population), such condition will only be satisfied if $sp - tq = 0$; with appropriate transformations we obtain $sp - tq = 0$, $sp = tq$, $s(1 - q) = tq$, $s - sq = tq$, $s = tq + sq = q(t + s)$, $q = \dfrac{s}{s + t}$

and, correspondingly

$$p = 1 - q = 1 - \frac{s}{s + t} = \frac{s + t - s}{s + t} = \frac{t}{s + t}$$

These equilibrium frequencies are stable because if p is greater than the equilibrium frequency, i.e.,

$$p > \frac{t}{s + t}$$

then $sp > tq$, and the value of Δq will be positive (because the value in parentheses in the numerator will be positive), and q will increase in frequency. On the other hand, if

$$p < \frac{t}{s + t}$$

then $sp < tq$, and the numerator of the expression for Δq will be negative, leading to a decrease in the frequency of q, until the equilibrium frequencies are reached.

References

Allison, A. C. 1954. Protection afforded by sickle cell trait against subtertian malarial infection. *British Medical Journal* 1:290–292.

Ayala, F. J. 1965. Evolution of fitness in experimental populations of *Drosophila serrata*. *Science* 150:903–905.

———. 1965. Sibling species of the *Drosophila serrata* group. *Evolution* 19: 538–545.

———. 1975. Genetic differentiation during the speciation process. *Evolutionary Biology* 8:1–78.

Ayala, F. J. and W. W. Anderson. 1973. Evidence of natural selection in molecular evolution. *Nature New Biology* 241:274–276.

Ayala, F. J. and M. L. Tracey. 1974. Genetic differentiation within and between species of the *Drosophila willistoni* group. *Proceedings of the National Academy of Sciences*, U.S.A. 71:999–1003.

Ayala, F. J., M. L. Tracey, L. G. Barr, and J. G. Ehrenfeld. 1974. Genetic and reproductive differentiation of *Drosophila equinoxialis caribbensis*. *Evolution* 28:24–41.

Ayala, F. J., M. L. Tracey, D. Hedgecock, and R. C. Richmond. 1974. Genetic differentiation during the speciation process in *Drosophila*. *Evolution* 28:576–592.

Ayala, F. J., J. W. Valentine, and G. S. Zumwalt. 1975. An electrophoretic study of the Antarctic zooplankter *Euphausia superba*. *Limnology and Oceanography* 20:635–640.

Bachmann, K., O. B. Goin, and C. J. Goin. 1972. Nuclear DNA amounts in vertebrates. In *Evolution of Genetic Systems*, H. H. Smith, ed., *Brookhaven Symposium Biology* 23:419–450.

Bakker, R. T. 1975. Experimental and fossil evidence for the evolution of tetrapod bioenergetics. In *Perspectives of Biophysical Ecology*, D. Gates and R. Schmerl, eds., Springer-Verlag, New York, pp. 365–395.

Berry, R. J. 1978. Genetic variation in wild house mice: where natural selection and history meet. *American Scientist* 66:52–60.

Bock, W. J. 1970. Microevolutionary sequences as a fundamental concept in macroevolutionary models. *Evolution* 24:704–722.

Brashaw, A. D. 1971. Plant evolution in extreme environments. In *Ecological Genetics and Evolution*, R. Creed, ed., Blackwell, Oxford, pp. 20–50.

Bretsky, S. 1970. Phenetic and phylogenetic classifications of the Lucinidae (Mollusca, Bivalvia). *Bulletin of the Geological Institute University of Upsala* new ser. 2:5–23.

Brown, W. L. and E. O. Wilson. 1956. Character displacement. *Systematic Zoology* 5:49–64.

Bruce, E. J. and F. J. Ayala. 1979. Phylogenetic relationships between man and the apes: Electrophoretic evidence. *Evolution* 33.

Bush, G. L. 1975. Modes of animal speciation. *Annual Review of Ecology and Systematics* 6:339–364.

Cain, A. J. and P. M. Sheppard. 1954. Natural selection in *Cepaea*. *Genetics* 39:89–116.

Carson, H. L. 1971. Speciation and the founder principle. *Stadler Symposium* 3:51–70.

Carson, H. L., D. E. Hardy, H. T. Spieth, and W. S. Stone. 1970. The evolutionary biology of the Hawaiian Drosophilidae. In *Essays in Evolution and Genetics in Honor of Th. Dobzhansky*, M. K. Hecht and W. C. Steere, eds., Appleton-Century-Crofts, New York, pp. 437–543.

Clark, R. B. 1964. *Dynamics in Metazoan Evolution*. Clarendon Press, Oxford.

Clarke, C. A. and P. M. Sheppard. 1960. The evolution of mimicry in the butterfly *Papilio dardanus*. *Heredity* 14:163–173.

Cloud, P. 1949. Some problems and patterns of evolution exemplified by fossil invertebrates. *Evolution* 2:322–350.

Connell, J. 1961. The influence of interspecific competition and other factors on the distribution of the barnacle *Chthamalus stellaus*. *Ecology* 42:710–723.

Conway Morris, S. 1976. A new Cambrian lophophorate from the Burgess shale of British Columbia. *Palaeontology* 19:199–122.

———. 1977. A new metazoan from the Cambrian Burgess shale of British Columbia. *Palaeontology* 20:623–640.

Crow. J. F. 1958. Some possibilities for measuring selection intensities in man. *Human Biology* 30:1–13.

Dawson, P. S. 1969. A conflict between Darwinian fitness and population fitness in *Tribolium* "competition" experiments. *Genetics* 62:413–419.

Dobzhansky. Th. 1970. *Genetics of the Evolutionary Process*. Columbia University Press, New York.

Dobzhansky. Th. and B. Spassky. 1969. Artificial and natural selection for two behavioral traits in *Drosophila pseudoobscura*. *Proceedings of the National Academy of Sciences*, U.S.A. 62:75–80.

Dobzhansky, Th. and O. Pavlovsky. 1957. An experimental study of interaction between genetic drift and natural selection. *Evolution* 11:311–319.

Doyle, J. A. 1977. Patterns of evolution in early angiosperms. In *Evolution as Illustrated by the Fossil Record*, A. Hallam, ed., Elsevier Publ. Co., Amsterdam, pp. 501–546.

Eldredge, N. and S. J. Gould. 1972. Punctuated equilibria: An alternative to phyletic gradualism. In *Models in Paleobiology*, T. J. M. Schopf, ed., Freeman, Cooper, San Francisco, pp. 82–115.

Fisher, W. L., P. U. Rodda, and J. W. Dietrich. 1964. Evolution of the *Athleta petrosa* stock (Eocene, Gastropoda) of Texas. *University Texas Publication* 6413.

Fitch, W. M. and E. Margoliash. 1967. Construction of phylogenetic trees. *Science* 155:279–284.

Fox, S. W. and Dose, K. 1972. *Molecular Evolution and the Origin of Life*. W. H. Freeman and Co., San Francisco.

Gingerich, P. D. 1977. Patterns of evolution in the mammalian fossil record. In *Patterns of Evolution as Illustrated by the Fossil Record*, A. Hallam, ed., Elsevier Publ. Co., Amsterdam, pp. 469–500.

Glaessner, M. F. and M. Wade. 1966. The late Precambrian fossils from Ediacara, South Australia. *Palaeontology* 9:599–628.

Gottlieb, L. D. 1973. Enzyme differentiation and phylogeny in *Clarkia franciscana*, *C. rubicunda*, and *C. amoena*. *Evolution* 27:205–214.

———. 1973. Genetic differentiation, sympatric speciation, and the origin of a diploid species of *Stephanomeria*. *American Journal of Botany* 60: 545–548.

Gould, S. J. 1977. *Ontogeny and Phylogeny*. Belknap Press of Harvard University Press, Cambridge, Massachusetts.

Gross, M. G. 1977. *Oceanography, A View of the Earth*. Prentice-Hall, Englewood Cliffs, New Jersey.

Haldane, J. B. S. 1949. Disease and evolution. *La Ricerca Scient.* 19 (suppl.): 68–76.

Hamilton, W. D. 1964. The genetic theory of social behavior. *Journal of Theoretical Biology* 1:1–16, 17–52.

Harding, J., R. W. Allard, and D. G. Smeltzer. 1966. Population studies in predominantly self-pollinated species. IX. Frequency-dependent selection in *Phaseolus lunatus*. *Proceedings of the National Academy of Sciences*, U.S.A. 56:99–104.

Harris, H. 1966. Enzyme polymorphisms in man. *Proceedings Royal Society Ser. B.* 164:298–310.

Hinegardner, R. 1976. Evolution of genome size. In *Molecular Evolution*, F. J. Ayala, ed., Sinauer, Sunderland, Massachusetts, pp. 179–199.

Johnson, F. M., C. G. Kanapi, R. H. Richardson, M. R. Wheeler, and W. S. Stone. 1966. An analysis of polymorphisms among isozyme loci in dark and light *Drosophila ananassae* strains from American and Western Samoa. *Proceedings of the National Academy of Sciences*, U.S.A. 56: 119–125.

Jowett, D. 1958. Population of *Agrostis* spp. tolerant to heavy metals. *Nature* 182:816–817.

Kettlewell, H. B. D. 1961. The phenomenon of industrial melanism in Lepidoptera. *Annual Review of Entomology* 6:245–262.

King, M. C. and A. C. Wilson. 1975. Evolution at two levels: Molecular similarities and biological differences between humans and chimpanzees. *Science* 188:107–116.

Kirk, D. 1968. Patterns of survival and reproduction in the United States: Implications for selection. *Proceedings of the National Academy of Sciences, U.S.A.* 59:662–670.

Kohne, D. E., J. A. Chiscon, and B. H. Hoyer. 1972. Evolution of primate DNA sequences. *Journal of Human Evolution* 1:627–644.

Lack, D. 1953. Darwin's finches. *Scientific American* 214: See also: *Darwin's Finches, An Essay on the General Biological Theory of Evolution.* 1961. Harper and Brothers, New York.

Lewis, H. 1966. Speciation in flowering plants. *Science* 152:167–172.

Lewontin, R. C. and J. L. Hubby. 1966. A molecular approach to the study of genic heterozygosity in natural populations. II. Amount of variation and degree of heterozygosity in natural populations of *Drosophila pseudoobscura*. *Genetics* 54:595–609.

MacArthur, R. H. 1955. Fluctuations of animal populations, and a measure of community stability. *Ecology* 36:533–536.

MacArthur, R. H. and E. O. Wilson. 1967. *The Theory of Island Biogeography.* Princeton Monographs in Population Biology 1.

Margulis, L. 1970. *Origin of Eukaryotic Cells.* Yale University Press, New Haven, Connecticut.

May, R. M. 1973. *Stability and Complexity in Model Ecosystems.* Princeton Monographs in Population Biology 6.

McDonald, J. F. and F. J. Ayala. 1974. Genetic response to environmental heterogeneity. *Nature* 250:572–574.

—— and ——. 1978. Genetic and biochemical basis of enzyme activity variation in natural populations. I. Alcohol dehydrogenase in *Drosophila melanogaster*. *Genetics* 89:371–388.

McDonald, J. F., G. K. Chambers, J. David, and F. J. Ayala. 1977. Adaptive response due to changes in gene regulation: A study with *Drosophila*. *Proceedings of the National Academy of Sciences, U.S.A.* 74:4562–4566.

Miller, S. J. 1953. A production of amino acids under possible primitive earth conditions. *Science* 117:528.

Moore, J. A. 1950. Further studies on *Rana pipiens* racial hybrids. *American Naturalist* 84:247–254.

Morgan, T. H. 1919. *The Physical Basis of Heredity.* Lippincott, Philadelphia.

Muller, H. J. 1950. Our load of mutations. *American Journal Human Genetics* 2:111–176.

——. 1963. Genetic progress by voluntarily conducted germinal choice. In *Man and His Future*, G. Wolstenholme, ed., pp. 247–262.

Nabours, R. K. 1937. Methoden und Ergebnisse bei der Züchtung von Tetriginae. *Abderhalden's Handb. Biol. Arbeitsmeth.* 9:1309–1365.

Nevo, E., Y. J. Kim, C. R. Shaw, and C. S. Thaeler. 1974. Genetic variation, selection and speciation in *Thomomys talpoides* pocket gophers. *Evolution* 28:1–23.

Nevo, E. and C. R. Shaw. 1972. Genetic variation in a subterranean mammal, *Spalax ehrenbergi. Biochemical Genetics* 7:235–241.

Newell, N. D. 1967. Revolutions in the history of life. *Geological Society of America, Special Papers* No. 89:63–91.

Paul, C. R. C. 1977. Evolution of primitive echinoderms. In *Patterns of Evolution as Illustrated by the Fossil Record*, A. Hallam, ed., Elsevier Publ. Co., Amsterdam, pp. 123–158.

Petit, C. and L. Ehrman. 1969. Sexual selection in Drosophila. *Evolutionary Biology* 3:177–223.

Prakash, S. and R. C. Lewontin. 1968. A molecular approach to the study of genic heterozygosity in natural populations. III. Direct evidence of co-adaptation in gene arrangements of Drosophila. *Proceedings* of the National Academy of Sciences, U.S.A. 59:398–405.

Raup, D. M., S. J. Gould, T. J. M. Schopf, and D. S. Simberloff. 1973. Stochastic models of phylogeny and the evolution of diversity. *Journal of Geology* 81:525–542.

Sanger, F. and E. O. P. Thompson. 1953. The amino acid sequence in the glycyl chain of insulin. *Biochem. Journal* 53:353–374.

Sarich, V. M. and A. C. Wilson. 1969. Immunological time scale for hominid evolution. *Science* 158:1200–1203.

Simpson, G. G. 1951. *Horses*. Oxford University Press, New York.

———. 1966. The biological nature of man. *Science* 152:472–478.

———. 1969. *Biology and Man*. Harcourt, Brace and World, New York.

Smith, A. G. and J. C. Briden. 1977. *Mesozoic and Conozoic Paleocontinental Maps*. Cambridge University Press.

Sparrow, A. H., H. J. Price, and A. G. Underbrink. 1972. A survey of DNA content per cell and per chromosome of prokaryotic and eukaryotic organisms: some evolutionary considerations. In *Evolution of Genetic Systems*, H. H. Smith , et al., eds., *Brookhaven Symposia* 23:451–495.

Stanley, J. M. 1975. A theory of evolution above the species level. *Proceedings of the National Academy of Sciences*, U.S.A. 72:646–650.

———. 1975. Clades versus clones in evolution: why we have sex. *Science* 190: 382–383.

Tarling, D. H. and M. P. Tarling. 1971. *Continental Drift*. G. Bell and Sons, London.

Thomson, K. S. 1977. The pattern of diversification among fishes. In *Patterns of Evolution as Illustrated by the Fossil Record*, A . Hallam, ed., Elsevier Publ. Co., Amsterdam, pp. 377–404.

Valentine, J. W. 1973. *Evolutionary Paleoecology of the Marine Biosphere*. Prentice-Hall, Englewood Cliffs, New Jersey.

Van Valen, L. 1973. A new evolutionary law. *Evolutionary Theory* 1:1–30.

Vaurie, C. 1951. Adaptive differences between two sympatric species of nut-hatches *(Sitta)*. *Proceedings X International Ornithological Congress* 1950:163–166.

Vine, F. J. 1969. Sea-floor spreading—new evidence. *Journal of Geological Education* 17:6–16.

Wallace, B. 1948. Studies on "sex-ratio" in *Drosophila pseudoobscura*. *Evolution* 2:189–217.

Watson, J. D. and F. H. Crick. 1953. A structure for deoxyribose nucleic acid. *Nature* 171:737.

———. 1953. Genetical implications of the structure of DNA. *Nature* 171:964.

White, M. J. D. 1968. Models of speciation. *Science* 159:1065–1970.

Whittaker, R. H. 1969. New concepts of kingdoms of organisms. *Science* 163: 150–160.

Williams, A. 1956. The calcareous shell of the Brachiopoda and its importance to their classification. *Biol. Review* 31:243–287.

Williams, G. C. 1975. *Sex and Evolution*. Princeton Monographs in Population Biology 8.

Wilson, A. C. 1976. Gene regulation in evolution. In *Molecular Evolution*, F. J. Ayala, ed., Sinauer, Sunderland, Massachusetts, pp. 225–234.

Wise, D. U. 1974. Continental margins, freeboard and the volumes of continents and oceans through time. In *The Geology of Continental Margins*, C. A. Burk and C. L. Drake, eds., Springer-Verlag, New York, Heidelberg, Berlin.

Wynne-Edwards, V. C. 1962. *Animal Dispersion in Relation to Social Behavior*. Oliver and Boyd, Edinburgh.

Glossary

Adaptive strategy. Usually, adaptation to a pattern of environmental variability. Also used for one of two or more alternative solutions to an adaptive problem.

Adaptive value. Same as Darwinian fitness.

Allele. One of two or more alternative forms of a gene (locus).

Allopatric. Having separate geographical ranges.

Allopolyploid. A polyploid arising when the chromosome number is doubled in an interspecific hybrid.

Altruism. Of individuals, behavior performed for the benefit of others; of genes, those that favor such behavior.

Amerous. Of coelomic cavities, not divided into segments or regions by septa.

Amniocentesis. Sampling of the amniotic fluid surrounding a fetus in order to test for certain genetic abnormalities.

Anagenesis. Progressive evolution.

Analogous. Of organs or genes, performing similar functions (though not necessarily descended from a common ancestor).

Aneuploidy. Change in the chromosome number, usually by an increase or decrease of one chromosome.

Autopolyploid. A polyploid arising when the number of chromosomes are doubled (in an individual that is not an interspecific hybrid).

Biomass. The weight of living matter measured in a standard manner.

Biosphere. The ecological system embracing all life on earth and the environment with which it interacts.

Bottleneck. Drastic decrease in the number of individuals of a population usually resulting in a reduction of genetic variability owing to genetic drift.

Character displacement. Divergence of two sympatric species with respect to a character owing to competition between the species.

Chromosome. A threadlike structure present in the nucleus of cells and containing genes in linear arrangement.

Chronospecies. Segments of a single lineage, between which there has been gradual evolutionary change so that the descendants are morphologically distinctive, but for which any subdivision within the gradational series would be arbitrary.

Clade. A single branch (which may be branched itself) of the tree of life.

Cladogenesis. The branching off of a new clade.

Coadaptation. The selection process by which harmoniously interacting genes accumulate in the gene pool of a population.

Coelom. A "true" body cavity in animals, filled with fluid, lined with mesoderm and communicating with the exterior by special ducts.

Coevolution. Evolution of two or more species resulting from biological interaction between them; affects plants and their herbivores, prey and their predators, hosts and their parasites.

Corridor. A migration route along which members of a biota may easily disperse.

Crossing over. Reciprocal exchange of homologous chromosome segments.

Deletion. Loss of a section of a chromosome (or of a gene).

Darwinian fitness. Reproductive efficiency of a genotype relative to other genotypes. Same as adaptive value and selective value.

Density-dependent factor. An ecological factor, usually a limiting factor (such as food), the effect of which increases with increasing population size.

Density-independent factor. An ecological factor (such as temperature), that acts independently of population size.

Deuterostomous. Animals that have their mouth formed during development at a site not previously marked by a pore, and usually share certain other features. This is characteristic of chordates and some of their close relatives.

Diploid.   Having two sets of chromosomes (or two genomes), usually one from each parent.

Diversity.   Of species or other taxa, usually the number present in a particular ecological unit. May also include a measure of the distribution of population sizes among the species.

Dominant.   Of genes, the allele that controls the phenotypic expression of a heterozygous locus.

Drift.   See genetic drift.

Duplication.   Presence of two copies of a chromosome segment (or of a gene), usually in the same chromosome.

Ecosystem.   A living system comprising all the organisms in an ecological unit (a community or province, for example) and the environment with which they interact.

Ectotherm.   An animal that approximately corresponds in body temperature to the temperature of its environment.

Electrophoresis.   Technique of separating molecules, particularly proteins and nucleic acids, according to their electric charge.

Emergent properties.   In a hierarchical system, properties that are manifest only on a more inclusive level of the hierarchy.

Endemism.   Of species or other taxa, a measure of the number that are restricted entirely to a particular ecological or geographic unit.

Endotherm.   An animal that has the ability to regulate body temperature.

Environmental grain.   The heterogeneity of the environment as it is "perceived" by a population. Populations that range freely through an environment perceive it as fine-grained, those that are limited to some small portion of it perceive it as coarse-grained.

Enzyme.   A protein that catalyzes a specific chemical reaction.

Eugenics.   The science and practice seeking to improve the hereditary constitution of a population, especially of mankind.

Eukaryote.   An organism whose cells have a distinct nucleus containing chromosomes.

F_1.   First filial generation. The progeny resulting from a parental cross, usually between homozygotes for different alleles.

F_2.   Second filial generation. The progeny resulting from mating F_1 individuals (or from selfing F_1 individuals in self-fertilizing organisms).

Filter route. A migration route along which only a small proportion of a biota can easily disperse.

Fission. Splitting of a chromosome into two.

Fitness. See Darwinian fitness.

Founder effect. Reduction in genetic variability owing to genetic drift when a new population is founded by a small group of individuals.

Frequency dependence. When the fitness of a genotype or phenotype depends on its frequency.

Gamete. A sexual, or reproductive, cell; in animals, a sperm or egg. A gamete contains only one set of genes and chromosomes, rather than two as are present in regular (somatic) cells.

Gel electrophoresis. Electrophoresis performed in a jellylike substrate, made of starch, polyacrylamide, or some other substance providing a homogeneous matrix.

Gene flow. Transfer of genes (by means of migration and interbreeding) from one to another population.

Gene locus (plural, gene loci). Position occupied by a gene in a chromosome; also a gene, when its particular allelic form is not specified.

Gene pool. The genetic constitution of a population.

Genetic coadaptation. See coadaptation.

Genetic code. Code that relates nucleotide sequences in nucleic acids to amino acid sequences in proteins.

Genetic distance. A measure of the genetic differentiation between two populations.

Genetic drift. Changes in gene frequencies owing to random fluctuations from generation to generation.

Genetic identity. A measure of the genetic similarity between two populations.

Genome. The total genetic material of an organism in its haploid state. Diploid individuals possess two genomes, ordinarily one from each parent.

Genotype. The genetic constitution of an individual (or of a character, locus, or other part thereof).

Geographic speciation. The rise of new species following the geographic isolation of two or more parts of the parent species.

Grade. The level of organization (as in body plan) of organisms.

Group selection. Selection due to the extinction of some populations and the propagation of other populations.

Haploid. A cell or organism that contains only one set of chromosomes.

Hardy-Weinberg law. A rule stating that, in the absence of evolutionary processes, gene and genotypic frequencies remain constant from generation to generation; the law moreover relates genotypic frequencies to gene frequencies.

Heterogametic. Resulting from the mating of dissimilar individuals.

Heterogametic sex. The sex (the males in mammals, insects, and other organisms, but the females in birds and butterflies) having two dissimilar sex chromosomes or only one sex chromosome.

Heterosis. "Hybrid vigor"; the higher fitness of heterozygotes.

Heterostyly. Of flowers, unequal lengths of styles and stamens to favor cross-fertilization.

Heterozygote. A genotype with two different forms of a gene (at a locus), or an individual bearing such a genotype.

Homogametic sex. The sex having two similar sex chromosomes.

Homologous. Of organs or genes, descended from a common ancestor (but not necessarily performing the same function). Of chromosomes, those that pair during meiosis; they usually resemble each other and contain genes coding for the same features.

Homozygote. A genotype with two identical alleles at a gene locus, or an individual bearing such a genotype.

Hybrid. An organism resulting from a cross between individuals belonging to dissimilar populations.

Inbreeding. Matings between consanguineous (closely related) individuals.

Independent assortment. Independent segregation of genes at meiosis. Genes in nonhomologous chromosomes assort independently.

Inversion. Rotation of a chromosomal segment.

Linkage. Presence of two or more loci on a single chromosome. Alleles at linked loci tend to be inherited together more often than separately.

Linkage disequilibrium. Nonrandom association of genes—the extent to which gametic frequencies of linked loci deviate from those expected by random combination of the alleles at the linked loci.

Meiosis. The sequence of two cell divisions that produce the sex cells or gametes and during which the number of chromosomes per cell is halved.

Metamerous. Of coelomic cavities, divided into regularized, repeated segments separated (at least primitively) by septa; most such segments contain a set of various internal organs.

Metazoa. Animals that include both coelenterates (jellyfish) and platyhelminths (flatworms) in their multicellular family tree; essentially all living animal phyla except sponges.

Migration. See gene flow.

Mitosis. Cell division producing daughter cells identical in genetic content to each other and to the parental cell.

Monophyletic. A taxon in which all the members have descended from a common ancestor that is included in the taxon or that is sometimes included in an ancestral taxon of the same rank.

Mosaic evolution. Evolutionary change in one or more body parts without a corresponding change in others.

Mutagen. A physical or chemical agent that increases the mutation rate.

Mutation. A heritable change in the genetic material; also, the effect of such change on the phenotype.

Mutation distance. The smallest number of mutations necessary to derive one DNA sequence from another.

Natural selection. Process, and result, of differential reproduction of alternative genetic variants.

Neoteny. The evolutionary process that results in the developmental retardation of some body features relative to reproduction, so that juvenile features in ancestors appear as adult features in descendants.

Niche. The ecological system of a population (or species) and the environment (including organisms) with which it interacts.

Nonhomologous. Of chromosomes, those that do not pair during meiosis.

Oligomerous. Of coelomic cavities, divided into two or (usually) three coelomic regions by septa; characteristic of chordates, echinoderms, and some other phyla.

Ontogeny. The developmental history of an organism from fertilized egg to adult.

Paradigm. A model; often applied to complex scientific hypotheses, which are conceptual models.

Peristalsis. Wave motion of a hollow muscular structure formed by a coordinated series of contractions.

Phenetic. In taxonomy, a classification formed by phenotypic characters without regard to lines of descent.

Phenotype. The appearance of an individual (or part thereof) resulting from its genetic makeup (its genotype) and its environmental history.

Pheromone. A chemical secretion that is used in communication within a species.

Phylogeny. The sequence of ancestor-descendant forms (including collateral branches if appropriate) created during the evolutionary history of a taxon.

Plate tectonics. The process of chiefly horizontal movements of the upper layer of the earth, including the continents and the ocean crust, as a result of heat flow from the interior.

Polymorphism. The presence of two or more distinctive genotypes or phenotypes within a population.

Polypeptide. A group of amino acids sequentially linked by peptide bonds. A protein consists of one or more polypeptides.

Polyphyletic. A taxon that includes individuals for which a common ancestor cannot be found within the taxon (or sometimes even within the immediately ancestral taxon of the same rank).

Polyploidy. Containing three or more sets of chromosomes.

Polytene chromosomes. Composed of numerous similar strands.

Postzygotic. After fertilization.

Prezygotic. Before fertilization.

Prokaryote. A unicellular organism that lacks a nucleus; bacteria and cyanobacteria ("blue-green algae") are prokaryotic.

Pseudoextinction. The disappearance of a lineage from the fossil record due to its evolution into a new form rather than to its termination.

Quantum speciation. The rise of new species rapidly, usually within small isolates with founder effects and genetic drift playing important roles.

Random genetic drift. See genetic drift.

Recapitulation. The relatively later onset of reproduction, or the relatively earlier appearance of nonreproductive features, during a life cycle. The term was once used for a theory that postulated the occurrence of the phylogenetic sequence of ancestral adult stages in the ontogeny of an organism, now disproven.

Recessive. An allele that is not expressed at a heterozygous locus.

Recombinant. A new phenotype or genotype arising through rearrangement of the genes via independent assortment or crossing over during meiosis.

Regulatory gene. A gene that governs or modifies the activity of another gene.

Reproductive isolating mechanisms. Biological properties of organisms that interfere with the interbreeding of organisms from different species.

Reproductive isolation. Inability to interbreed due to biological differences.

RIMs. Reproductive isolating mechanisms.

Saltational speciation. Same as quantum speciation.

Segregation. The separation of allele pairs from one another and their distribution to different gametes.

Selection coefficient. The difference in fitness between a particular genotype and another genotype chosen as a standard of reference.

Selective value. Same as Darwinian fitness.

Sex chromosomes. The chromosomes (usually represented as X and Y) that differ in males and females.

Species. A group of individuals able to interbreed producing fertile progeny.

Stasigenesis. Of lineages, a lack of significant evolutionary change or of branching over a long period of time.

Structural gene. A gene that codes for a protein.

Supergene. A DNA segment containing a number of closely linked genes that affect a single trait or an array of interrelated traits.

Sweepstakes route. A migration route along which no species may disperse easily; given enough time, however, some species will by chance manage to cross.

Sympatric. Having overlapping geographical ranges.

Taxon. A group of individuals classed together as a formal unit and ranked in a taxonomic category; Arthropoda (a phylum), Insecta (an order), and *Danaus plexippus* (a species of monarch butterfly) are examples.

Taxonomic category. The rank of a taxon in the hierarchy of classification; phylum, order, and species are examples.

76989-11005 Bio Ethics corrs. sequence into a messenger RNA molecule.

Transcription. In genetics, the conversion of the genetic information in a DNA sequence into a messenger RNA molecule.

Translation. In genetics, the conversion of the genetic information in messenger RNA into an amino acid sequence, which forms a polypeptide.

Translocation. When a chromosomal segment changes from one to another position in the chromosomes.

Zygote. The cell formed by the union (fertilization) of a male gamete and a female gamete, and from which the individual develops.

Index